Lecture Notes in Physics

Volume 869

W0234527

For further volumes:
www.springer.com/series/5304

The Lecture Notes in Physics

The series Lecture Notes in Physics (LNP), founded in 1969, reports new developments in physics research and teaching—quickly and informally, but with a high quality and the explicit aim to summarize and communicate current knowledge in an accessible way. Books published in this series are conceived as bridging material between advanced graduate textbooks and the forefront of research and to serve three purposes:

- to be a compact and modern up-to-date source of reference on a well-defined topic
- to serve as an accessible introduction to the field to postgraduate students and nonspecialist researchers from related areas
- to be a source of advanced teaching material for specialized seminars, courses and schools

Both monographs and multi-author volumes will be considered for publication. Edited volumes should, however, consist of a very limited number of contributions only. Proceedings will not be considered for LNP.

Volumes published in LNP are disseminated both in print and in electronic formats, the electronic archive being available at springerlink.com. The series content is indexed, abstracted and referenced by many abstracting and information services, bibliographic networks, subscription agencies, library networks, and consortia.

Proposals should be sent to a member of the Editorial Board, or directly to the managing editor at Springer:

Christian Caron
Springer Heidelberg
Physics Editorial Department I
Tiergartenstrasse 17
69121 Heidelberg/Germany
christian.caron@springer.com

Ion Geru · Dieter Suter

Resonance Effects of Excitons and Electrons

Basics and Applications

 Springer

Ion Geru
Institute of Chemistry
Academy of Sciences of Moldova
Chisinau
Moldova

Dieter Suter
Fakultät Physik
Universität Dortmund
Dortmund
Germany

ISSN 0075-8450   ISSN 1616-6361 (electronic)
Lecture Notes in Physics
ISBN 978-3-642-35806-7   ISBN 978-3-642-35807-4 (eBook)
DOI 10.1007/978-3-642-35807-4
Springer Heidelberg New York Dordrecht London

Library of Congress Control Number: 2013941482

Printed on acid-free paper

Springer is part of Springer Science+Business Media (www.springer.com)

In memory of E.K. Zavoisky, discoverer of the electron paramagnetic resonance phenomenon, on the eve of the 70th anniversary of this discovery

Foreword

Electrons constitute the major building material of matter. Without understanding their functionality, we cannot properly appreciate the inner workings of nature. Electrons also provide physicists, chemists, and biologists a fascinating playground with a virtually unlimited plethora of experimental possibilities, from microscopic tools for exploring molecular and biological systems to novel possibilities for implementing the most powerful computers conceivable, the quantum computers. This book offers to the interested and patient reader an understanding of the inherent fascination of electronic systems.

Electrons have the virtue of interacting in various ways with the ambient electric, magnetic, and acoustical phonon fields created by other particles. To fully appreciate their virtues, a full-blown quantum mechanical treatment is indispensable. The quantum theoretical approach fills a large part of the first seven chapters of the book. Their digestion might require some stamina from the reader, but will allow him/her to obtain nearly unlimited insight into features and effects to be expected in localized and delocalized electronic systems.

The reader will become acquainted with the importance of excitonic states and the phenomena of paramagnetic, paraelectric, and zero-field resonance, for example, in magnetic semiconductors. The reader will also learn about transitions between exciton subbands, the resonant absorption of hypersound in condensed matter, and the fascinating possibilities of double resonance methods.

Electrons and electronic paramagnetism never exist in isolation; they interact with nuclear spins. The latter provide further insight by their localized character and can be used as observers of events in the electronic realm. Spin relaxation is a phenomenon that reveals much insight on structures and on excitonic motions, for example, in diluted magnetic semiconductors. Deep saturation effects are another feature of unsteady quantum systems.

Electrons and excitons also open fascinating possibilities for the implementation of quantum information processors, alluded to in Chap. 8. Chapter 9 reveals several possibilities for designing quantum information processors. So far, quantum computers of practical utility are still in the realm of future dreams. But this book convincingly demonstrates that quantum computing possibilities are close to being implemented.

In summary, this book contains much knowledge that could be of future importance for the invention and practical design of useful electronic and excitonic devices.

1 January 2013   Richard R. Ernst
                                                              Nobel Prize in Chemistry 1991

Preface

The discovery of electron paramagnetic resonance (EPR) at Kazan State University by Soviet physicist Yevgeny Zavoisky in 1944 [1] represented the beginning of broad studies in the field of the energetic structure and dynamics of spin and pseudospin systems. In the short period of time that has passed since this fundamental discovery, the narrow area of research samples has been significantly enlarged to include almost all types of compounds containing paramagnetic centers [2].

The importance of Zavoisky's discovery is its real universal application in the study of the energy spectra of paramagnetic centers and the shapes of EPR lines, which are conditioned by spin–spin, spin–phonon, and other interactions in different systems: ionic [3–5] and molecular [6] crystals, coordination compounds in solid [4–7] and liquid [8] crystals, free radicals [6, 9, 10], chemical solvents [3, 11], color centers [12, 13], semiconductors [14–16], magnetic semiconductors [17–20], glasses (dielectric [3] and semiconductor [21, 22]), ordinary [14, 23, 24] and superconducting [25–27] metals and alloys, and spin-assigned biological systems [28], etc. This discovery also has a methodological importance, because the method used by Zavoisky to detect magnetic resonance based on changes in the field of electromagnetic radiation (instead of matter) during the interaction between the substance and the field has been widely applied to detect many other low-frequency resonances in condensed matter. The following list of phenomena observed and studied by the resonance method after the EPR discovery is a conclusive proof of the great importance of EPR, not only as a new physical phenomenon, but also as a new method for research: nuclear magnetic resonance (NMR) [29–34], nuclear quadrupole resonance (NQR) [35], acoustic paramagnetic resonance (APR) [3, 36–40], paraelectric resonance (PER) [41, 42], cyclotron resonance [43–45], spin resonance of conduction electrons in metals [23] and semiconductors [46–48], mixed resonance [44, 49], ferromagnetic resonance (FMR) [50–52], antiferromagnetic resonance (AFMR) [50], Pound–Overhauser double resonance [14], electron-nuclear double resonance (ENDOR) [4, 5, 53, 54], nuclear–nuclear double resonance [55–57], electron–electron double resonance (ELDOR) [58], electron–nuclear double magnetoacoustic resonance (acoustical ENDOR) [59], and more.

Some resonance phenomena among those listed above, due to the importance of the research results obtained by using these methods and their valuable practical applications, have triggered the development of new research directions in solid state physics. Thus, the phenomenon of APR [36], discovered in 1952 by Al'tshuler, later became the starting point for the development of direct methods for the research of spin interactions in paramagnetic centers and nuclei with phonons in condensed matter, including both magneto-dispersed and magneto-condensed systems with magnetic ordering. The ensemble of these methods forms modern quantum acoustics, with several independent research directions.

The reciprocal influence of one domain of spectroscopy on others as well as the unitary physical character of similar phenomena can be extended further. For example, it is easy to see that the coupled magnetoelastic waves in ferromagnetic materials, which are generated as a result of interactions of elastic waves with spin waves in conditions of magnetoacoustic resonance [50], resemble, to a wide extent, polaritons in the excitonic domain of the spectrum and phonon polaritons [60].

Another typical example of the branching of a research direction is represented by the phenomenon of NMR, discovered in 1946 independently by Purcell, Torrey, and Pound [61] and by Bloch, Hansen, and Packard [62]. Because the local magnetic fields on the nuclei of magnetic (and in some cases, nonmagnetic) ions conditioned by the hyperfine interaction in ferro- and antiferromagnetic materials are two to three orders of magnitude higher than those for these systems in the paramagnetic phase, as soon as radio spectrometers with frequencies in the range of 10^2–10^3 MHz were built (13 years after the discovery of NMR in paramagnetic materials), an independent research direction of great importance—NMR in magneto-ordered crystals [63–65]—was developed. However, the new peculiarities of magneto-ordered phase have led to the discovery of new phenomena, such as the effect of amplification of NMR signals, electron–nuclear double ferromagnetic and electron–nuclear double antiferromagnetic resonances [66], a new type of electron–nuclear double resonance that is determined by a substantial change in the rate of nuclear transversal relaxation at parametric excitation of electronic spin waves in ferrites [67], and so on.

The further development of the methods themselves and the realization of substantial new ideas are important as well. A good illustration of this aspect of the process in radio spectroscopy development is represented by the sharp increase in information given by NMR methods in studying condensed matter by using multipulse sequences and the successful experimental realization of "coherent averaging" methods in spin spaces [68, 69].

Radio-frequency spectroscopic methods can be applied under specific conditions to study degenerate exciton states in crystals. Regarding their real application, magnetic microwave spectroscopy of the Frenkel excitons in molecular crystals has been developing for a long time. Meanwhile, the application of these methods to study the Wannier–Mott excitons in semiconductors is a topic of future research.

The lack of hyperfine structure in the paramagnetic resonance spectra of triplet excitons due to the effect of translational movement [70] allows us to firmly differentiate experimentally the exciton states in molecular crystals from the localized triplet excited states of molecules. The small radius of the Frenkel exciton and, accordingly, the narrow exciton band which differs insignificantly from the enlarged level, as well as the sufficiently large lifetime of metastable triplet states in molecular crystals (noticeably larger than the period of a microwave) mean that the study of EPR spectra of Frenkel excitons differs insignificantly from the results obtained for EPR spectra of paramagnetic centers. In this way, naturally, the investigation of EPR of triplet Frenkel excitons in molecular crystals actually began in the 1960s [58].

Things look completely different in the study of Wannier–Mott excitons in semiconductors by methods of radio spectroscopy. The first experimental works in this field were performed substantially later [71–74]. It is thus reasonable to elucidate the research possibilities of exciton states in semiconductors by means of radio and microwave spectroscopy at small and large levels of optical excitation in crystals.

During the last decades, many of the physical systems that we discuss in this book, as well as the magnetic resonance techniques that were developed to study them, have been investigated from a very different angle: they may be used as "quantum bits" or qubits, for storing and processing information. The field of quantum information (or, more generally quantum technology) was born from the realization that information is not only an abstract concept, but that it is intimately connected to its physical representation [75]. In particular, it was conjectured [76] and proved [77] that quantum mechanical systems can process information in a way that makes them qualitatively more powerful than classical systems. As a result, problems for which no efficient solution appears to exist for classical computers may be solved efficiently by information processing devices that use the laws of quantum mechanics [78, 79]. The basic building blocks of quantum information processing are called quantum logical gates, in analogy to the logical gate operations of classical computers.

The realization of this potential requires two main ingredients: a physical system for storing the quantum information and external controls for implementing the quantum logical gate operations. The basic elements for storing the information are quantum mechanical two-level systems: the qubits. Since spins $1/2$ are the only physical systems whose Hilbert space is exactly two dimensional, they represent ideal blueprints for qubits. Similarly, all required quantum gate operations can be implemented by pulses of resonant radio-frequency or microwave fields interacting with the spins, in combination with free precession of the spins under an internal Hamiltonian that couples them to each other. As a result, the basic principles of quantum information processing have all been demonstrated on the basis of physical systems and spectroscopic methods described in this book.

In Chaps. 1–7 we discuss Wannier–Mott excitons, free and localized biexcitons, paramagnetic centers and nuclei which interact with the high-density excitons, as well as the influence of paramagnetic centers of high concentration on low-density excitons. The aim of these investigations is to obtain, by means of radio spectroscopy methods, research data on the hyperfine interaction of nuclear spins with excitons for small levels of optical excitation in crystals, to elucidate the mechanisms of influence of free excitons and free and localized biexcitons on the magnetic properties of semiconductors at large levels of their optical excitation, and also to obtain the relations regarding the inverse influence of magneto-condensed systems of impurity centers on free excitons. The methods of radio and microwave spectroscopy then have a special significance in those cases when the exciton states cannot be studied by means of optical methods (e.g., optically forbidden transitions in the exciton states, or extremely low concentrations of excitons).

Chapter 1 reviews the individual and collective properties of excitons in semiconductors studied by means of optical spectroscopy. Special attention is given to some fundamental difficulties in experimentally supporting the theory of Bose–Einstein condensation of excitons by means of optical spectroscopy experiments. These difficulties can be overcome by using the methods of radio spectroscopy.

In Chap. 2 we present the research on exciton paramagnetic and exciton paraelectric resonances. The possibilities of generation by excitons of coherent magnons (in magnetic semiconductors) and coherent electromagnetic radiation in submillimeter and infrared diapasons are analyzed.

In Chap. 3 we study the acoustic resonance of excitons and biexcitons. The theory of phonon masers based on exciton transitions is developed.

Chapter 4 is devoted to double resonances. Electron–nuclear double resonance and electron–nuclear magnetoacoustic double resonance of paramagnetic centers, hole–nuclear double resonance of the biexcitons captured by isoelectronic traps in GaP:N crystals, transfer of radio-frequency coherence in the cycle of optical pumping of excitons and, thereby, excitonic optical radio-frequency double resonance are presented.

In Chap. 5 we study excitons and biexcitons by means of NMR spectroscopy. The effects of excitons and orthobiexcitons on the relaxation rate of nuclear spin are discussed. The Knight excitonic shift of NMR lines for a degenerate and nondegenerate exciton gas is studied. The possibility of expanding the multipulse methods of high-resolution NMR spectroscopy to the optical spectroscopy of excitons is suggested based on an example of a four-pulse sequence of WAHUHA (WAugh, HUber, HAeberlen) type.

In Chap. 6 we study the interaction of excitons with paramagnetic centers: exchange exciton scattering on deep paramagnetic centers, indirect exchange interactions between paramagnetic centers through excitons, giant spin splitting of the exciton band in the exchange field formed by paramagnetic centers of high concentration (first being partially spin-polarized by using an external magnetic field) and, thereby, anomalous high increases in magneto–optical effects in the exciton band.

Chapter 7 is devoted to the effects of strong saturation of exciton and spin transitions. Nonstationary exciton states in the field of an intense hypersonic wave, nonstationary spin states (for an arbitrary spin value) in a constant magnetic field and low-frequency magnetic field of high amplitude, and the time-reversal symmetry for systems having a quasi-energy spectrum are studied.

In Chap. 8 we give a brief overview of the basics of quantum information processing. Taking the quantum mechanical nature of the physical system into account allows one to build qualitatively more powerful computers than is possible with the already extremely powerful classical devices. The fundamental notions are quantum superposition and the parallel processing of superposition states by unitary control operations.

Chapter 9 discusses specific examples where spins in solids have been used to demonstrate important concepts of quantum information processing. This includes very different systems like semiconductor quantum dots, dielectric solids containing rare-earth ions, or defects, such as the nitrogen–vacancy center in diamond.

Chapters 1–7 and Sect. 8.11 were written by I. Geru, while Chaps. 8 and 9 were written by D. Suter.

One of the authors (I. Geru) wishes to thank Professor S. Moskalenko and Professor S. Ryabchenko for their scientific collaboration during different periods of time.

Contents

Chapter 1
Excitons and Biexcitons in Semiconductors

This chapter briefly deals with information on the physics of excitons. In Sects. 1.1 and 1.2, and partially in Sect. 1.5, the individual properties of the excitons are considered. The exciton states are classified on the basis of microscopic theory and group theory. The characteristic properties of the exciton influenced by the division of its movements, the relative movement of the electron–hole and the translational motion of the center of gravity of the electron–hole pair, as well as the spin structure and longitudinal–transverse splitting of the exciton by Coulomb far action are considered in the framework of a simple two-band model for a semiconductor.

The most impressive successes have been achieved in the physics of high-density excitons, and this field continues to develop intensively (Chap. 4). The main results in this domain are briefly reflected in Sect. 1.3. We note that the Lenard–Dyson theorem is significant in understanding the stability of the ground energy state of a system with an arbitrary number of charged particles interacting according to Coulomb's law.

In Sect. 1.5 the role of impurities and their capture of one or two excitons is discussed. In particular, we note the capture of excitons by isoelectron traps with the formation of a bound exciton or a localized exciton molecule.

1.1 The Electronic Structure of Excitons

The concept of excitons as specific excited states of crystals was first introduced in 1931 by Frenkel [80] in regard to molecular crystals. In the framework of the theory of excitons the processes of the absorption of light by intrinsic (without impurities) crystals without the accompaniment of photoconductivity and the mechanism of transformation of light into heat can be easily explained. The representations of excitons were further developed by Wannier [81] and Mott [82], and it was shown that the presence of excitons is a characteristic feature of the spectrum of elementary excitations of any nonmetallic crystal. Some 30–35 years after the basic work of Frenkel, the theory of excitons turned into a developed topic in solid state physics.

I. Geru, D. Suter, *Resonance Effects of Excitons and Electrons*,
Lecture Notes in Physics 869, DOI 10.1007/978-3-642-35807-4_1,
© Springer-Verlag Berlin Heidelberg 2013

General monographs on the theory of excitons [83–87], and later books [88, 89] and other works were published. In a rather detailed form they set forth the achievements of that period, which could be called "the period of all-round investigation of individual exciton properties."

The first clear confirmation of Frenkel's work [80] on the existence of excitons in crystals was determined experimentally in the work of Gross and coauthors [90] in 1952 and in Hayashi's and Katsuki's work [91] on the rich discrete structure of the long-wave edge of the absorption spectrum of the Cu_2O crystal.

The concrete structure of excitons is different for different types of crystals and, according to their electron structure, three types of excitons are distinguished: small-radius excitons (Frenkel's excitons), large-radius excitons (Wannier–Mott excitons), and intermediate-radius excitons.

Frenkel's exciton corresponds to the case when the electron and hole formed during the excitation of the crystal by a quantum of light are localized on the same node of the crystal lattice and they move together on the lattice. Although the translational symmetry of the crystal lattice requires that each exciton correspond to a certain wave vector $\mathbf{K}$ in the space of the reciprocal lattice ($\mathbf{K}$ plays the role of the quantum number, determining the irreducible representation of the translational group of the symmetry of the crystal), the concept of quasi-particles for the electron and hole with certain wave vectors at such a strong localization loses value, and the description in the node representation is more convenient. Frenkel's exciton is formed, as a rule, in molecular crystals between molecules on which the weak van der Waals forces act as a sequence in which the wave functions of electrons on neighboring molecules are slightly covered, and there is not a big probability of the transition of the excited electron from one molecule to another. For this reason, the width of the exciton band is an order less than that of the electron band for these crystals, but the effective mass is ten times bigger than that of the band charge carriers [92].

Another model of the exciton with a radius of the order of the constant lattice is applied in the alkali halide crystals with ionic bonds but small dielectric constant. This is called the "charge transition model," and it was investigated in a more detailed way by Overhauser [93]. On the basis of the NaCl crystal, in which every Cl^- ion is surrounded by six Na^+ ions localized on an axis of type $\langle 100 \rangle$, one electron in the transition model of the charge during the absorption of light passes from the filled $2p$ shell of the Cl^- ion to the free $3s$ shell of the Na^+ ion. The model is collectivized on the basis of the nearest configuration sphere. This kind of binding state of the electron and hole which is translated as a whole on the lattice corresponds to an exciton of intermediate radius.

In semiconductors with ionic bonds but with big dielectric constants, e.g., Cu_2O, CuCl, CuBr types and semiconductors A_2B_6, A_3B_5, A_3B_6, as well as in semiconductors with covalent bonds of Ge, Si types and others, the large-radius exciton[1]

---

[1]Excitons are also revealed in biological media, where they play a big role, for example, in the processes of photosynthesis [86, 94–96]. Excitons may be present in superconductors, formed from quasi-particles with an energetic level that lies in the interior of the energy gap [86]. In normal metals excitons usually have a very short lifetime and cannot be detected experimentally [86].

model is realized. The electron and hole in this model participate simultaneously in a quick relative movement and in the translational movement of the coupled electron–hole pair as a whole (the hydrogenlike model of Wannier–Mott excitons). We shall limit ourselves by considering the excitons in semiconductors.

The binding energy of the Wannier–Mott exciton Ry^{ex} and its effective radius a_{ex} may be estimated according to Bohr's formula for the hydrogen atom, taking into consideration that the effective masses of the electron m_e and hole m_h are distinguished from the mass of the free electron m_0, and that the Coulomb interaction between the electron and hole is weakened by the presence of the dielectric constant of the medium ε [97]:

$$Ry^{ex} = Ry \mu m_0^{-1} \varepsilon^{-2}, \qquad a_{ex} = a_B \varepsilon m_0 \mu^{-1}. \tag{1.1}$$

Here $\mu = m_e m_h m_{ex}^{-1}$ is the reduced effective mass of the exciton, $m_{ex} = m_e + m_h$ is the exciton effective mass, and Ry and a_B are Rydberg's constant and Bohr's radius for the hydrogen atom. For $\mu \sim (1-0, 1)m_0$ and $\varepsilon \sim 10$, which apply for the semiconductors such as germanium, silicon, and the $A_3 B_5$ and $A_2 B_6$ groups, we get $Ry^{ex} \sim 10^{-2}$–10^{-1} eV and $a_{ex} \sim 10^{-6}$–10^{-7} cm. Thus, the exciton binding is two to three orders of magnitude smaller than the characteristic atom energies, but their radii are one to two orders of magnitude greater than the interatomic distances in the crystals.

The division of direct and indirect excitons depends on the band structure of the energetic spectrum of the semiconductor. Excitons for which the bottom of the exciton band is at $\mathbf{K}_{ex} = 0$ are called direct excitons. If the minimum of the energy of electrons (or holes) is achieved at wave vectors $\mathbf{K}_{oe}$ (respectively, $\mathbf{K}_{oh}$) that are not equal to zero, then the exciton formed by the electron–hole pair, which will have the minimum of energy at $\mathbf{K}_{oex} = \mathbf{K}_{oe} + \mathbf{K}_{oh}$, is called an indirect exciton. The electron and the hole in the exciton may be recombined by the radiation of light quanta and other quasi-particles. For a direct exciton whose wave vector is small, the radiative recombination takes place without phonon participation or the participation of another exciton with probability $\sim a_{ex}^{-3}$. For indirect excitons this process is forbidden by the law of conservation of kinetic quasi-momentum and can take place only with the participation of a phonon, impurity center, or another exciton. Therefore, the speed of the radiative recombination of indirect excitons is two to three orders of magnitude smaller than that for the direct ones. In pure semiconductors at low temperatures ($k_0 T \ll Ry^{ex}$), the lifetime of direct excitons is 10^{-7}–10^{-9} s, while for indirect excitons it is significantly longer: $\sim 10^{-4}$–10^{-6} s [97]. The time of the free run of the exciton during diffusion on the phonons at low temperatures is of the same order as that for the free electrons and holes, i.e., 10^{-8}–10^{-10} s at $T \lesssim 4$ K. Knowing this time and the lifetime of the exciton we can find the distance covered by the exciton during its lifetime (the diffusion length). In pure semiconductors the diffusion length is on the order of 10^{-1} cm for indirect excitons and 10^{-2}–10^{-3} cm for direct excitons [97].

For large-radius excitons, as is seen from (1.1), the crystal may be substituted by a continuous dielectric medium, and the band structure influences the energetic spectrum of the exciton not only through μ and ε, but also through the nondiagonal components of the tensor of the effective masses of the electron and hole, the resulting spin (in the general case, the total angular momentum) of the electron and the hole in the ground state, and the symmetry properties of the crystal. The characteristic features of the electronic structure of the exciton can be elucidated on the basis of a simple model for the semiconductor with two energy bands (the conduction and the valence bands), the degeneration of which in the absence of spin–orbital interaction is characterized by the spin and the orbital quantum numbers, but in a more general case by the irreducible representations of the point group of crystal symmetry for direct excitons and the group of the wave vector for indirect excitons. Below we give the calculation of the energy spectrum of the exciton according to [98].

We shall consider a crystal whose atoms have filled valence shells, with empty levels lying higher. Let us assume that there are N nodes of the lattice in the crystal. The nuclei of the atoms occupying these nodes and the internal electron shells can be considered as cores with charge $+Ze$. (The polarization of the core and the displacement of the nuclei from the position of equilibrium are not considered here.) Then due to the condition of electric neutrality in the crystal there exist ZN valence electrons, the Hamiltonian of which in the coordinate representation has the form

$$H = -\sum_{i=1}^{ZN} \frac{\hbar^2}{2m_0} \Delta_i + \sum_{i=1}^{ZN} V(\mathbf{r}_i) + \frac{1}{2} \sum_{i,j=1(i\neq j)}^{ZN} \frac{e^2}{|\mathbf{r}_i - \mathbf{r}_j|}, \qquad (1.2)$$

where

$$V(\mathbf{r}_i) = \sum_{j=1} U\big(|\mathbf{r}_i - \mathbf{R}_j|\big).$$

Here $U(\mathbf{r}_i - \mathbf{R}_j)$ is the potential of the core, found in the node $\mathbf{R}_j$, but $V(\mathbf{r}_i)$ is the basic periodic potential of the lattice cores.

The complete periodic potential in which the band carrier is moving and which determines its energy spectrum and the real structure of the band, is created in a "self-agreed" way by all the electrons together with the core of atoms. This is that effective potential $V_{\text{eff}}(\mathbf{r})$ in which a separate electron is moving in the field of all the other electrons and cores of the lattice. The wave function $\Psi_{n\mathbf{k}}(\mathbf{r})$ of the band carrier satisfies Schrödinger's equation,

$$\left[-\frac{\hbar^2}{2m_0} \Delta + V_{\text{eff}}(\mathbf{r}) \right] \Psi_{n\mathbf{k}}(\mathbf{r}) = \varepsilon_{n\mathbf{k}} \Psi_{n\mathbf{k}}(\mathbf{r}),$$

which we will present in the form

$$\left[-\frac{\hbar^2}{2m_0} \Delta + V(\mathbf{r}) \right] \Psi_{n\mathbf{k}}(\mathbf{r}) = \varepsilon_{n\mathbf{k}} \Psi_{n\mathbf{k}}(\mathbf{r}) + \big[V(\mathbf{r}) - V_{\text{eff}}(\mathbf{r}) \big] \Psi_{n\mathbf{k}}(\mathbf{r}).$$

The discrete index "n" of the Bloch function $\Psi_{n\mathbf{k}}(\mathbf{r})$ in the case of two spin degenerate bands v and c has the corresponding values $(v, \pm 1/2)$ and $(c, \pm 1/2)$. Here v means the valence band, and c is the conduction band.

In the representation of the second quantization [99, 100], the Hamiltonian (1.2), with the basis of a one-electron wave function $\Psi_{n\mathbf{k}}(\mathbf{r})$, is expressed in the form

$$H = \sum_{n\mathbf{k}\sigma} \varepsilon_{n\mathbf{k}} a^+_{n\mathbf{k}\sigma} a_{n\mathbf{k}\sigma} + \sum_{n\mathbf{k},n'\sigma} W_{n\mathbf{k},n'\mathbf{k}} a^+_{n\mathbf{k}\sigma} a_{n'\mathbf{k}\sigma}$$

$$+ \frac{1}{2} \sum_{n_1\mathbf{k}_1 n_2\mathbf{k}_2 n'_1\mathbf{k}'_1 n'_2\mathbf{k}'_2 \sigma_1 \sigma_2} F\left(n_1\mathbf{k}_1, n_2\mathbf{k}_2; n'_1\mathbf{k}'_1, n'_2\mathbf{k}'_2\right)$$

$$\times a^+_{n_1\mathbf{k}_1\sigma_1} a^+_{n_2\mathbf{k}_2\sigma_2} a_{n'_2\mathbf{k}'_2\sigma_2} a_{n'_1\mathbf{k}'_1\sigma_1}. \tag{1.3}$$

Here $a^+_{n\mathbf{k}\sigma}$ and $a_{n\mathbf{k}\sigma}$ are the operators of the creation and destruction of the electron in the state $|n\mathbf{k}\sigma\rangle$, σ is the projection of the electron spin, and

$$W_{n\mathbf{k},n'\mathbf{k}'} = \int \Psi^*_{n\mathbf{k}}(\mathbf{r})(V - V_{\text{eff}})\Psi_{n'\mathbf{k}'}(\mathbf{r})\,\mathrm{d}\tau,$$

$$F\left(f_1, f_2; f'_1, f'_2\right) = \int \Psi^*_{f_1}(1)\Psi^*_{f_2}(2)\frac{e^2}{r_{12}}\Psi_{f'_1}(1)\Psi_{f'_2}(2)\,\mathrm{d}\tau_1\,\mathrm{d}\tau_2, \tag{1.4}$$

where $f \equiv (n, \mathbf{k})$. Calculations of the matrix elements using Bloch functions are found in many works, for example, in [101, 102]. For the calculation of the energy of exciton formation in a simple two-band model, it is sufficient to preserve only the diagonal terms in the Hamiltonian (1.3) that contain a similar number of creation and destruction operators of the electrons in each of the bands [98]. Here the terms pertain to $n_1 = n'_1 = c$ and $n_2 = n'_2 = c$; $n_1 = n'_1 = v$ and $n_2 = n'_2 = v$; $n_1 = n'_1 = c$ and $n_2 = n'_2 = v$; $n_1 = n'_1 = v$ and $n_2 = n'_2 = c$. Among the remaining possibilities it is necessary to select only those for which the spin indices at wave functions depending on the same variable are similar. The same relates to the spin indices of the operators.

The Hamiltonian (1.3) may be simplified by using the properties of the symmetry of the coefficients (1.4),

$$F\left(v\mathbf{k}_1, c\mathbf{k}_2; c\mathbf{k}'_1, v\mathbf{k}'_2\right) = F\left(c\mathbf{k}_2, v\mathbf{k}_1; v\mathbf{k}'_2, c\mathbf{k}'_1\right), \tag{1.5}$$

and the redesignation of the indices $\mathbf{k}_1 \rightleftarrows \mathbf{k}_2$ and $\mathbf{k}'_1 \rightleftarrows \mathbf{k}'_2$ under the summation sign, and by discarding the terms containing operators of the types $a^+_c a^+_c a_v a_v$, $a^+_v a^+_v a_c a_c$, $a^+_c a^+_c a_c a_v$, and $a^+_c a^+_v a_v a_c$, which do not preserve a number of quasi-particles of the given type c and v.

After the transition to the creation and destruction operators of the hole in the valence band

$$b^+_{\mathbf{k}\sigma} \equiv a_{v,-\mathbf{k},-\sigma}, \qquad b_{\mathbf{k}\sigma} \equiv a^+_{v,-\mathbf{k},-\sigma},$$

the Hamiltonian (1.3) may be written in the electron–hole representation taking into account that, due to translational symmetry, the conservation law of impulses applies. In this case the Hamiltonian will not have a normal-ordered form, and in order for the operators in the electron–hole representation to become normally ordered, it is necessary to use the Fermi commutation relations. The terms which do not contain operators in the normal-ordered Hamiltonian can be added, and they determine the energy E_0 of the ground state of the crystal:

$$E_0 = 2 \sum_{\mathbf{k}} \varepsilon_v(\mathbf{k}) + 2 \sum_{\mathbf{k}} W_{v\mathbf{k}} + 2 \sum_{\mathbf{k}_1 \mathbf{k}_2} F(v\mathbf{k}_1, v\mathbf{k}_2; v\mathbf{k}_1, v\mathbf{k}_2)$$
$$- \sum_{\mathbf{k}_1 \mathbf{k}_2} F(v\mathbf{k}_1, v\mathbf{k}_2; v\mathbf{k}_2, v\mathbf{k}_1).$$

As a result, we see that E_0 contains the Coulomb and exchange energy of interaction of the valence electrons. Using the properties of symmetry

$$F(\mathbf{k}_1, \mathbf{k}_2; \mathbf{k}_1, \mathbf{k}_2) = F(\mathbf{k}_2, \mathbf{k}_1; \mathbf{k}_2, \mathbf{k}_1),$$

one may write two expressions:

$$H_1 = \sum_{\mathbf{k}\sigma} \left[W_{c\mathbf{k}} + 2 \sum_{\mathbf{k}'} F(c\mathbf{k}, v\mathbf{k}'; c\mathbf{k}, v\mathbf{k}') \right.$$
$$\left. - \sum_{\mathbf{k}'} F(c\mathbf{k}, v\mathbf{k}'; v\mathbf{k}', c\mathbf{k}) \right] a^+_{\mathbf{k}\sigma} a_{\mathbf{k}\sigma}$$

and

$$H_2 = \sum_{\mathbf{k}\sigma} \left[W_{v\mathbf{k}} + 2 \sum_{\mathbf{k}'} F(v\mathbf{k}, v\mathbf{k}'; v\mathbf{k}, v\mathbf{k}') \right.$$
$$\left. - \sum_{\mathbf{k}'} F(v\mathbf{k}, v\mathbf{k}'; v\mathbf{k}', v\mathbf{k}) \right] b^+_{\mathbf{k}\sigma} b_{\mathbf{k}\sigma},$$

which correspond to the differences of the energies of the unrenormalized quasi-particles and band carriers $W_{i\mathbf{k}}$ ($i = c, v$), as well as Coulomb and exchange interaction of band carriers with the valence electrons. These expressions may be changed to zero by self-consistent selection of the periodic potential $V_{\mathrm{eff}}(\mathbf{r})$ and wave functions of band carriers, after which the Hamiltonian may be written without a constant term E_0:

$$\tilde{H} = \sum_{\mathbf{k}} \varepsilon_e(\mathbf{k}) a^+_{\mathbf{k}\sigma} a_{\mathbf{k}\sigma} + \sum_{\mathbf{k}} \varepsilon_h(\mathbf{k}) b^+_{\mathbf{k}\sigma} b_{\mathbf{k}\sigma}$$
$$+ \frac{1}{2} \sum_{\mathbf{k}_1 \mathbf{k}_2 \mathbf{k}'_1 \mathbf{k}'_2 \sigma_1 \sigma_2 (\mathbf{k}_1 + \mathbf{k}_2 = \mathbf{k}'_1 + \mathbf{k}'_2)} F(c\mathbf{k}_1, c\mathbf{k}_2; c\mathbf{k}'_1, c\mathbf{k}'_2) a^+_{\mathbf{k}_1 \sigma_1} a^+_{\mathbf{k}_2 \sigma_2} a_{\mathbf{k}'_2 \sigma_2} a_{\mathbf{k}'_1 \sigma_1}$$
$$+ \frac{1}{2} \sum_{\mathbf{k}_1 \mathbf{k}_2 \mathbf{k}'_1 \mathbf{k}'_2 \sigma_1 \sigma_2 (\mathbf{k}_1 + \mathbf{k}_2 = \mathbf{k}'_1 + \mathbf{k}'_2)} F(v\mathbf{k}'_1, v\mathbf{k}'_2; v\mathbf{k}_1, v\mathbf{k}_2) b^+_{\mathbf{k}_1 \sigma_1} b^+_{\mathbf{k}_2 \sigma_2} b_{\mathbf{k}'_2 \sigma_2} b_{\mathbf{k}'_1 \sigma_1}$$

$$+ \sum_{\mathbf{k}_1\mathbf{k}_2\mathbf{k}_1'\mathbf{k}_2'\sigma_1\sigma_2(\mathbf{k}_1+\mathbf{k}_2=\mathbf{k}_1'+\mathbf{k}_2')} F\left(c\mathbf{k}_1, v\mathbf{k}_2; c\mathbf{k}_1', v\mathbf{k}_2'\right)$$

$$\times a_{\mathbf{k}_1\sigma_1}^+ b_{-\mathbf{k}_2',-\sigma_2}^+ b_{-\mathbf{k}_2,-\sigma_2} a_{\mathbf{k}_1'\sigma_1}$$

$$+ \sum_{\mathbf{k}_1\mathbf{k}_2\mathbf{k}_1'\mathbf{k}_2'\sigma_1\sigma_2(\mathbf{k}_1+\mathbf{k}_2=\mathbf{k}_1'+\mathbf{k}_2')} F\left(c\mathbf{k}_1, v\mathbf{k}_2; v\mathbf{k}_1', c\mathbf{k}_2'\right)$$

$$\times a_{\mathbf{k}_1\sigma_1}^+ b_{-\mathbf{k}_1',-\sigma_1}^+ b_{-\mathbf{k}_2,-\sigma_2} a_{\mathbf{k}_2'\sigma_2}. \tag{1.6}$$

Here

$$\varepsilon_e(\mathbf{k}) = \varepsilon_c(\mathbf{k}), \qquad \varepsilon_h(\mathbf{k}) = -\varepsilon_v(\mathbf{k}).$$

For large-radius exciton states and a low-density exciton gas, the first three matrix elements of (1.6) are equal among themselves and are determined by the expression [98]

$$F\left(c\mathbf{k}_1, c\mathbf{k}_2; c\mathbf{k}_1', c\mathbf{k}_2'\right) = F\left(v\mathbf{k}_1, v\mathbf{k}_2; v\mathbf{k}_1', v\mathbf{k}_2'\right)$$

$$= F\left(c\mathbf{k}_1, v\mathbf{k}_2; c\mathbf{k}_1', v\mathbf{k}_2'\right) = V_{\mathbf{k}} = \frac{1}{\mathcal{V}} \frac{4\pi e^2}{\mathbf{k}^2}, \tag{1.7}$$

where $|\mathbf{k}| = |\mathbf{k}_1 - \mathbf{k}_1'| = |\mathbf{k}_2 - \mathbf{k}_2'|$ and $\mathcal{V}$ is the volume of the crystal.

The fourth matrix element is expressed through the terms of dipole–dipole interaction, and for small transfer of impulses for small $\mathbf{k}$ (when $\mathbf{k} \to 0$, see below) it has the form [98]

$$F\left(c\mathbf{k}_1, v\mathbf{k}_2; v\mathbf{k}_1', c\mathbf{k}_2'\right) = \frac{1}{\mathcal{N}}\left[V(\boldsymbol{\xi}; \mathbf{d}_{cv}) + A_{cv}\right]\delta_{\mathbf{k}_1+\mathbf{k}_2, \mathbf{k}_1'+\mathbf{k}_2'}, \tag{1.8}$$

where

$$V(\boldsymbol{\xi}, \mathbf{d}_{cv}) = \lim_{\varkappa \to 0} \frac{1}{\mathcal{V}_0} \int e^{i\varkappa\boldsymbol{\xi}\mathbf{R}}\left[\frac{|\mathbf{d}_{cv}|^2}{R^3} - \frac{3(\mathbf{d}_{cv}\mathbf{R})(\mathbf{R}\mathbf{d}_{cv}^*)}{R^5}\right]d\mathbf{R},$$

$\mathcal{N}$ is the full number of elementary cells in the crystal, and $\mathcal{V}_0 = \mathcal{V}/\mathcal{N}$ is the volume of the crystal elementary cell.

In (1.4) and (1.5) the vector of the electric dipole moment of transition $\mathbf{d}_{cv} = e\mathbf{X}_{cv} = e\mathbf{X}_{vc}^*$, the orth $\boldsymbol{\xi}$ of the wave vector $\mathbf{k}$ ($\mathbf{k} = \varkappa\boldsymbol{\xi}$), and the vector $\mathbf{R} = \mathbf{r}_l - \mathbf{r}_m - (\boldsymbol{\rho}_l - \boldsymbol{\rho}_m)$ are introduced. Here $\mathbf{r}_i$ ($i = l, m$) is the radius vector of the ith electron counted from an arbitrary lattice site and $\boldsymbol{\rho}_i$ ($i = l, m$) is the radius vector of the same electron counted from the ith lattice site. The magnitude A_{cv} from (1.8) is determined by the expression

$$A_{cv} = \frac{1}{\mathcal{V}_0} \int d\boldsymbol{\rho}_m \frac{1}{\mathcal{V}_0} \int d\boldsymbol{\rho}_l \frac{e^2 e^{i\mathbf{k}(\boldsymbol{\rho}_l - \boldsymbol{\rho}_m)}}{|\boldsymbol{\rho}_l - \boldsymbol{\rho}_m|} U_{n_1\mathbf{k}_1}^*(\boldsymbol{\rho}_m) U_{n_1'\mathbf{k}_1'}(\boldsymbol{\rho}_m)$$

$$\times U_{n_2\mathbf{k}_2}^*(\boldsymbol{\rho}_l) U_{n_2'\mathbf{k}_2'}(\boldsymbol{\rho}_l),$$

where ρ_l and ρ_m vary in the limits of the elementary cell and e is the electron charge. Here $U_{nk}(\boldsymbol{\rho})$ is the periodic part of the Bloch function

$$\Psi_{nk}(\boldsymbol{\rho}) = \frac{1}{\sqrt{\mathcal{V}}} U_{nk}(\boldsymbol{\rho}) e^{ik\rho}.$$

The functions $U_{nk}(\boldsymbol{\rho})$ are normalized to the volume of an elementary cell of the crystal

$$\frac{1}{\mathcal{V}_0} \int U_{nk}^* U_{nk}\, d\tau = 1 \tag{1.9}$$

and satisfy the translational symmetry condition

$$U_{nk}(\boldsymbol{\rho} + \mathbf{R}) = U_{nk}(\boldsymbol{\rho}).$$

The value $V(\boldsymbol{\xi}; \mathbf{d}_{cv})$ depends on the orientation of vectors $\boldsymbol{\xi}$ and $\mathbf{d}_{cv}$. This dependence is the sequence of nonanalytical dipole–dipole interactions known as Coulomb far action:

$$V(\boldsymbol{\xi} \parallel \mathbf{d}_{cv}) = \frac{8\pi |\mathbf{d}_{cv}|^2}{\mathcal{V}_0} \sqrt{\frac{\pi}{2}} \int_0^\infty \frac{J_{5/2}(y)}{y^{3/2}}\, dy = \frac{8\pi |\mathbf{d}_{cv}|^2}{3\mathcal{V}_0},$$

$$V(\boldsymbol{\xi} \perp \mathbf{d}_{cv}) = -\frac{1}{2} V(\boldsymbol{\xi} \parallel \mathbf{d}_{cv}) = -\frac{4\pi |\mathbf{d}_{cv}|^2}{3\mathcal{V}_0}. \tag{1.10}$$

Here $J_{5/2}(y)$ designates the Bessel function of the real argument [103].[2]

The far-acting Coulomb interaction of the electron and hole, which appears during the calculation of the exchange interaction of the electrons in the valence and the conduction bands can be considered the result of the virtual recombination and generation of the excitons [105, 106]. Therefore, in different works Coulomb far action is also called the exchange, resonance, or annihilation interaction [44].

Let us insert expressions (1.7) and (1.8) in (1.6) and make the following substitution of the indices of summation: $\mathbf{k}_1 = \mathbf{q}$, $\mathbf{k}_2 = \mathbf{q}'$, $\mathbf{k}_1' = \mathbf{q} - \mathbf{k}$, $\mathbf{k}_2' = \mathbf{q}' + \mathbf{k}$. Then the Hamiltonian (1.6) will have the final form [98]

$$\tilde{H} = \sum_{k\sigma} \varepsilon_e(\mathbf{k}) a_{k\sigma}^+ a_{k\sigma} + \sum_{k\sigma} \varepsilon_h(\mathbf{k}) b_{k\sigma}^+ b_{k\sigma}$$

$$+ \frac{1}{2} \sum_{qq'k\sigma\sigma'} V_k \big\{ a_{q\sigma}^+ a_{q'\sigma'}^+ a_{q'+k,\sigma'} a_{q-k,\sigma}$$

$$+ b_{q\sigma}^+ b_{q'\sigma'}^+ b_{q'+k,\sigma'} b_{q-k,\sigma} - 2 a_{q\sigma}^+ b_{q'\sigma'}^+ b_{q'+k,\sigma'} a_{q-k,\sigma} \big\}$$

$$+ \frac{1}{N} \sum_{qq'k\sigma\sigma'} \big[A_{cv} + V(\mathbf{k}; \mathbf{d}_{cv}) \big] a_{q\sigma}^+ b_{-q+k,-\sigma}^+ b_{-q',-\sigma'} a_{q'+k,\sigma'}.$$

---

[2] Another method of deducing the correlations (1.10) on the basis of equations from electrostatics is given in [104].

This is the initial Hamiltonian for investigating the energetic spectrum and different properties of Wannier–Mott excitons.

The wave function of the exciton in the second quantization representation is determined by the expression

$$|n\mathbf{K}, SM_S\rangle = \frac{1}{\sqrt{\mathcal{V}}} \sum_{\mathbf{q}\sigma\sigma'} \varphi_n(\mathbf{q}) a^+_{\alpha\mathbf{K}+\mathbf{q},\sigma} b_{\beta\mathbf{K}-\mathbf{q},\sigma'} C^{SM_S}_{\sigma\sigma'} |0\rangle,$$

where $\alpha = m_e/m_{ex}$, $\beta = m_h/m_{ex}$, $m_{ex} = m_e + m_h$, and $\mathbf{K}$ is the wave vector of the center of gravity of the exciton, $C^{SM_S}_{\sigma\sigma'}$ are spin coefficients, depending both on the projections of the electron and hole spins in the exciton, and on the summary exciton spin S and its projection M_S on the quantization axis ($\mathbf{S} = \mathbf{S}_e + \mathbf{S}_h$); $\varphi_n(\mathbf{q})$ is the wave function describing the relative electron–hole movement with a set of quantum numbers, designated by a single symbol n, which satisfies the correlation

$$\frac{1}{\mathcal{V}} \sum_{\mathbf{q}} |\varphi_n(\mathbf{q})|^2 = 1.$$

For crystals with simple (not orbitally degenerate) energy bands, it is possible to build four different spin functions for the exciton; one corresponds to the paraexciton ($S = 0$), but the other three correspond to the orthoexciton ($S = 1$). The average Hamiltonian $\tilde{H}$ from (1.6) and the application of these functions results in the following equation [98]:

$$\left[E_{n\mathbf{K}S} - 2\delta(S)\tilde{V}\left(\mathbf{K}, \mathbf{d}^{(n)}_{ex}\right)\right] \frac{1}{\mathcal{V}} \sum_{\mathbf{q}} |\varphi_n(\mathbf{q})|^2$$

$$= \frac{1}{\mathcal{V}} \sum_{\mathbf{q}} |\varphi_n(\mathbf{q})|^2 \left[\varepsilon_c(\alpha\mathbf{K} + \mathbf{q}) + \varepsilon_h(\beta\mathbf{K} - \mathbf{q})\right] - \frac{1}{\mathcal{V}} \sum_{\mathbf{q}\mathbf{q}'} \varphi^*_n(\mathbf{q}) \varphi_n(\mathbf{q}') V_{\mathbf{q}-\mathbf{q}'},$$

$$\tag{1.11}$$

where

$$\tilde{V}\left(\mathbf{K}; \mathbf{d}^{(n)}_{ex}\right) = \mathcal{V}_0 \left[A_{cv} + V(\mathbf{K}, \mathbf{d}_{cv})\right] |\varphi_n(\mathbf{r} = 0)|^2,$$

$$\delta(S) = \begin{cases} 1 & \text{at } S = 0, \\ 0 & \text{at } S = 1, M_S = 0. \end{cases}$$

In (1.11) $\mathbf{d}^{(n)}_{ex}$ is the electric dipole moment of transition from the ground state of the crystal to the nth exciton state, and by $E_{n\mathbf{K}S}$ we designate the energy of the exciton in the state $|n\mathbf{K}S\rangle$.

After the variation of expression (1.11) on $\varphi^*(\mathbf{q})$ and the transition from an impulse representation to a coordinate one, we obtain the following Schrödinger equation:

$$\left[\hat{\varepsilon}_e(\alpha\mathbf{K} - i\nabla) + \hat{\varepsilon}_h(\beta\mathbf{K} + i\nabla)\right]\varphi_n(\mathbf{r}) - \frac{e^2}{r}\varphi_n(\mathbf{r})$$

$$= \left[E_{n\mathbf{K}S} - 2\delta(S)\tilde{V}\left(\mathbf{K}; \mathbf{d}_{ex}^{(n)}\right)\right]\varphi_n(\mathbf{r}).$$

Expanding the operators $\hat{\varepsilon}_e$ and $\hat{\varepsilon}_h$ in a series on $(\alpha\mathbf{K} - i\nabla)$ and $(\beta\mathbf{K} + i\nabla)$ and neglecting terms of third and higher order, we obtain the equation

$$\left(\frac{\hbar^2 K^2}{2m_{ex}} + 2\delta(S)\tilde{V}\left(\mathbf{K}; \mathbf{d}_{ex}^{(n)}\right) - E_{n\mathbf{K}S}\right)\varphi_n(\mathbf{r})$$

$$- \frac{\hbar^2}{2\mu}\Delta\varphi_n(\mathbf{r}) - \frac{e^2}{r}\varphi_n(r) = 0, \qquad (1.12)$$

where μ is the reduced exciton mass.

Equation (1.12) corresponds to the simplest two-band model. However, one should consider that, when an electron and a hole form in a crystal, polarization of the atom shells takes place. Thus, the electron wave function must be built in the form of a linear combination of Bloch functions including not only the lowest conduction band, but also higher bands. Similarly, the hole wave function must contain the superposition of deeper hole bands. As was shown in the monograph [86], the calculation in this situation involves the introduction of the Coulomb potential, as well as the terms of the exchange and dipole–dipole interactions and the dielectric constant ε, which in general depend on $\mathbf{r}$.

The energy spectrum of the hydrogen-like exciton accounting for translational movement, Coulomb far action, and the resulting spin, after we solve (1.12), has the form

$$E_{n\mathbf{K}S} = E_g + \frac{\hbar^2\mathbf{K}^2}{2m_{ex}} + 2\delta(S)\tilde{V}\left(\mathbf{K}; \mathbf{d}_{ex}^{(n)}\right) - \frac{\text{Ry}^{ex}}{n^2}; \quad n = 1, 2, 3, \ldots,$$

where E_g is the energy gap and Ry^{ex} is the Rydberg effective constant (see (1.1)) equal to the ionization potential of the exciton, if we neglect Coulomb far action. For the same value of the principal quantum number n there are three levels (bands) of energy:

$$S = 1: \qquad {}^3E_n(\mathbf{K}) = E_g + \frac{\hbar^2\mathbf{K}^2}{2m_{ex}} - \frac{\text{Ry}^{ex}}{n^2},$$

$$S = 0, \mathbf{d}_{ex}^{(n)} \perp \mathbf{K}: \quad {}^1E_{n\perp}(\mathbf{K}) = {}^3E_n(\mathbf{K}) - \frac{8\pi}{3V_0\varepsilon}\left|\mathbf{d}_{ex}^{(n)}\right|^2 + 2A_{ex},$$

$$S = 0, \mathbf{d}_{ex}^{(n)} \parallel \mathbf{K}: \quad {}^1E_{n\parallel}(\mathbf{K}) = {}^3E_n(\mathbf{K}) + \frac{16\pi}{3V_0\varepsilon}\left|\mathbf{d}_{ex}^{(n)}\right|^2 + 2A_{ex},$$

where

$$A_{ex} = A_{cv} \cdot V_0\left|\varphi_n(\mathbf{r} = 0)\right|^2.$$

The longitudinal–transverse splitting of the exciton is determined by the formula

$$^1E_{n\parallel}(\mathbf{K}) - {}^1E_{n\perp}(\mathbf{K}) = \frac{8\pi}{\mathcal{V}_0\varepsilon}|\mathbf{d}_{\mathrm{ex}}^{(n)}|^2,$$

and decreases with the growth of the number n as $1/n^3$, because $|\mathbf{d}_{\mathrm{ex}}^{(n)}|^2 \sim |\varphi_n(0)|^2 \sim 1/n^3$. According to a similar law, it appears that the oscillator force of the transition from the ground state of the crystal to the exciton one also decreases. According to the phenomenological theory [85, 107] (see also Ref. [108]), in the annihilation interaction responsible for splitting electrons into longitudinal and transverse excitons, ε is the dielectric constant of the frequency of the exciton transition minus the contribution in ε connected with the excitation of the exciton itself, i.e., the optic dielectric constant ε_∞.

Thus, in the framework of the simplest two-band model it is possible to explain the main characteristic peculiarities of the electronic structure of the direct exciton: division of the movements in the exciton into the relative movement of the electron–hole and the translational movement of the center of gravity of the electron–hole pair, the spin structure of the exciton, and longitudinal–transverse splitting of the exciton by Coulomb far action. The indirect exciton may be considered analogically. However, calculating the real structure of crystals during the investigation of direct and indirect excitons leads to significant difficulties.

In the approximation of the effective mass method, exciton formation, accounting for the real structure of the crystal, is described as a result of the interaction of two electrons, one of which is in the conduction band near the extreme $\mathbf{K}_c$ and the other in the valence band with the extreme $\mathbf{K}_v$ (for a direct exciton, $\mathbf{K}_c = \mathbf{K}_v = \mathbf{K}_0$). Incidentally, each of the bands has a corresponding degeneration at the extreme point [44]. Thus, for the orbitally degenerate exciton energy bands the energy spectrum becomes complicated and depends on the constants of the spin–orbital interaction in the bands. The effects of the exchange interaction in the degenerated bands are also complicated if the structure of the crystal is taken into account.

In [108] the operator of the exchange interaction of the electron and hole in the exciton was obtained for an arbitrary degeneration degree of the bands which has a local character and is determined only by the short-acting part of the Coulomb potential. The exchange splitting of the exciton according to the order of magnitude is $(a_0/a_{\mathrm{ex}})^3$ times less than the corresponding atomic splitting (a_0 is the lattice constant). In that work it was also shown that the calculation of the exchange interaction between the electrons of the valence and conduction bands in the **kp** approximation (i.e., the calculation of only the far-acting part of the electron–hole interaction) in the case of free electrons does not lead to exchange splitting, but to splitting into longitudinal and transverse excitons (annihilation interaction). In contrast, in the case of bound excitons, as shown in [109], the annihilation interaction contributes to the exchange splitting. Let us note that, apart from the usual display of the exchange splitting of the exciton states, in the case of a large exchange interaction (untypical for semiconductors), interference of the discrete exciton state with a continuous spectrum of band carriers is possible (Fano's antiresonance effect [110]), if the ex-

citon binding energy is less than that of spin–orbital splitting of the valence band
[111].

Exchange splitting of the exciton lines in semiconductors was first observed
experimentally in 1960 [112]. As a result, the exchange interaction in Wannier–
Mott excitons has been well studied experimentally [112–123] and theoretically
[44, 86, 105, 108, 109, 124, 125].

1.2 Classification of the Exciton States

Symmetry theory has wide applications in solid state physics. In particular, it allows
one to conduct a strict classification of the exciton states in a crystal and to establish
important peculiarities of the structure of the exciton energy spectrum. The methods
of investigation of the symmetry properties of different systems are based on group
theory and are described in a number of monographs [44, 126–138]. In fact, all the
physical applications of group theory are based on the theory of representations.

Irreducible representations play an especially important role in the physical applications, since the wave functions being transformed according to these representations correspond to the same energy eigenvalues. The exciton is not an exception to these rules. The exciton states must also be transformed according to
certain irreducible representations of the symmetry group of the crystal (the symmetry point group for the direct-band exciton and the group of the wave vector $\mathcal{G}_{\mathbf{K}}$
for the indirect-band exciton). This section considers the symmetry properties of
the ground state of the crystal and the appointed type of the excited states that are
used to construct the exciton wave functions. Note, first of all, that the invariance
of the ground state of the crystal relative to the transformations of its symmetry
group follows from the physical considerations. This conclusion is proved by direct
calculations as well [98].

The symmetry transformation of the Hamiltonian is determined as a linear transformation of the coordinates of the particles, making this Hamiltonian invariant.
Let $\mathbf{r}_i$ and $\mathbf{R}_j$ be the coordinates of the electrons and nuclei of the crystal forming
the space of the configurations. We shall consider only those transformations that
change the coordinates of the electrons or, in other words, that change the coordinates of both the electrons and the nuclei at the same time, after which they return
the nuclei to the initial positions in the nodes of the crystalline lattice. In the case
of simultaneous "turnover" of both the electrons and the nuclei that designate the
turnover of the crystal as a whole, it is impossible to reveal any symmetry properties characteristic to the crystal itself. If the nuclei are shifted from their positions
of equilibrium, then the symmetry operation will give their shifts to other nuclei,
together with all the degrees of freedom, including the spin degree of freedom—but
the nuclei themselves usually return to their initial positions.

This definition of the symmetry operation is applied in [98] to find the symmetry properties of the exciton wave functions. It was also applied by one of the
authors during the investigation of hyperfine and spin–phonon interaction in local

electron centers by the method of invariants [139]. Note that for regular representations this definition apparently is close to the sense of the mathematical definition of the Q-symmetry transformation [140] (or $\bar{P}$-symmetry [141]) consisting in the isometric reflection $g^{(p)} = pg$ of the preliminary "indexed" figure $F^{(N)}$ on itself ($N = \{1, 2, \ldots, m\}$), in which the transformation g acts on the points of the figure and on the indices on the previously given law, which doesn't depend on the points, but where p is the compensating substitution of the indices that supplies the mapping of the figure $F^{(N)}$ on itself. This substitution belongs to the fixed group of substitutions of indices $1, 2, \ldots, m$, designating m qualities of a general nature, that were attributed before the transformation at least once to all of the points of the figure $F^{(N)}$.

The wave function of the exciton with wave vector $\mathbf{K}$ and the function of the relative electron–hole movement in the impulse representation $\varphi_{\Gamma_n m}(\mathbf{q})$ has the form [98]

$$\Psi_{\mathbf{K}, \Gamma_n, m, \Gamma_c, i, \Gamma_v, j} = \sum_{\mathbf{q}} \varphi_{\Gamma_n m}(\mathbf{q}) \Psi(\alpha \mathbf{K} + \mathbf{q}, \Gamma_c, i; \beta \mathbf{K} - \mathbf{q}, \Gamma_v, j). \tag{1.13}$$

Under the actions of the symmetry operation, the exciton wave functions are transformed in the following way:

$$G\Psi_{\mathbf{K}, \Gamma_n, m, \Gamma_c, i, \Gamma_v, j} = \sum_{m'} G^{\Gamma_n}_{m'm} \sum_{i'} G^{\Gamma_c}_{i'i} \sum_{j'} G^{\Gamma_v^*}_{j'j} \Psi_{G\mathbf{K}, \Gamma_n, m', \Gamma_c, i', \Gamma_v, j'},$$

$$G \in \mathcal{G}, \tag{1.14}$$

where $\mathcal{G}$ is one of the 32 symmetry point groups (corresponding to the crystalline classes of the crystals whose exciton properties are investigated) for direct excitons or the group $\mathcal{G}_{\mathbf{K}}$ of the wave vector $\mathbf{K}$ ($G\mathbf{K} = \mathbf{K}$) for indirect excitons.

According to formula (1.14), the representation Γ (on the basis of which the exciton wave function (1.13) is transformed) is equal to the direct product of the representations Γ_n, Γ_c, and Γ_v^* on which the function of the relative electron–hole movement in the exciton $\varphi_{\Gamma_n m}(\mathbf{q})$ and the wave function of the excited state of the crystal's corresponding electron–hole pair with wave vectors $\mathbf{k}_e$ and $\mathbf{k}_h$, $\psi(\Gamma_c, \mathbf{k}_e, i; \Gamma_v, \mathbf{k}_h, j)$ are transformed. The function $\psi(\Gamma_c, \mathbf{k}_e, i; \Gamma_v, \mathbf{k}_h, j)$ is similar to the product of two one-electron Bloch functions $\psi_{\Gamma_c, \mathbf{k}_e, i}(\mathbf{r}_1) \cdot \psi^*_{\Gamma_v, \mathbf{k}_h, j}(\mathbf{r}_2)$. Thus, the exciton wave functions are transformed according to the representation

$$\Gamma = \Gamma_n \otimes \Gamma_c \otimes \Gamma_v^*.$$

Fulfilling the decomposition of the reducible representation Γ on the irreducible representations of the symmetry group $\mathcal{G}$, it is possible to classify the exciton states by the symmetry types depending on the symmetry of the conduction and valence bands in the actual points of the Brillouin zone and on the symmetry function of the relative movement of the electron–hole in the exciton.

Knowing the wave functions of the excitons allows us to investigate their individual and collective properties. Herewith, the Lenard–Dyson theorem is important in understanding the energetic stability (H-stability) of a substance [142, 143] (see also [144, 145]) and in turn understanding the collective effects in a system of high-density excitons.

1.3 Lenard–Dyson Theorem

Let us consider the stability of the ground state of a multiple system following the works [142, 143]. The problem of the stability of matter has been posed by Fisher and Ruelle [146]. The mathematical model for "matter" is a system of point particles, finite in number, obeying the laws of nonrelativistic quantum mechanics, and interacting with each other only by Coulomb forces. The word "stability" is used to mean that there exists a lower bound for the energy per particle which is independent of the state and the size of the system. The problem of stability for single atoms was solved quantitatively in 1925 in the framework of wave mechanics. It became clear that an atom with a nuclear charge Ze and Z electrons of charge $-e$ could not have an energy state lower than $-Z^3$Ry, where Ry $= me^4/2\hbar^2$ is the natural atomic energy unit (Rydberg's constant) formed from the fundamental constants m, e, and $\hbar$. However, bulk matter consists of a very large number of positively and negatively charged particles, attracting and repulsing each other by the Coulomb force. The effects of the Coulomb force include such diverse phenomena as chemical binding, metallic cohesion, van der Waals forces, superconductivity, and superfluidity. The stability of matter does not depend on non-Coulombian forces (e.g., nuclear forces, magnetic dipole interactions, retardation and relativistic effects, and radiative corrections). These forces contribute to very small corrections of the binding energies of atoms and molecules.

According to [142], we can introduce the following definition of stability. Let the Hamiltonian operator of $N \geq 2$ charged particles be

$$H_N = \sum_{j=1}^{N}\left(-\frac{\hbar^2}{2m_j}\Delta_j\right) + \sum_{1 \leq i < j \leq N}\frac{e_i e_j}{|\mathbf{r}_i - \mathbf{r}_j|}.$$

Here the standard notation is used; the charges e_j may have either sign. We write

$$E_{\min}(N, e, m) = \mathrm{Inf}\langle\psi, H_N\psi\rangle,$$

where the minimization operation is taken with respect to all N-particle wave functions $\psi = \psi(\mathbf{r}_1, \mathbf{r}_2, \ldots, \mathbf{r}_N)$ normalized according to $\langle\psi|\psi\rangle = 1$, with all values of the masses satisfying

$$0 < m_j < m, \tag{1.15}$$

and all values of the charges satisfying

$$-e \leq e_j \leq e. \tag{1.16}$$

If there is a numerical constant A such that, for all N,

$$E_{\min} > -AN\mathrm{Ry} \quad \left(\mathrm{Ry} = m_{\mathrm{e}}^4/2\hbar^2\right), \tag{1.17}$$

we consider the system to be stable.

In this definition the statistics of the particles is not mentioned. The complete statement of stability and instability involves a specification of the particles. In that case the constant A may depend on the number and kind (in the sense of statistics) of different particle species.

Theorem *Let N negatively charged particles belong to q different fermion species. Let their masses and charges be subject to (1.15) and (1.16), respectively. Let an arbitrary number of positively charged particles be subject to the sole restriction (1.16) on their charges, their statistics and their masses being arbitrary. Then*

$$E_{\min} > -Aq^{2/3}N\mathrm{Ry}, \tag{1.18}$$

where A is an absolute constant.

Note that the constants A appearing in (1.17) and (1.18) are not the same. Unfortunately, the proof of this theorem is lengthy and complicated [143]; therefore, it is not given here.

It is a remarkable fact that Lenard and Dyson's proof of the stability of matter, after a few artificial tricks, comes down in the end to an estimate of the binding energy of a single electron in a periodic Coulomb potential. This appearance of the periodic potential at the kernel of the proof is not accidental. The ground states of most forms of matter are crystals in which electrons are actually moving in periodic Coulomb potentials. According to Lenard and Dyson [143], the essence of the proof of the stability of matter should be a demonstration that an aperiodic arrangement of the particles cannot give greater binding than the periodic arrangement. If this dominance of the periodic potential could be proved directly, then we would have a vastly simpler and more satisfactory proof of the Lenard–Dyson theorem than the one presented in [143].

The Lenard–Dyson theorem has a fundamental significance, as it solves the problem of the stability of a substance in macroscopic volumes, consisting of a large number of positively and negatively charged particles interacting according to Coulomb's law. The effects conditioned by Coulomb interaction are very fine and extremely diverse. Dyson and Lenard showed that the property of saturation is a general property characteristic to all Coulomb effects, so the binding energy falling on one particle always remains limited. One of the proofs of these general conclusions is the variety of possible collective effects and new states in a system of high-density excitons.

1.4 Effects of Exciton–Exciton Interaction

An exciton gas of low density exhibits individual exciton properties. With increasing exciton concentration n_{ex}, the average distance between excitons decreases. During rather high levels of crystal excitation, which are usually achieved by means of a laser source or a powerful electron beam, this distance may be compared with the radius of the exciton ($n_{ex}a_{ex}^3 \lesssim 1$) or an even smaller value ($n_{ex}a_{ex}^3 \gtrsim 1$). In this situation the Fermi nature of the electron and hole is displayed; they form an exciton, which is not a boson anymore, and now the properties of the system are determined by the interaction between all electrons and holes.

The exciton creation operator $A_{K,n}^+$ is determined through Fermi operators of the electron a_p^+ and hole b_p^+ creation according to the formula [147]

$$A_{K,n}^+ = \frac{1}{\sqrt{V}} \sum_{pp'} \delta_{K,p+p'} \varphi_n(q) a_p^+ b_{p'}^+, \tag{1.19}$$

where $q = \beta p - \alpha p'$, with $\alpha = m_e/(m_e + m_h)$ and $\beta = m_h/(m_e + m_h)$, and where

$$\varphi_n(q) = \int dr\, \varphi_n(r) e^{-iqr}$$

is the Fourier transform of the function of the relative motion of the electron–hole in the exciton.

Taking into consideration formula (1.19) and the similar formula for $A_{K,n} = (A_{K,n}^+)^+$, we have the following commutation relations for the operators of exciton creation and destruction:

$$[A_{K,n}, A_{K',n'}] = 0,$$

$$[A_{K,n}^+, A_{K',n'}^+] = 0,$$

$$[A_{K,n}, A_{K,n}^+] = \delta_{KK'}\delta_{nn'} - \frac{1}{V} \sum_l \varphi_{n'}^*(\alpha K' - l')\varphi_n(\alpha K - l) a_{K-l}^+ a_{K'-l}$$

$$- \frac{1}{V} \sum_l \varphi_{n'}^*(1 - \beta K')\varphi_n(1 - \beta K) b_{K-l}^+ b_{K'-l} \tag{1.20}$$

$$= \delta_{KK'}\delta_{nn'} + O(n_{ex}a_{ex}^3).$$

From the third relation of commutation in (1.20), the excitons will act as bosons as long as $n_{ex}a_{ex}^3 \ll 1$, i.e., if the interexciton spacing is larger compared with the exciton Bohr radius a_{ex}.

The simplest display of the interaction between excitons is their heating and the discharge of a quasi-equilibrated exciton thermostat. The increase in heating of an exciton subsystem with the growth of the level of excitation was observed in the emission spectra of the crystals CdS [148, 149], CdSe [150], Cu_2O [151], and others. A convincing experimental proof of the fact that the distribution of hot excitons on the kinetic quasi-momenta is quasi-equilibrated is found when the temperature of the phonons is controlled using the form of the band I_1, conditioned by the radiative recombination of the excitons bounded on the neutral acceptors, while the temperature of the excitons is determined from the spectra of exciton–phonon luminescence. The violation of the hot equilibrium between the excitons and phonons is influenced by the fact that, at high exciton densities, the scattering of excitons on acoustic phonons (which is the main mechanism of the longitudinal relaxation of excitons near the bottom of the band when $n_{ex}a_{ex}^3 \ll 1$) ceases to be the most rapid process of relaxation versus the energy near the bottom of the band.

Exciton–exciton interaction causes new bands of luminescence: $P(1)$, $P(2)$, $P(3), \ldots, P(\infty)$ that have been observed in crystals at high exciton concentrations. The P-band was observed for the first time in the crystal CdS in the process of its excitation by an electron beam [152, 153] and in the crystals CdS, CdSe, and ZnO during intensive laser excitation. This band is shifted in the long-wave part relative to the zero-phonon radiation line of the free exciton by a magnitude roughly equal to the exciton binding energy. It is influenced by the radiative recombination of two direct free excitons due to exciton–exciton interaction, as a result of which one of the excitons is annihilated and gives a part of its energy to the formation of another exciton. If the remaining exciton passes into state n of the relative motion, then the corresponding band is $P(n)$. The band of radiation observed in [152–154] is interpreted as the band $P(\infty)$. The simultaneous presence of bands $P(2)$, $P(3)$, $P(4)$, and $P(\infty)$ was revealed for exciton series A of CdS and CdSe crystals [155].

The rapid development of the physics of high-density excitons, along with the effective application of lasers for excitation of the crystals, does not permit us, in a short discussion, to enumerate the enormous quantity of obtained and forthcoming scientific results in optic spectroscopy—not to mention their analysis and systematization. Therefore, we shall limit ourselves to the general state of things in this domain and indicate only the results of a general character that are sufficiently time-tested. A detailed analysis is given in the review articles [156–161], collections of articles [162–165], and monographs [84–89, 166–172].

With the growth of the exciton concentration, depending on the parameters of the crystal, the exciton system (at low temperatures) evolves over one of the following paths: the formation of molecules containing two excitons (biexcitons) [173, 174], the formation of electron–hole drops and electron–hole metallic liquid (condensation in the coordinate space) [97, 156, 175, 176], Bose–Einstein condensation of excitons and biexcitons (condensation in the space of kinetic quasi-momenta) [171, 177–182], the appearance of a condensate–photon mode similar to a coherent giant polariton of macroscopic amplitude [171, 183], and the formation of coherent excitons and photons [184, 185]. The last states appear during Bose–Einstein condensation of dipole-active excitons and photons in the resonance cavity or in the stationary field of the external resonance laser radiation [185]. Coherent polaritons also appear under the action of ultrashort impulses of resonance laser irradiation [186, 187].

The exciton molecule, the existence of which was forecast independently by Moskalenko [173] and Lampert [174], may be realized in the presence of the attractive forces between excitons. It represents by itself a rather friable formation of four fermions that move relative to each other and simultaneously participate in the translational movement of the excitons as a whole. From this point of view, the biexciton is similar to the hydrogen molecule, but unlike the hydrogen molecule, it has a short lifetime with respect to the processes of radiative and nonradiative annihilation.

When calculating the biexciton binding energy there are essential difficulties that are affected not only by the four-particle component of the biexciton itself, but also by the necessity of calculating the complex structure of the charge carrier energy bands, the anisotropy of their effective masses, the influence of the vibrations of the crystalline lattice, etc. For simple bands and isotropic effective masses of the charge carriers, the exciton binding energy is a monotonously decreasing function of the parameter $\sigma = m_e/m_h$ in the domain $0 \leq \sigma \leq 1$, where $\mathrm{Ry}^M = 0.298\,\mathrm{Ry}^{ex}$ for $\sigma = 0$ and $\mathrm{Ry}^M = 0.273\,\mathrm{Ry}^{ex}$ for $\sigma = 1$ [188, 189]. The conclusion that Ry^M must be a convex and monotonously changing function on σ in the interval $0 \leq \sigma \leq 1$ came from Adamovsky and others [190]. These authors, guided by work established by Wehner [191] on the basis of Feynman's theory for the energy of the ground state boundary condition

$$\left.\frac{\partial \mathrm{Ry}^M(\sigma)}{\partial \sigma}\right|_{\sigma=1} = 0,$$

obtained the inequalities

$$\frac{\partial \mathrm{Ry}^M(\sigma)}{\partial \sigma} \geq 0, \qquad \frac{\partial^2 \mathrm{Ry}^M(\sigma)}{\partial \sigma^2} \leq 0.$$

Thus, it was proved that, if the biexciton is stable at $\sigma = 1$, then it must be stable at all other values of σ.

Let us mention that the ratio of the energy of dissociation of a biexciton to the energy of exciton binding for $\sigma = 0$ is a value of the same order as the ratio of the energy of dissociation of a hydrogen molecule to the binding energy of a hydrogen atom. This inexact coincidence is affected by the difference of $m_e/m_h = 0$ from $m_e/m_p = 5.46 \cdot 10^{-4}$. At the same time, at $\sigma = 1$ this relation is much less, in connection with the effect of loosening of the action of zero vibrations on the molecular bond. For the biexciton, due to the lack of heavy particles, this effect is much more essential than in the usual molecules.

Due to the Pauli principle, the wave function of the biexciton must be antisymmetric relative to the permutation of the identical Fermi particle, and this puts certain limits on the accepted orientation of the electron and hole spins in the biexciton. In the initial works on the calculation of the biexciton binding energy, the biexciton was assumed to be formed from two singlet excitons in the state of relative movement of $1s$ type, during which the wave function of the biexciton satisfies Pauli's principle automatically.

In the works [166, 192, 193] the dependence of the energy of interaction between excitons on the orientation of the spins of electrons and holes in the model of two simple spin-degenerate bands in the absence of spin–orbit interaction was investigated. It was shown that the character of the interaction between excitons essentially depends on the symmetry of the spin wave functions relative to the transposition of the identical particles. If the distance Δ between the position of the levels of ortho- and paraexcitons (singlet–triplet splitting, caused by the intraexciton electron–hole exchange interaction) is bigger than the energy of interaction W between excitons ($\Delta > W$), then in the approximation $\mathbf{K}_1 = \mathbf{K}_2 = \mathbf{K}'_1 = \mathbf{K}'_2$ ($\mathbf{K}_i$ and $\mathbf{K}'_i$ are the ith exciton wave vectors before and after the elementary act of interaction, respectively) the Fourier transform of the energy of interaction of two paraexcitons, as well as that of an ortho- and a paraexciton, is positive, while in case of two orthoexcitons it is negative. Thus, the biexciton is formed from two orthoexcitons. In the opposite case ($\Delta < W$) the interaction between excitons breaks the exchange bond of the electron and hole spins in each exciton, and the precise wave functions of the interacting excitons [194] will be 16 functions, from which nine symmetric functions with respect to the simultaneous transposition of the spins of the electrons and holes (the values of the resulting spins are $S = 2, 1, 0$). There are six antisymmetric wave functions: three functions symmetric with respect to transposition of the electron spins and antisymmetric with respect to transposition of the hole spins and other three functions with the opposite symmetric properties. There is only one antisymmetric function with respect to the transposition of spins of electrons and holes, for which the resulting spin is $S = 0$. Calculations in the framework of the same model [166, 192, 193] showed that the Fourier transform of the energy of interaction of two excitons is negative only at antiparallel orientation of both electron spins and spins of the holes, and this orientation of the spins is preferable during the formation of biexcitons in this case.

If the Fourier transform of the exciton interaction energy is positive and the biexcitons are not formed, then with the growth of the exciton concentration, a short-wave shift of the exciton level takes place [182, 195]. The concentration shift of the exciton level accounting for different orientations of the electron and hole spins of the interacting excitons was considered in [168, 196]. Both blue [197–199] and red [200, 201] shifts of the exciton lines of absorption with the growth of the level of optical excitation have been observed experimentally in different crystals.

The authors of the works [166, 168, 202, 203], on the basis of the generalized Morse potential, and depending on the parameter σ, determined the location of the excited levels of the biexciton and showed that its vibrational and rotational states exist only at very small values of σ. This allowed one to find approximate expressions for the energy of the vibrational and rotational biexciton levels using the analogy between the biexciton and the hydrogen molecule [204, 205]. The first excited (rotational) state of the biexciton was also considered in [206, 207].

The calculation of the anisotropy of the electron and hole effective masses according to [208–210] does not lead to an essential change of the biexciton binding energy, but according to Ref. [211] this change may be essential. In [211] it was confirmed that for a large anisotropy the biexciton binding energy according to the order of magnitude may become comparable with the exciton binding energy, and the upper limit of the biexciton binding energy for types of anisotropy like anisotropic holes–anisotropic electrons was established. Analogical results were obtained during simultaneous calculation of the anisotropy of the effective masses of the electron and hole [211]. In [212] three general theorems of Gell-Mann–Feynman type for multiparticle stationary nonadiabatic systems were proved, setting up the qualitative peculiarities of the dependence of the binding energy of the nonadiabatic system on the parameter that appears in the kinetic energy operator of the Hamiltonian (for the biexciton this parameter is σ).

The exciton–phonon interaction has an essential influence on the processes of excitons binding into biexcitons, which for acoustic phonons leads to an effective attraction between excitons, but for optical phonons leads to their repulsion [213]. The contribution of the exciton–phonon interaction to the biexciton binding energy was studied for cases of similar [214] and different [215] masses of the carriers. In [216, 217] the probability of binding two excitons into a biexciton was calculated for interaction with acoustic and optic phonons under radiation of infrared light quanta and due to exciton–exciton collisions.

In a number of crystals, polariton effects have an essential influence on the biexcitons [218–222]. We note also the important role of the Fano antiresonance effect in biexciton spectroscopy in semiconductors [223].

As a whole, after the experimental discovery of biexcitons in the works of Nikitine and co-workers [224, 225], and the Japanese authors [226], it was established that the most favorable conditions for the formation of biexcitons are in the direct-band semiconductors. The most noncontradictory data on the existence of biexcitons have been accumulated for the CuCl and CuBr, although there are many works dedicated to biexcitons in other direct-band semiconductors. In indirect-band semiconductors, due to anisotropy and multivalley effects, there are essentially fewer possibilities for the formation of biexcitons. In connection with this, the detection by Kulakovsky, Timofeev, Edelstein [227] and Gourley and Wolfe [228] of the co-existence of electron–hole drops, excitons, and biexcitons in uniaxially deformed silicon is of special interest. There also exists information about the possibilities of the existence of biexcitons in unstrained indirect-band crystals [229]. The formation of a complex system from several free excitons [230] is not excluded.

The formation of electron–hole drops (EHDs) and an electron–hole liquid (EHL) of Fermi type in semiconductors at $na_{ex} \geq 1$ was discovered by Keldysh in 1968 [175]. At low temperatures, $k_0 T \ll \text{Ry}^{ex}$, the nonequilibrium charge carriers during a short period of time, 10^{-12}–10^{-11} s, are bound in excitons. At concentrations $n_{ex} \sim a_{ex}^{-3}$ the energy interaction between excitons becomes the same as the energy of the interaction of particles in the exciton, which in turn decreases itself due to shielding effects. Therefore, the behavior of a system of electrons and holes interacting according to Coulomb's law during an increase of the carrier concentration is similar to the behavior of a substance during an increase of pressure or density [97]. The EHL is similar to the gas compression at an increase of the density up to a degree where all the particles are maintained by internal forces at atomic distances from each other. The difference consists in the fact that the characteristic scale of the lengths and energies in this system are determined, not by the atomic parameters $a_B \sim 10^{-8}$ cm and Ry ~ 10 eV, but by the corresponding parameters of the separate excitons a_{ex} and Ry^{ex}. Therefore, the density of EHDs and EHL corresponds to the concentrations of excitons $n_{ex} \sim 10^{16}$–10^{18} cm^{-3}, in agreement with the results of experimental investigation of new radiation in semiconductors at intensive optical pumpings, caused by the existence of EHDs.

A new line of radiation which at first appeared at high levels of optical excitation was discovered by Haynes [231] in the silicon crystal, and it was attributed to radiative recombination from the biexciton state. Later this line was studied by Asnin and Rogachev [232]. Pokrovsky and Svistunova [233] detected new lines of radiation in germanium and indicated the existence of a sharp threshold for these lines with the change of temperature at a constant power of excitation. Asnin and Rogachev [234] brought proofs of the presence of the new phase in Ge based on the optical absorption. Vavilov, Zayats, and Murzin [235] observed the resonance absorption of far-IR radiation by metallic EHDs of small size in Ge and determined the concentration of the electron–hole pairs in these drops ($n_{e-h} = 2 \cdot 10^{17}$ cm^{-3}). Bagaev, Galkina, Gogolin, and Keldysh [236] showed that EHDs move in the crystal under the action of a pressure gradient.

Thus, after the theoretical prediction by Keldysh of the EHD state, the metallic phase in the crystals Ge and Si was observed, the density and binding energy of particles in this phase were determined [237], big charge pulsations during the dissociation of the drops in a strong electric field of $n-p$ transition with inverted displacement [238, 239] were detected, the light scattering on EHL [240] was investigated, the influence of the flow of nonequilibrium phonons (phonon wind) on EHDs [241] was established, etc. The state of the EHL was detected in the crystals Ge and Si, as well as in CdS [242], GaP [243], SiC [244], and in other semiconductors.

The theoretical and experimental investigations in this domain are reflected, respectively, in Keldysh's [156] and Pokrovsky's [157] reviews. A number of problems on the theory of the early phase of development have also been considered in the works [245–249]. A detailed analysis of the investigations in the field of EHL in semiconductors is given in the book by Rice, Hansel, Phillips, and Thomas [250].

If the idea of the existence of EHL in semiconductors was almost immediately confirmed experimentally, then, in quite another way, the investigation of the phenomenon of exciton condensation in the space of kinetic quasi-momenta was developed. After the possibility of Bose–Einstein condensation (BEC) of excitons was forecast independently by Moskalenko [177] and Blatt, Boer, and Brandt [178] in 1962, in 1970 Akopyan, Gross, and Razbirin's experimental work [251] was published. In this work for the first time the idea of BEC was used in the interpretation of the narrow line which appeared with the growth of the level of optic excitation in the domain of the LO-phonon replica of the exciton line in the radiation spectrum of the CdSe crystal. However, in this and many other later works (for example, some in which the crystals CdSe [252–254], CuCl [255–257], and CuBr [258] were investigated), the existence of the spontaneous BEC of excitons or biexcitons was not proved definitive experimentally.

The history of these experimental findings is full of dramatic situations. Thus, in Ref. [254] experimental data proving the existence of BEC of biexcitons in the crystal CdSe were given. Having been convinced of the fact that the temperature of biexcitons goes above the critical temperature earlier than the concentration of biexcitons achieves its critical value necessary for Bose condensation during excitation of the crystal by nanosecond impulses of a nitrogen laser, the authors of this work used light impulses from a neodymium picosecond laser. Exciting the crystal by a two-photon pump at a temperature of 1.8 K, they obtained a luminescence spectrum consisting of an extremely narrow line a width less than the slot of the spectrograph (0.17 meV) and a wide line with the maximum situated in the long-wave part of the narrow peak. The delta-form peak was attributed to the radiation from a Bose condensate of biexcitons in the state with $\mathbf{K} = 0$, but the wide band was considered to be affected by the overcondensate biexcitons, which corresponded well with the theory of BEC. However, very soon in [259] theoretical proofs of the impossibility of BEC of biexcitons in CdSe under the experimental conditions used in [254] were given. It appeared that the considered narrow peak was influenced by the decay of the biexciton in which one exciton is annihilated, but the other occurs in the polariton state, and in the radiation spectrum of the exciton molecule the polariton doublet is displayed, similarly to the way in which this happens in the crystal CuCl [255].

As these examples show, for a more effective recognition of the phenomenon of BEC of excitons and biexcitons, along with the optical property it is desirable to also have other characteristic properties of this phenomenon. In this book the possibility of using new (radio-frequency) indications of the phenomenon of BEC of excitons will be shown in Chap. 5.

Brodin and Matsko [260–262] investigated the possibility of BEC of excitons in the presence of the polariton effect. They used an experimental methodology which allowed them to show the contribution to the radiation, conditioned by luminescence and by Raman scattering of the light. The delta peak of luminescence in the crystal ZnTe that they observed at low temperatures was interpreted as the one conditioned by the BEC of polaritons in the upper branch in the state with $\mathbf{K} = 0$.

The theoretical aspects of the BEC phenomenon in exciton and biexciton systems were considered in many works [160, 166, 171, 177–182, 185, 187]. Elesin and Kopaev first studied the possibility of the induced BEC of excitons in a field of laser radiation [185]. The experimentally observed process of the direct creation of coherent biexcitons in a state with $\mathbf{K} = 0$ in the CuCl crystal by means of coherent photons coming from the opposite side [263] is probably an example of a similar phenomenon for biexcitons.

To conclude this section, we shall discuss the problem of crystallization of excitons and biexcitons. The possibility of exciton crystallization was considered for the first time by Keldysh [97], who showed that in the phase of the EHL, due to large zero vibrations, the spatial arrangement of the crystallization type cannot appear at any exciton temperatures and densities with $m_e = m_h$ ($\sigma = 1$), because the localization of excitons at distances of order a_{ex} in the hypothetical lattice led to the ionization of the exciton itself. In this regard the exciton system is distinguished from the usual substances, except for liquid helium, which remains liquid up to $T = 0$ at not very high pressures, namely due to large zero vibrations.

Nikitine [264] also considered the possibility of the spatial arrangement of the excitations. Using the analogy between a hydrogen atom and an exciton at $\sigma \ll 1$, and the existence of solid hydrogen in the form of the molecular crystal CRH_2, the author of [264] admits the possibility of crystallization of biexcitons in a "solid" biexciton lattice CR_{ex_2}. The estimations show that in CuCl and CuBr crystals a face-centered cubic lattice from parabiexcitons with the binding energy on a pair of particles exceeding the energy of zero vibrations can exist. According to [264] the formation of the hypothetical lattice from orthobiexcitons in these crystals is not excluded.

1.5 Excitons Captured by Isoelectronic Impurities

In the radiation spectra of semiconductors, along with the radiation of free exciton lines, one observes lines influenced by the radiative recombination of excitons captured on the impurity centers. Lampert [174] studied for the first time the possibility of the capture of excitons by small impurity centers with the formation of exciton–impurity complexes (EICs) in 1958, but in 1960 Haynes [265] found EICs in silicon. Later along with the exciton–impurity complexes on neutral centers in many crystals, excitons captured by the charged centers and isoelectron traps were also detected. The lines of radiation of all these EICs usually represent optical bands which are much more intensive than the lines of radiation of the free excitons.

Let us consider the binding of excitons on the isoelectron traps, following the work of Faulkner [266]. The binding of the exciton on the isoelectron acceptor takes place in such a way that at first the electron from the conduction band is attracted by the short-acting potential of the impurity, as a result of which a negatively charged system is formed, which further binds to it the hole from the valence band by long-range Coulomb interaction. Isoelectron donors act in the opposite direction, first binding the hole with the short-range potential of the impurity and already binding the electron by long-range Coulomb interaction.

The processes of binding of excitons on isoelectronic centers may be described much more simply in the framework of the one band-one node Slater–Koster model, which admits an analytical solution and contains only one adjusting parameter [266]. If the interaction between electrons in the crystal and isoelectronic impurity is described by the potential $V(\mathbf{r})$, then the matrix elements of the type $\langle n\mathbf{k}|V|n'\mathbf{k}'\rangle$ are calculated on the Bloch functions $|n\mathbf{k}\rangle$, which are contained in the Schrödinger's equation for the wave function of the binding state. This takes into consideration the interaction of the electron and hole with the impurity and Coulombic attraction of the electron and hole. This may all be represented by means of Wannier functions in the form

$$\langle n\mathbf{k}|V|n'\mathbf{k}'\rangle = \frac{\mathcal{V}_0}{(2\pi)^3}\sum_{\mathbf{R}\mathbf{R}'}\exp(\mathrm{i}\mathbf{k}\mathbf{R})\langle n\mathbf{R}|V|n'\mathbf{R}'\rangle\exp(-\mathrm{i}\mathbf{k}'\mathbf{R}'), \qquad (1.21)$$

where $\mathbf{R}$ is the vector of the direct lattice, and $\mathcal{V}_0$ is the volume of the elementary cell.

The Slater–Koster model consists in the fact that, for the short-range potential of the impurity (sharply decreasing at distances essentially smaller in comparison with the constant lattice), the main contribution in expression (1.21) is given by the terms with $\mathbf{R} = \mathbf{R}'$ (if potential $V(\mathbf{r})$ is centered at the beginning of the coordinates) and, also, that only one conduction band c and only one valence band v are considered:

$$\langle n\mathbf{k}|V|n'\mathbf{k}'\rangle = J_i \frac{V_0}{(2\pi)^3},$$

where

$$J_i = \langle n\mathbf{R} = 0|V|n\mathbf{R} = 0\rangle, \quad i = c, v.$$

In the case of exciton binding with an isoelectron acceptor the electron part of the wave function in the Slater–Koster approximation satisfies the equation

$$\left[E - \epsilon_c(\mathbf{k})\right]\Psi_e(\mathbf{k}) = J_c V_0 (2\pi)^{-3} \int d^3q\, \theta(\mathbf{q})\Psi_e(\mathbf{q}), \tag{1.22}$$

where the binding energy E and the kinetic energy $\epsilon_c(\mathbf{k})$ of the electron are counted from the bottom of the conduction band ($E < 0$). Here the function $\theta(\mathbf{q})$ is equal to unity, if the wave vector $\mathbf{q}$ lies in the first Brillouin zone, and is equal to zero if $\mathbf{q}$ goes out of the limits of the reduced Brillouin zone. From expression (1.22) one immediately obtains

$$1 + J_c \frac{V_0}{(2\pi)^3} \int d^3q\, \theta(\mathbf{q}) \frac{1}{\epsilon_c(\mathbf{q}) - E} = 0, \tag{1.23}$$

from which we can see that the approximation of the method of effective mass is not a good one for this problem. If we use the square law of dispersion for $\epsilon_c(\mathbf{k})$ and expand the integration in (1.23) on all $\mathbf{q}$-space, then the integral from (1.23) diverges. However, expression (1.23) may be transformed into the form

$$1 + J_c \left\langle \frac{1}{\epsilon_c} \right\rangle = -J_c \frac{V_0}{(2\pi)^3} \int d^3q\, \theta(\mathbf{q}) \frac{E}{\epsilon_c(\mathbf{q})(\epsilon(\mathbf{q}) - E)}, \tag{1.24}$$

where

$$\left\langle \frac{1}{\epsilon_c} \right\rangle = \frac{V_0}{(2\pi)^3} \int d^3q\, \theta(\mathbf{q}) \frac{1}{\epsilon_c(\mathbf{q})}.$$

Integration of the right-hand side of (1.24) may be spread to all $\mathbf{q}$-space; now the approximation of the method of effective mass can be applied, only if $|E|$ is small. Taking these remarks into account, we can obtain the following expression for the electron wave function on the basis of equation (1.22):

$$\Psi_e(\mathbf{k}) = \frac{(2m_e^*|E_1|)^{1/4}}{\sqrt{3}(2\pi m_e^*)} \cdot \frac{1}{\epsilon_c(\mathbf{k}) + |E_1|},$$

where m_e^* is the effective mass of the electron. The parameter $|E_1|$ of the wave function $\Psi_e(\mathbf{k})$ is determined by the expression

$$\frac{3V_0}{2\pi} m_e^* \left(2m_e^*|E_1|\right)^{1/2} = \frac{1 + J_c\langle\epsilon\rangle}{J_c}.$$

Similarly, we find the wave function of the hole $\Psi_h(\mathbf{K})$ (with the corresponding parameter $|E_2|$), and then for the integral cross section of the optical absorption in the indirect-band semiconductor we obtain

$$
\int \sigma(\omega)\, \mathrm{d}(\hbar\omega) = \frac{16}{3}(4\pi)^2\left(\frac{e^2}{\hbar c}\right) \cdot \frac{1}{n\hbar\omega_0}
$$

$$
\times |M|^2\left(\frac{E_\mathrm{g}+\Delta}{\Delta}\right)^2 \cdot \left(\frac{m_\mathrm{h}^*}{m_\mathrm{e}^*}\right)^{3/2} |E_1|^{1/2}|E_2|^{3/2},
$$

where $\hbar\omega_0 = E_\mathrm{g} - |E|$, E_g is the "indirect" gap, m_h^* is the effective mass of the hole, n is the refraction index of the crystal, $|M|$ is the matrix element of the optical transition, and Δ for semiconductors of type GaP with minima of the conduction band in the points Γ and X of the Brillouin zone is determined by the expression

$$
\Delta = \epsilon_c(\Gamma) - \epsilon_c(X).
$$

The states of single excitons bound on isoelectronic impurities in the crystals GaP:N, GaP:Bi, ZnTe:O, CdS:Te are investigated in a number of works [266–270]. The zero-phonon transitions in the luminescence decay of these states in the crystals, the valence bands of which are formed from p-type electron states, cause two radiation lines ($\mathcal{A}$ and $\mathcal{B}$), appearing at optical transitions from the states of the bound exciton with total angular momentum $J_t = 1$ and, respectively, $J_t = 2$ in the ground state of the crystal. This experimental fact reflects the presence of a twofold spin degeneration of the conduction band and a fourfold degeneration of the upper valence subband, split as a result of the large spin–orbital interaction. The lower twofold degenerate valence subband does not participate in the optical transition. In a crystalline field of cubic symmetry the level of the bound exciton with $J_t = 2$ is split into the components Γ_5 and Γ_3, while the states with $J_t = 1$ are transformed according to the representation Γ_4. The crystalline splitting of the $\mathcal{B}$ line is small in comparison with its width; therefore, the doublet structure of the $\mathcal{B}$ line situated from the long-wave part of the $\mathcal{A}$ line is not observed experimentally. Moreover, due to the selection rules for the angular momentum at the optical transition, line $\mathcal{B}$ is less intensive compared with the $\mathcal{A}$ line.

Faulkner, Merz, and Dean [271, 272] discovered new lines of radiation, situated on the long-wave part of line $\mathcal{A}$, in the photoluminescence spectra of GaP:N crystal at low temperatures ($T \leq 4.2$ K) and high levels of optical excitation. With an increase of pumping power, the intensity of these lines grows as the square of the intensity of lines $\mathcal{A}$–$\mathcal{B}$ up to the saturation of these lines. The authors of the works [271, 272] assumed that the new spectral lines of photoluminescence ($\mathcal{A}^*$ and $\mathcal{B}^*$) are influenced by radiational disintegration of the exciton molecule, formed by two excitons, bound on one isoelectron trap (the localized biexciton). The localized exciton molecule contains two electrons and two holes. Pauli's principle of interdiction requires antiparallel orientation for the electron spins. The angular momenta $3/2$ of the holes can be combined only in a way that forms a biexciton with resulting angular momentum $J_t = 0$ or $J_t = 2$. The level with $J_t = 2$ of the bound exciton molecule is split by the crystalline field into levels Γ_5 and Γ_3 with an interval of 0.16 meV between them. Level Γ_1 ($J_t = 0$) is situated 0.17 meV higher than the center of gravity of the levels with $J_t = 2$.

The high-energy doublet ($\mathcal{B}^*$) appears at transitions from biexciton doublet's components with $J_t = 2$ to doublet's components with $J_t = 2$ ($\mathcal{B}$) of the single bound exciton. The transitions ($J_t = 0$) $\rightarrow$ ($J_t = 2$) are forbidden electric dipole transitions. The low-energy triplet ($\mathcal{A}^*$) appears at transitions from biexciton doublet's components with $J_t = 2$ and biexciton level $J_t = 0$ to level $J_t = 1$ of the single bound exciton.

Unlike bound excitons on neutral donors, for which the exciton binding energy has a magnitude of 0.1 order of the donor binding energy, for a localized biexciton on an isoelectronic center, the binding energy of the second exciton is nearly as large as the binding energy of the first exciton, characteristic for the isoelectron binding. The difference between the binding energies to the isoelectron center of the first and second (in the presence of the first) excitons of the localized biexciton in the crystal GaP:N is only 1.88 meV, in spite of the fact that the binding energy with the nitrogen center of the single exciton has a magnitude of $\sim$11 meV. Investigations of the Zeeman effect on the $\mathcal{A}^*$ and $\mathcal{B}^*$ lines in GaP:N [271, 272] confirmed the model of the bound biexciton on the isoelectron acceptor in this crystal.

In 1970 Kaminsky and Pokrovsky discovered new collective states of nonequilibrium charge carriers, bound on impurity Bohr atoms in silicon—multiparticle exciton–impurity complexes (MEICs) [273]. At helium temperatures and moderate levels of optical excitation they observed a series of narrow lines situated on the long-wave side of the radiation line of the bound exciton in the recombination radiation spectra. The spectral widths of these lines are less than the average energy of the heat vibrations of the lattice, and with the growth of the level of excitation new lines appear in a sequence in which the following lines are situated in the much longer wavelength part of the spectrum. At high pumpings the discrete spectrum of the new radiation changes into a continuous one. The new lines of radiation appear in recombination of electrons and holes in MEICs which are formed as a result of the consecutive binding of many excitons on the impurity atoms. The work [273] began the development of the physics of MEICs. The results obtained during the ten years since that time are discussed in [274].

Chapter 2
Exciton Paramagnetic, Paraelectric, and Zero-Field Resonances

The Coulomb binding of electrons and holes into excitons causes peculiarities in their resonance absorption of the energy of a microwave electromagnetic field. In addition to information about the structure of the conduction and valence bands, obtained by means of spin resonance of the conduction electrons (holes), exciton paramagnetic resonance can yield information about the exciton energy spectrum. The possible existence of paramagnetic resonance on excitons in semiconductors was noticed by Deigen and Pekar [70], but after their pioneer work no successful investigations were carried out for a long time, either theoretically or experimentally.[1]

In a low-density exciton gas ($n_{ex}a_{ex} \ll 1$), which is what we consider in this chapter, the width of the exciton paramagnetic resonance line cannot be influenced by the exciton–exciton interaction. During the exciton–exciton collisions, processes with reciprocal spin reorientation of interrelating excitons are possible, affected, in particular, by the exchange electron scattering of one exciton on the hole of another one due to the mechanism of long-range Coulomb interactions (annihilation interaction of an electron and a hole of different excitons). However, for a low exciton concentration the contribution of this mechanism is lower compared with the contribution to the value of transverse exciton spin relaxation from dephasing, with the reorientation of exciton spins during their intraband spin-dependent scattering on phonons. In undoped crystals longitudinal exciton spin relaxation is also determined by the interaction of spin excitons with phonons.

Here and further the concept of the spin exciton is not used in a literal sense. In the model of simple energy bands, as was mentioned in Sect. 1.1, the exciton states can be split into para states and ortho states. In all other cases the projection of the total angular momentum of the electron and hole in the exciton is not a good quantum number. In this case it turns out that if we build a Hamiltonian of the interaction of the localized magnetic moment (or the external magnetic field) with

---

[1] We are talking about direct detection of the paramagnetic resonance of excitons. Indirect (optical) detection of exciton paramagnetic resonance, which has been realized experimentally in the crystal GaSe, will be discussed in Chap. 5.

I. Geru, D. Suter, *Resonance Effects of Excitons and Electrons*,
Lecture Notes in Physics 869, DOI 10.1007/978-3-642-35807-4_2,
© Springer-Verlag Berlin Heidelberg 2013

the excitons in terms of exciton operators of creation and destruction (see Chap. 4), then we can see from this Hamiltonian that the turnover of the spins of the paramagnetic centers leads to transitions between different exciton bands with a change in both the quantum numbers of relative electron–hole motion and the exciton wave vector. An external alternating magnetic field of the respective polarization causes similar transitions. On this basis, in those cases when the exciton states are not characterized by a definite spin projection, we shall conditionally apply the concept of exciton spin and, respectively, the concept about the processes of the exciton spin–lattice relaxation if magnetic dipole transitions are possible between the exciton states.

The utilization of the method of correlation functions and graphic techniques of Konstantinov–Perel' [275] (which was used for a study of electron paramagnetic resonance (EPR) line shape by Aminov [276]) allows us to consider different mechanisms of exciton spin–lattice relaxation. Because the valence bands of most semiconductors are built from the p and d states (in rare-earth semiconductors from the f states) which correspond to different-from-zero values of the orbital angular momenta of the valence electrons, the modulation of spin–orbital interaction by lattice vibrations in the valence band causes the effective mechanism of the exciton spin–lattice relaxation.

This chapter discusses the contribution to the exciton spin–lattice relaxation of the dynamic contact hyperfine interaction between electron components of the free excitons (s electrons of the conduction band, which are coupled with the holes from the valence band during the formation of free excitons) with the nuclear spins in the nodes of the crystalline lattice, although the low efficiency of this relaxation mechanism is pre-evident. What is interesting here is the fact that, despite the complete averaging of the hyperfine interaction during the motion of the individual exciton [70], for the system of excitons there exists a different-from-zero contribution to the linewidth of the exciton paramagnetic resonance due to the hyperfine interaction of all electrons and holes coupled in excitons with the crystal nuclei.

With regard to paraelectric resonance, despite the case of localized paraelectric centers or the so-called "noncentral ions" having the ability to tunnel between some equilibrium positions, making inhibited but not translational movements, in the case of excitons the electric dipole moments are translated through the crystal. Paraelectric resonance of excitons, despite sufficient evidence of its existence, has previously not been considered at all either in molecular crystals or in semiconductors. An interesting feature of paraelectric resonance of excitons in semiconductors is the lack of the isotopic shift effect of the paraelectric resonance lines for the intraserial exciton transitions and the presence of a large isotopic shift of the lines of interserial paraelectric resonance.

Finally, this chapter suggests and theoretically justifies the idea of using a large set of initial splittings of exciton levels in crystalline fields of different semiconductors to generate coherent electromagnetic radiation in the submillimeter and far-infrared (IR) ranges at intra- and interserial exciton transitions and also to generate coherent magnons by excitons in magnetic semiconductors.

The main results of this chapter are published in [170, 277–282].

2.1 Paramagnetic Resonance of Small-Radius Triplet Excitons

We shall consider the features of the paramagnetic resonance spectra of small-radius excitons using the example of triplet excitons in organic crystals containing aromatic molecules such as naphthalene [283]. The aromatic molecules in the triplet state are characterized by initial spin level splitting. Removal of the triple spin degeneration may be complete or partial, depending on the symmetry of the molecule and the nature of the molecular wave function of the triplet state. In aromatic molecules, the initial splitting of spin levels is conditioned by the dipole–dipole interaction between electron spins

$$\mathcal{H}_D = g_e^2 \mu_B^2 \sum_{i<j} \left[(\mathbf{S}_i \mathbf{S}_j) r_{ij}^{-3} - 3(\mathbf{r}_{ij}\mathbf{S}_i)(\mathbf{r}_{ij}\mathbf{S}_j) r_{ij}^{-5} \right], \tag{2.1}$$

where the sum is taken over all electrons of the molecule, g_e is the electron g-factor, μ_B is Bohr's magneton, $\mathbf{S}_i$ is the spin operator of the ith electron, and $r_{ij} = |\mathbf{r}_i - \mathbf{r}_j|$, where $\mathbf{r}$ is the electron coordinate.

After averaging operator $\mathcal{H}_D$ on the coordinate part of the wave function, we may represent it as

$$\mathcal{H}_D = D \left(S_z^2 - \frac{1}{3}\mathbf{S}^2 \right) + E \left(S_x^2 - S_y^2 \right), \tag{2.2}$$

where D and E are constants of the spin Hamiltonian.

Since the energy of the spin–spin interaction is often compared in magnitude to the Zeeman energy corresponding to the operator

$$\mathcal{H}_z = -g_e \mu_B \sum_i \mathbf{S}_i \mathbf{H}$$

(for naphthalene, for example, $D = 0.1006$ cm^{-1} and $E = 0.0138$ cm^{-1}), the EPR spectra are highly anisotropic. In particular, the anisotropic EPR spectrum of a naphthalene solid solution in durene contains four lines. Each pair of lines is caused by two differently oriented locations of the naphthalene molecules in durene (naphthalene and durene crystals belong to the monoclinic system and have two molecules in the elementary cell). Four lines merge into two at certain orientations of the external magnetic field (as evidenced by the elements of crystal symmetry).

The EPR spectrum of the lowest triplet excited state in pure aromatic single crystals is qualitatively different from the EPR spectrum of the excited triplet states of the molecules from which the crystal is built. This difference is particularly important if (1) there are two or more molecules in the elementary cell that are not connected by the inversion center, and (2) the Davydov splitting of the triplet state is greater than the D and E constants. Thus, for example, the paramagnetic resonance spectrum of the triplet excitons in naphthalene consists of two lines due to dipole–dipole interaction (2.1) for any orientations of the external magnetic field.

Also, since the triplet exciton is spread over many molecules, the hyperfine structure of the EPR spectra, as has already been indicated, is washed away due to the protons, in contrast to the proton hyperfine structure observed at isolated molecules.

The EPR spectrum from two lines does not necessarily indicate the existence of coherent exciton waves. When the wave of excitation undergoes strong scattering by lattice vibrations, then the propagation of this excitation may be regarded as a random process or as diffusion. If in addition $\tau^{-1} \gg |D|, |E|$, then the EPR spectrum is averaged on all nonequivalent positions of molecules in the elementary cell. Here τ is the characteristic time of the disordered movement between two neighbors in the elementary cell that is not transferred from one to the other at translation.

Singlet and triplet excitons in the molecular crystals at low levels of optical excitation and without their interaction with the lattice vibrations can be considered on the basis of elementary perturbation theory. If $|s\rangle$ is the wave function of the crystal in which the molecule s is in the first excited state, but all other $N - 1$ molecules (N is the number of molecules in the crystal) are in the ground state, then the energy of the singlet excitons is the solutions of the $N \times N$ secular determinant

$$\left| \langle s | \mathcal{H}_0 | s' \rangle - \langle s | s' \rangle E \right| = 0, \tag{2.3}$$

where $\mathcal{H}_0$ contains the kinetic and Coulomb electron energy. If $|t\rangle$ is the lowest triplet exciton state, then all the matrix elements of the type $\langle t | \mathcal{H}_0 | t' \rangle$ in (2.3) are reduced to two-electron exchange integrals (containing the orbitals of different molecules) that are small. For the triplet excitons the effects of the intermolecular charge transfer can contribute to the width of the absorption bands. Taking into account these effects, it is necessary to replace $\mathcal{H}_0$ by

$$\mathcal{H}_1 = \mathcal{H}_0 + \sum_{t,t'} R_{tt'} |t\rangle\langle t'|,$$

where

$$R_{tt'} = \sum_{\mathbf{K}} (E_{\mathbf{K}m} - E_0)^{-1} \langle \mathbf{K}, m | \mathcal{H}_0 | t' \rangle \langle \mathbf{K}, m | \mathcal{H}_0 | t \rangle.$$

Here, $|\mathbf{K}, m\rangle$ is the ionized state taking into consideration the intermolecular charge transfer ($\mathbf{K}$ is the wave vector and m is the band index). Then the energies of the triplet excitons are the solutions of the secular determinant

$$\left| \langle t' | \mathcal{H} | t' \rangle - \langle t | t' \rangle E \right| = 0. \tag{2.4}$$

Let λ denote the location of the molecule in the nth elementary cell with the number of elementary cells $n = 1, 2, 3, \ldots$. Let $|n\lambda M\rangle$ be the antisymmetric wave function of the crystal in which the molecule $n\lambda$ is excited in the lowest triplet electron state with the spin projection M and the rest of the molecules are in the ground state. Then, the wave functions diagonalizing the secular determinant (2.4) have the form [283]

$$|\mathbf{K}\lambda M\rangle = \frac{1}{N} \sum_n |n\lambda M\rangle \exp\{2\pi i \mathbf{k} \mathbf{R}_{n\lambda}\},$$

where $\mathbf{R}_{n\lambda}$ is the coordinate of the molecule $n\lambda$. Taking into consideration the nonequivalences of the molecule locations in the elementary cell ($\lambda \neq \lambda'$), the exciton wave functions diagonalizing the determinant (2.4) have the form

$$|\mathbf{K}_l M\rangle = \sum_\lambda C_{l\lambda} |\mathbf{K}\lambda M\rangle.$$

Here l is the index of the triplet exciton band. The relationship between the coefficients $C_{l\lambda}$ can be found in specific cases on the basis of symmetry properties.

In order to calculate the parameters of the spin Hamiltonian describing the paramagnetic resonance spectrum of the excitons, we shall divide the Brillouin zone into two regions P and Q, which relate, respectively, to the cases of nondegenerate and degenerate exciton bands. In region P, for which the wave vectors of the excitons correspond to the nondegenerate exciton bands, without considering the spin we can use a 3×3 determinant to find the spin energy levels:

$$\left| \langle \mathbf{K}_l M | \mathcal{H}_D + \mathcal{H}_Z | \mathbf{K}_l M' \rangle - E\delta_{MM'} \right| = 0. \tag{2.5}$$

Since the operator $\mathcal{H}_D + \mathcal{H}_Z$ is invariant under the transformations of the crystal translation group, it is sufficient to consider the matrix elements between states with the same $\mathbf{K}$.

In the Q region, where there is a degeneracy (not considering the spin) of the exciton bands or a quasi-degeneracy (when the distance between bands is of the order of the spin–spin interaction constants), it is necessary to replace (2.5) by

$$\left| \langle \mathbf{K}_l M | \mathcal{H}_D + \mathcal{H}_Z | \mathbf{K}'_l M' \rangle - E\delta_{MM'}\delta_{ll'} \right| = 0,$$

where l is now the number of degenerate (quasi-degenerate) exciton bands.

The excitons with the wave vectors $\mathbf{K}$ from the P region of the Brillouin zone have an exciton paramagnetic resonance spectrum that is not dependent on $\mathbf{K}$, whereas the paramagnetic resonance spectrum of the excitons with the wave vectors $\mathbf{K}$ from the Q region strongly depends on $\mathbf{K}$.

The degeneracy of exciton bands can be affected by the symmetry properties or may be accidental. The Hamiltonians $\mathcal{H}_0$ and $\mathcal{H}$ are invariant to the symmetry operations of the crystal space group and to the time-reversal transformation. In aromatic crystals the degeneracy of the exciton bands due to time-reversal symmetry has been revealed (benzene may also have an accidental degeneracy) [283].

The matrix element of the spin–spin interaction operator may be presented as follows:

$$\langle \mathbf{K}_l M | \mathcal{H}_D | \mathbf{K}_l M' \rangle = \sum_\lambda |C_{l\lambda}|^2 \langle n\lambda M | \mathcal{H}_D | n\lambda M' \rangle$$
$$+ \sum_\lambda F(\mathbf{K}_l \lambda M M') + \sum_{\lambda < \lambda'} G(\mathbf{K}_l \lambda \lambda' M M'), \tag{2.6}$$

where

$$F\left(\mathbf{K}_l \lambda M M'\right) = \sum_{t(t\neq n)} |C_{l\lambda}|^2 \langle n\lambda M | \mathcal{H}_D | t\lambda M'\rangle \exp\{2\pi i \mathbf{K}(\mathbf{R}_{t\lambda} - \mathbf{R}_{n\lambda})\},$$

$$G\left(\mathbf{K}_l \lambda\lambda' M M'\right) = \sum_{t} C_{l\lambda}^* C_{l\lambda'} \langle n\lambda M | \mathcal{H}_D | t\lambda' M'\rangle \exp\{2\pi i \mathbf{K}(\mathbf{R}_{t\lambda'} - \mathbf{R}_{n\lambda})\}$$

$$+ C_{l\lambda} C_{l\lambda'}^* \langle t\lambda M | \mathcal{H}_D | n\lambda' M'\rangle \exp\{2\pi i \mathbf{K}(\mathbf{R}_{n\lambda} - \mathbf{R}_{t\lambda'})\}.$$

The numerical estimation for the **K**-dependent terms from (2.6) in the case of naphthalene gives [283]

$$0 < \varepsilon\left(\mathbf{K}_l, \lambda' M M'\right) < 4.8 \cdot 10^{-4} \ \text{cm}^{-1},$$

where

$$\varepsilon\left(\mathbf{K}_l, \lambda' M M'\right) = \left| \sum_{\lambda} F\left(\mathbf{K}_l, \lambda M M'\right) - \sum_{\lambda<\lambda'} G\left(\mathbf{K}_l, \lambda\lambda' M M'\right) \right|.$$

This value is much smaller than the energy of the spin–spin interaction; therefore, the matrix element (2.6) may be represented with a high degree of accuracy as

$$\langle \mathbf{K}_l M | \mathcal{H}_D | \mathbf{K}_l M'\rangle \sim \frac{1}{2}\{\langle M | \mathcal{H}_S^{(\mathrm{I})} | M'\rangle + \langle M | \mathcal{H}_S^{(\mathrm{II})} | M'\rangle\}. \tag{2.7}$$

Here $\mathcal{H}_S$ is the spin Hamiltonian of rhombic symmetry determined by (2.2), in which the axes x and y correspond to the axis of molecular elongation and compression. The upper indices (I) and (II) of $\mathcal{H}_S$ in (2.7) represent two nonequivalent positions of the molecule in the elementary cell.

In order to simplify the calculation of the spin energy levels, it is convenient to select the crystal screw axis as a quantization axis. In this coordinate system the spin energy levels obtained by solving the secular equation (2.5) in the case of the naphthalene crystal are the eigenvalues of the spin Hamiltonian

$$\mathcal{H}_S = -0.00588\left(S_z^2 - \frac{1}{3}\mathbf{S}^2\right) - 0.00345\left(S_x^2 - S_y^2\right)$$

$$+ 0.0332(S_x S_y + S_y S_x),$$

where the energy is expressed in cm^{-1}.

In this way one can obtain the spin Hamiltonians for the triplet excitons in any molecular crystal.

2.2 Spin-Dependent Intraband Scattering of Triplet Wannier–Mott Excitons on Phonons

The experimental technique used for the excitation of semiconductors by optical means or by an electron beam has reached the level necessary to create a concentration of excitons that allows the observation of the resonance paramagnetic absorption effect of microwaves by excitons. The value of the necessary exciton concentration was estimated in the work [70]. According to this estimation, the stationary concentration of excitons created in the crystal during optical band–band transitions is determined by the formula

$$n_{ex} = \tau_{ex} \Phi_{\varkappa} / \hbar\omega, \tag{2.8}$$

if we assume that exciton de-excitation occurs only due to the processes of radiative recombination, that is, if nonradiative de-excitation processes of excitons are not important. Here Φ is the light flow per unit of illuminated area and per unit of time, $\varkappa$ is the coefficient of the exciton absorption of light in the crystal, ω is the frequency of the absorbed light, and τ_{ex} is the average lifetime of the exciton (at $\Phi = 1$ W/cm^2, $\varkappa \sim 10^5$, $\hbar\omega \sim 1$ eV, and $\tau_{ex} \sim 10^{-8}c$,[2] according to (2.8) we obtain $n_{ex} \simeq 10^{16}$ cm^{-3}).

Electron spin resonance on excitons is discussed in [70]. In this process a microwave quantum is actually absorbed by the electron with reorientation of its spin. After absorbing microwave quanta the excitons have a longer lifetime with respect to luminescence than ordinary excitons, as the projection of the electron spin is changed with the electron spin resonance, and the total spin of the excited crystal becomes different from zero, unlike the zero spin of the crystal in the ground state. This apparently corresponds to the case when the Zeeman energy is much greater than the energy of exchange splitting of the exciton. Another case proposed for consideration in the work [284] is the absorption of the energy of a microwave magnetic field at the transitions between the Zeeman components of the energy spectrum of the triplet exciton.

Spin–lattice relaxation of excitons, as already mentioned in this chapter, is determined by different mechanisms of interaction of the electrons and holes (coupled into excitons) with phonons. As explained at the beginning of this chapter, we shall consider the modulation by lattice vibrations of the contact hyperfine interaction of the electrons coupled into excitons with the magnetic moments in the nodes of the crystal lattice.

Another reason for this consideration is that hyperfine interaction, on one hand, is in a number of cases a dominant factor determining the width of the EPR line (local electron and hole centers in alkali halide crystals), and, on the other hand, it does not contribute to the linewidth of the single free exciton in the same crystals.

The Hamiltonian of the contact hyperfine interaction between the spins of intrinsic electrons of the crystal and the nuclear spins in the presence of an external constant magnetic field has the form (it is assumed that hyperfine splitting is much

---

[2] At low temperatures ($k_0 T < \mathrm{Ry}^{ex}$) in sufficiently pure semiconductors, it was established according to extensive experimental data that the lifetime of direct excitons actually has a value on the order taken in [70] (10^{-9} s $\leq \tau_{ex} \leq 10^{-7}$ s [97]).

less than Zeeman splitting) [14]

$$\mathcal{H}_{IS} = \frac{8\pi}{3}\gamma_e\gamma_N\hbar^2 \sum_i \sum_n S_{zi} I_{zn} \delta(\mathbf{r}_i - \mathbf{R}_n),\qquad(2.9)$$

where γ_e and γ_N are the gyromagnetic ratios for the electron and nucleus, $\mathbf{r}_i$ and $\mathbf{R}_n$ are radius vectors of the ith electron and the nth nucleus, respectively (it is assumed that all the magnetic nuclei in the crystal are the same), and S_{zi} and I_{zn} are the spin projection operators of the ith electron and the nth nucleus. We shall pass in (2.9) to the representation of the second quantization, taking into account that in the node representation the state of a single electron in a crystal is described by a set of quantum numbers $\mathbf{f}, \lambda, \sigma$:

$$\mathcal{H}_{IS} = \frac{8\pi}{3}\gamma_e\gamma_N\hbar^2 \sum_n \sum_{\mathbf{f},\mathbf{f}'} \sum_{\lambda,\lambda'} \psi_\lambda^*(\mathbf{R}_n - \mathbf{f})\psi_{\lambda'}(\mathbf{R}_n - \mathbf{f}')$$

$$\times \left(a_{\mathbf{f}\lambda\uparrow}^+ a_{\mathbf{f}'\lambda'\uparrow} - a_{\mathbf{f}\lambda\downarrow}^+ a_{\mathbf{f}'\lambda'\downarrow}\right) I_{zn}.\qquad(2.10)$$

Here $\Psi_\lambda(\mathbf{R}_n - \mathbf{f})$ is the wave function of the electron in an atom that is situated in a crystal node $\mathbf{R}_n$, $\mathbf{f}$ is the vector defining the electron position relative to its intrinsic nucleus, λ is the discrete quantum number characterizing the non-spin electron state, σ is the spin projection ($\sigma = \downarrow, \uparrow$), and $a_{\mathbf{f}\lambda\sigma}^+$ and $a_{\mathbf{f}\lambda\sigma}$ are the operators of creation and destruction of the electron in the $|\mathbf{f}\lambda\sigma\rangle$ state.

In the future we will discuss the electron scattering process on the phonons in the presence of contact hyperfine interaction but without changing the spin projections of the nuclei. Therefore, deriving the Hamiltonian of the interaction of electron spins with phonons from which the Hamiltonian of the spin-dependent intraband scattering of excitons on the phonons can be obtained, we shall replace the operator $\mathbf{I}_z$ by its eigenvalue M_I.

In order to account for the modulation effects of the hyperfine interaction by lattice vibrations, we shall represent the radius vector of the nth nucleus in the form $\mathbf{R}_n = \mathbf{R}_n^0 + \mathbf{U}_n$, where $\mathbf{R}_n^0$ determines the nucleus equilibrium position in the nth lattice node, and $\mathbf{U}_n$ is the displacement vector of the nucleus from this position due to thermal vibrations. Let us expand the operator (2.10) in a series according to the shift of the nuclei from their equilibrium positions (limited to linear terms) and express $\mathbf{U}_n$ in terms of normal coordinates of the crystal $q_{\varkappa_\alpha}$ ($\varkappa$ is the wave vector, α is the branch of vibrations) taking into account the cyclical conditions. Then we obtain the Hamiltonian of the spin-dependent interaction of electrons with vibrations of atoms in nodes of the crystalline lattice.

In the continuum approximation that we use, the electron interacts only with the longitudinal acoustic vibrations:

$$(\mathbf{e}_{\varkappa_l}, \varkappa_l) = \varkappa_l, \qquad (\mathbf{e}_{\varkappa_t}, \varkappa_t) = 0,\qquad(2.11)$$

where $\varkappa_l(\varkappa_t)$ and $\mathbf{e}_{\varkappa_l}(\mathbf{e}_{\varkappa_t})$ are the wave vector and the unit vector of polarization for longitudinal (transversal) acoustic vibrations, respectively.

Let us pass from the normal coordinates of the crystal $\mathbf{q}_{\varkappa_l}$ to the phonon creation and destruction operators $B_{\varkappa_l}^+$ and $B_{\varkappa_l}$ and, also, from the electron operators $a_{\mathbf{f}\lambda\sigma}^+$ and $a_{\mathbf{f}\lambda\sigma}$ to the exciton operators $A_{\mathbf{K}M_S}^+$ and $A_{\mathbf{K}M_S}$ ($\mathbf{K}$ and M_S are the wave vector and the spin projection of the triplet exciton) by means of orthonormalized wave functions of the orthoexciton recorded in the node representation:

$$\psi_{\mathbf{K}v,1} = \frac{1}{\sqrt{N}} \sum_{\mathbf{g},\mathbf{h}} e^{i(\mathbf{K},\alpha\mathbf{g}+\beta\mathbf{h})} \varphi_v(\mathbf{g}-\mathbf{h}) a_{\mathbf{g}\uparrow}^+ b_{\mathbf{h}\uparrow}^+ |0\rangle,$$

$$\psi_{\mathbf{K}v,0} = \frac{1}{\sqrt{2N}} \sum_{\mathbf{g},\mathbf{h}} e^{i(\mathbf{K},\alpha\mathbf{g}+\beta\mathbf{h})} \varphi_v(\mathbf{g}-\mathbf{h}) \left(a_{\mathbf{g}\uparrow}^+ b_{\mathbf{h}\downarrow}^+ + a_{\mathbf{g}\downarrow}^+ b_{\mathbf{h}\uparrow}^+ \right) |0\rangle, \qquad (2.12)$$

$$\psi_{\mathbf{K}v,-1} = \frac{1}{\sqrt{N}} \sum_{\mathbf{g},\mathbf{h}} e^{i(\mathbf{K},\alpha\mathbf{g}+\beta\mathbf{h})} \varphi_v(\mathbf{g}-\mathbf{h}) a_{\mathbf{g}\downarrow}^+ b_{\mathbf{h}\downarrow}^+ |0\rangle,$$

where $|0\rangle$ is the wave function of the vacuum state, the sum is taken over the lattice nodes, $\varphi_v(\mathbf{g}-\mathbf{h})$ is the hydrogen-like wave function describing the relative motion of the electron and hole in the exciton (v is a set of quantum numbers characterizing the state of the relative motion), $e^{i(\mathbf{K},\alpha\mathbf{g}+\beta\mathbf{h})}$ is the translational part of the wave function (α and β are the ratios of the effective masses of the electrons and holes to the exciton mass), and $a_{\mathbf{g}\sigma}^+$ and $b_{\mathbf{h}\sigma'}^+$ are the creation operators of the electron and hole with the spin projections σ and σ' in the corresponding lattice nodes $\mathbf{g}$ and $\mathbf{h}$.

Note that the spin indices of the operators $a_{\mathbf{g}\sigma}^+$ and $b_{\mathbf{h}\sigma'}^+$ from (2.12) correspond to two different spinor bases,[3] belonging to different quasi-particles (electrons and holes). Thus, the spin operators of the holes S_z^h, $S_\pm^h = S_x^h \pm i S_y^h$ differ from the corresponding electron spin operators S_z^e, $S_\pm^e$ only by replacing operators a^+, a by b^+, b:

$$S_z^{(p)} = \frac{1}{2} \sum_{\mathbf{f}\lambda} \left(\alpha_{\mathbf{f}\lambda\uparrow}^+ \alpha_{\mathbf{f}\lambda\uparrow} - \alpha_{\mathbf{f}\lambda\downarrow}^+ \alpha_{\mathbf{f}\lambda\downarrow} \right),$$

$$S_+^{(p)} = \sum_{\mathbf{f}\lambda} \alpha_{\mathbf{f}\lambda\uparrow}^+ \alpha_{\mathbf{f}\lambda\downarrow}, \qquad (2.13)$$

$$S_-^{(p)} = \sum_{\mathbf{f}\lambda} \alpha_{\mathbf{f}\lambda\downarrow}^+ \alpha_{\mathbf{f}\lambda\uparrow},$$

where

$$\alpha_{\mathbf{f}\lambda\sigma}^{(p)} \left(\alpha_{\mathbf{f}\lambda\sigma}^{+(p)} \right) = \begin{cases} a_{\mathbf{f}\lambda\sigma} \left(a_{\mathbf{f}\lambda\sigma}^+ \right), & p = e, \\ b_{\mathbf{f}\lambda\sigma} \left(b_{\mathbf{f}\lambda\sigma}^+ \right), & p = h. \end{cases} \qquad (2.14)$$

It can be seen that the wave functions (2.12) are indeed the eigenfunctions of the operators $S_z^e + S_z^h$ and $(\mathbf{S}_e + \mathbf{S}_h)^2$, built on the basis of the operators (2.13),

---

[3]For the simple energy bands of the free carriers that we consider in this section, these bases coincide accurately to the designations.

(2.14). Therefore, the use of these spin operators and their corresponding wave functions (2.12) seems more convenient in comparison with the case when the complex-conjugated electron spinor basis is applied to characterize the spin part of the wave function of the hole from the exciton. In this latter case, for example, the paraexciton and orthoexciton with zero spin projection are characterized by symmetric and, respectively, antisymmetric spin wave functions with respect to transposition of the spin variables [98, 166, 168]. In contrast, the wave functions (2.12) correspond to the usual rules of addition of the angular momenta of the electrons and holes (whose operators are given in natural spinor bases), the singlet state, as usual, being antisymmetric, and the triplet one being symmetric with respect to the permutations of the electron and hole spin coordinates.

The spin operators of the holes from (2.13) (the case $p = h$) are obtained by taking into account the transition rules for the hole creation and destruction operators according to Bir and Pikus [44]:

$$b^+_{f\lambda\sigma} = a_{\hat{\mathcal{K}}(f\lambda\sigma)}, \qquad b_{f\lambda\sigma} = a^+_{\hat{\mathcal{K}}(f\lambda\sigma)},$$

where $\hat{\mathcal{K}}$ is the time-reversal operator.

After the introduction of exciton creation and destruction operators $A^+_{KM_S}$ and A_{KM_S}, the Hamiltonian of the spin-dependent interaction of triplet excitons with phonons, describing the intraband exciton–phonon scattering, will take the form

$$\mathcal{H}_1 = \frac{1}{\sqrt{N}} \sum_{K,K'} \theta(K, K'; S, I) \left(A^+_{K,1} A_{K',1} - A^+_{K,-1} A_{K',-1} \right)$$

$$\times \left(B_{K-K'} + B^+_{K'-K} \right), \tag{2.15}$$

where N is the number of atoms in the crystal, and $B^+_{K'-K}$ and $B_{K-K'}$ are the creation and destruction operators of the longitudinal acoustic phonons with wave vectors $q = K' - K$ and $-q$, respectively.

The constant of the spin-dependent exciton–phonon interaction $\theta(K, K'; S, I)$ in (2.15) is determined by the formula

$$\theta(K, K'; S, I) = -\frac{\sqrt{2}\pi}{3} \cdot \frac{M_I \mu_B \mu_N}{I S V_0} \sqrt{\frac{\hbar|K - K'|}{m_a v_l}} \frac{1}{\sqrt{N}} \sum_{gg'h} \varphi^*_{1S}(g - h)\varphi_{1S}(g' - h)$$

$$\times \exp\{-i\alpha Kg + i\alpha K'g' + i\beta(K' - K)h\}$$

$$\times [(1 + i)\mathcal{I}_1 + (1 - i)\mathcal{I}_2], \tag{2.16}$$

where

$$\mathcal{I}_1 = \int d^3R \, \psi^*_{2S}(R - g)\psi_{2S}(R - g') \exp\{i(K - K')R\},$$

$$\mathcal{I}_2 = \int d^3R \, \psi^*_{2S}(R - g)\psi_{2S}(R - g') \exp\{i(K' - K)R\}.$$

In (2.16) m_a is the mass of the atom, v_l is the speed of the longitudinal acoustic phonons, and $\mathcal{V}_0$ is the volume of the elementary cell. The summation over $\mathbf{g}$, $\mathbf{g}'$, and $\mathbf{h}$ in (2.16) is performed on all lattice nodes. Here, for simplicity, we consider a model of a crystal for which the ground state corresponds to the hydrogen-like wave functions of $1s$-type valence electrons, which, after optical excitation by π-polarized light, are transferred into a $2s$-type state with the formation of $\Gamma_6 \otimes \Gamma_6 \otimes \Gamma_1$-excitons.

Following the normalization conditions for the wave functions of the orthoexciton (2.12), the functions of relative motion of the electron–hole in the exciton $\varphi_{1S}(\mathbf{g} - \mathbf{h})$ and $\varphi_{1S}(\mathbf{g}' - \mathbf{h})$ from (2.16) are normalized to the volume of the elementary cell of the crystal $\mathcal{V}_0$.

According to (2.16), at $\mathbf{K} = \mathbf{K}'$ the constant of spin-dependent exciton–phonon interaction $\theta(\mathbf{K}, \mathbf{K}', S, I)$ becomes zero. It can be shown that the absence of bonding between long-wave phonons ($\varkappa_l = \mathbf{K} - \mathbf{K}' \to 0$) and the spins of the conduction electrons, holes, or triplet excitons is a general property of such spin–phonon interactions, regardless of the concrete mechanism of phonon bonding with the spins of the specified elementary excitations.

2.3 Contribution of Hyperfine Interaction to Exciton Paramagnetic Resonance Linewidth

The contribution of the exciton–phonon spin-dependent interaction to the linewidth of the exciton paramagnetic resonance may be estimated using the correlation functions method or on the basis of the graphical techniques of Konstantinov and Perel' [275] developed for paramagnetic resonance by Aminov [276, 285]. The general scheme of the spin–lattice relaxation by these methods (we shall use the first of them) is applied to any system regardless of the specific types of relaxation transitions.[4] Therefore, before finalizing the explicit form of the operator $\mathcal{H}_1$ of exciton–phonon interaction, we can use the formulas of this section to describe the exciton–phonon relaxation without using an explicit form of the bonding constant $\theta(\mathbf{K}, \mathbf{K}', S, I)$ from (2.16). In particular, the contact hyperfine exciton interaction and the exciton exchange interaction with the paramagnetic centers can be treated in a unified way.

We shall present the Hamiltonian of the exciton–phonon system in the form

$$\mathcal{H} = \mathcal{H}_0 + \mathcal{H}_1, \qquad (2.17)$$

where

$$\mathcal{H}_0 = \sum_{\mathbf{K}, M_S} E_{M_S}(\mathbf{K}) A^+_{\mathbf{K}, M_S} A_{\mathbf{K}, M_S} + \sum_{\varkappa_\alpha} \hbar \omega_{\varkappa_\alpha} B^+_{\varkappa_\alpha} B_{\varkappa_\alpha} \qquad (2.18)$$

---

[4]This is true if we are interested in the behavior of a system for not very small time scales after the switching of the external excitation when the details of its initial state have already become nonessential [286].

is the Hamiltonian of the free excitons and phonons, and the operator $\mathcal{H}_1$ was determined above (formula (2.15)). Further, we shall introduce the $\mathcal{H}_t$ operator for the interaction of triplet excitons with the alternative magnetic field. The x-component of the magnetic moment of triplet excitons is determined by

$$
\mathcal{M}_x^{ex} = \frac{1}{2\sqrt{2}} \mu_B (g_h - g_e) \sum_{\mathbf{K}} \left(A_{\mathbf{K},1}^+ A_{\mathbf{K},0} + A_{\mathbf{K},0}^+ A_{\mathbf{K},1} \right.
$$
$$
\left. + A_{\mathbf{K},-1}^+ A_{\mathbf{K},0} + A_{\mathbf{K},0}^+ A_{\mathbf{K},-1} \right), \tag{2.19}
$$

where μ_B is the Bohr magneton, and g_e and g_h are the effective g-factors of the electron and hole. Then for linear polarization of the microwave field, the operator $\mathcal{H}_t$ has the form

$$
\mathcal{H}_t = -2\mathcal{M}_x^{ex} H_1 e^{\varepsilon t} \cos \omega t,
$$

where the operator $\mathcal{M}_x^{ex}$ is determined by the formula (2.19), H_1 is the amplitude of the magnetic component of the microwave, and ε is the adiabatic parameter of the connection of the electromagnetic field of frequency ω.

The $\mathcal{H}_t$ operator causes quantum transitions between the orthoexciton states, from which we shall select the two lowest states $\Psi_{\mathbf{K};-1}(n = 1, l = 0, M_s = -1)$ and $\Psi_{\mathbf{K};0}(n = 1, l = 0, M_s = 0)$ for consideration. The total transition probability per time unit between these states summed on the final states and averaged on the initial ones (for the given value of the exciton wave vector $\mathbf{K}$) is

$$
P(\mathbf{K}, \omega) = \frac{(g_e - g_h)^2}{32\hbar^2} (\mu_B H_1)^2 \int_{-\infty}^{\infty} dt\, e^{i\omega t}
$$
$$
\times \left\langle A_{\mathbf{K},-1}^+(t) A_{\mathbf{K},0}(t) A_{\mathbf{K},0}^+(0) A_{\mathbf{K},-1}(0) \right\rangle_{\mathcal{H}}. \tag{2.20}
$$

Here the statistical averaging is performed by means of the Hamiltonian $\mathcal{H}$ from (2.17); $A_{\mathbf{K},-1}^+(t)$ and $A_{\mathbf{K},0}(t)$ are the operators of creation and destruction of excitons with the wave vector $\mathbf{K}$ and corresponding to the spin projections $M_s = -1$ and $M_S = 0$ in Heisenberg's representation. In deriving (2.20) we used the integral representation of Dirac's δ-function.

According to the definition, we shall introduce the following correlation function:

$$
F_{\mathbf{K}}(t) = e^{i\omega_0 t} \left\langle A_{\mathbf{K},-1}^+(t) A_{\mathbf{K},0}(t) A_{\mathbf{K},0}^+(0) A_{\mathbf{K},-1}(0) \right\rangle_{\mathcal{H}}, \tag{2.21}
$$

where $\hbar\omega_0 = (E_{\mathbf{K},0} - E_{\mathbf{K},-1})$ is the exciton Zeeman splitting. Then for transition probability $P(\mathbf{K}, \omega)$, instead of (2.20), we obtain

$$
P(\mathbf{K}, \omega) = \frac{(g_e - g_h)^2}{32\hbar^2} (\mu_B H_1)^2 \int_{-\infty}^{\infty} F_{\mathbf{K}}(t) e^{i(\omega - \omega_0)t}\, dt. \tag{2.22}
$$

In formula (2.21) the statistical average for the correlation function $F_{\mathbf{K}}(t)$

$$
\langle \dots \rangle_{\mathcal{H}} = \frac{\langle \mathcal{U}(\beta) \dots \rangle_{\mathcal{H}_0}}{\langle \mathcal{U}(\beta) \rangle_{\mathcal{H}_0}} \tag{2.23}
$$

is performed by means of the evolution operator

$$\mathcal{U}(\beta) = T \exp\left\{-\int_0^\beta d\lambda\, \mathcal{H}_1(\lambda_1)\right\},$$

where $\hat{T}$ is the Dyson time-regulating operator, $\mathcal{H}_1(\lambda)$ is operator $\mathcal{H}_1$ in the representation of interaction, $\beta = (k_0 T)^{-1}$, k_0 is Boltzmann's constant, and T is temperature. The averaging on the right side of (2.23) is performed on the Hamiltonian of zero approximation from (2.18), where the known series from thermodynamic perturbation theory is used for the evolution operator $\mathcal{U}(\beta)$ [287, 288].

Formula (2.21) is initial during the calculation of the spectral moments of the nth order of the line of exciton paramagnetic resonance under consideration by nth differentiating of the correlation function

$$(-\mathrm{i})^n M_n(\mathbf{K}) = \left.\frac{\partial^n}{\partial t^n} F_{\mathbf{k}}(t)\right|_{t=0}. \tag{2.24}$$

By means of formulas (2.24) and (2.21) and the equations of motion for the operators $A_{\mathbf{q},-1}^+(t) A_{\mathbf{K},0}(t)$, $B_{\mathbf{q}-\mathbf{K}}(t)$ and $B_{\mathbf{q}-\mathbf{K}}^+(t)$ in the Heisenberg representation we find the following expression for the second moment of the absorption line:

$$
\begin{aligned}
M_2(\mathbf{K}) = \frac{1}{\hbar^2 N} &\left\{\sum_{\mathbf{q}\mathbf{q}'} \theta(\mathbf{q},\mathbf{q}'; S, I)\theta(\mathbf{q}',\mathbf{q}; S, I)\right.\\
&\times \left\langle A_{\mathbf{q}',-1}^+ A_{\mathbf{K},0}\left(B_{\mathbf{q}'-\mathbf{q}} + B_{\mathbf{q}-\mathbf{q}'}^+\right)\left(B_{\mathbf{q}-\mathbf{K}} + B_{-\mathbf{q}+\mathbf{K}}^+\right) A_{\mathbf{K},0}^+ A_{\mathbf{K},-1}\right\rangle_{\mathcal{H}} \\
&+ \sum_{\mathbf{q}} (T_{\mathbf{K}} - T_{\mathbf{q}} + \hbar\omega_{\mathbf{q}-\mathbf{K}})\theta(\mathbf{q},\mathbf{K}; S, I) \\
&\times \left\langle A_{\mathbf{q},-1}^+ A_{\mathbf{K},0} B_{\mathbf{q}-\mathbf{K}} A_{\mathbf{K},0}^+ A_{\mathbf{K},-1}\right\rangle_{\mathcal{H}} \\
&+ \sum_{\mathbf{q}} (T_{\mathbf{K}} - T_{\mathbf{q}} - \hbar\omega_{\mathbf{K}-\mathbf{q}})\theta(\mathbf{q},\mathbf{K}; S, I) \\
&\left.\times \left\langle A_{\mathbf{q},-1}^+ A_{\mathbf{K},0} B_{\mathbf{K}-\mathbf{q}}^+ A_{\mathbf{K},0}^+ A_{\mathbf{K},-1}\right\rangle_{\mathcal{H}}\right\},
\end{aligned}
\tag{2.25}
$$

where $T_{\mathbf{K}}$ and $T_{\mathbf{q}}$ are the kinetic energies of excitons with the wave vectors $\mathbf{K}$ and $\mathbf{q}$ in the band characterized by the spin projection $M_S = -1$, and $\theta(\mathbf{K}, \mathbf{K}'; S, I)$ is determined by (2.16).

If in the evolution operator $\mathcal{U}(\beta)$ from (2.3.8a) we are limited by the first item (unit) during its representation in the form

$$\mathcal{U}(\beta) = 1 + \sum_{(n=1)}^{\infty} \frac{(-1)^n}{n!} \int_0^\beta d\lambda_n \int_0^\beta d\lambda_{n-1} \cdots \int_0^\beta d\lambda_1$$

$$\times \hat{T}\{\mathcal{H}_1(\lambda_1)\mathcal{H}_1(\lambda_2)\cdots\mathcal{H}_1(\lambda_n)\},$$

that corresponds to the averaging in (2.25) by means of the Hamiltonian $\mathcal{H}_0$, then for the contribution of the hyperfine interaction to the second moment of the exciton paramagnetic resonance line we obtain

$$M_2(\mathbf{K}) = \frac{1}{\hbar^2 N} \sum_{\mathbf{q}} \theta(\mathbf{q}, \mathbf{K}; S, I)\theta(\mathbf{K}, \mathbf{q}; S, I)n_{\mathbf{K},-1}$$

$$\times (n_{\mathbf{K},0} + 1)(2m_{\mathbf{K}-\mathbf{q}} + 1), \tag{2.26}$$

where $n_{\mathbf{K},M_S}$ and $m_{\mathbf{K}-\mathbf{q}}$ are, respectively, the average numbers of filling excitons with wave vector $\mathbf{K}$ and spin projection M_S and phonons with wave vectors $\mathbf{K} - \mathbf{q}$. Note that this formula and the more general equation (2.25) are true only if the energy of the thermal vibrations of the lattice is less than the Zeeman splitting of the exciton, and spin transitions $0 \rightarrow 1$ in the exciton are not essential compared with transitions $-1 \rightarrow 0$.

Formulas (2.26) and (2.25) are applied if the inverted time of the exciton relaxation by the phonons τ_{rel}^{-1} exceeds the transition probability $P(\mathbf{K}, \omega)$ from (2.22), which in turn must be greater than the inverted time of the exciton life τ_{ex}^{-1} $(\tau_{\text{ex}}^{-1} < P(\mathbf{K}, \omega) < \tau_{\text{rel}}^{-1})$.[5] From formula (2.26) we can see that with decreasing exciton concentration the contribution of hyperfine interaction to the width of the resonance line decreases, and at low exciton occupation numbers it falls practically to zero, in accordance with the results of the work [70] for separate free excitons.

2.4 Generation of Coherent Electromagnetic Radiation at Intra- and Interseries Exciton Transitions

The great diversity of optic quantum generators satisfies different scientific and applied purposes to a considerable degree. However, besides the variety of applied optically active media and ways of excitation of the coherent radiation, due to the mastering of submillimeter and far-infrared (far-IR) ranges we come across many difficulties, both in creating sources with a rebuilt frequency of radiation in a wide interval and in selecting the active media for laser generation.

In the far-IR and submillimeter ranges, to create a volumetrical negative differential conduction usually an inverted distribution of the hot carriers of the current in the impulse space is used [289]. The inversion is created due to the accumulation of carriers in a region of closed trajectories [290], where the electron energy is less than the energy of the optical phonon. In particular, in p–Ge crystals in sufficiently strongly crossed electric $\mathbf{E}$ and magnetic $\mathbf{H}$ fields there is a superpopulation in the subband of light holes, which is preserved even at comparatively high concentrations of impurities ($n_{\text{imp}} \sim 5 \cdot 10^{15}$ cm^{-3}) and lattice temperatures from liquid

---

[5]For a more general expression of (2.25) including the case of nonequilibrium thermodynamic states, the first of these inequalities can probably be weakened.

helium to liquid nitrogen temperatures. This causes a negative absorption coefficient of the far-IR radiation [291]. The far-IR radiation is generated at direct optical transitions from the band of light holes in p–Ge to the band of heavy holes under conditions of the inversion of the distribution function of light holes relative to the heavy ones [292]. The creation of an inverted function of the free carrier distribution is possible, not only in external crossed $\mathbf{E}$ and $\mathbf{H}$ fields, but also in their absence with the capture of electrons (holes) by shallow impurities in semiconductors [293].

The possibility of resonance parametric generation of submillimeter waves of frequency $\omega_3 = \omega_2 - \omega_1$ from the region of the lattice absorption at high levels of semiconductor excitation by bichromatic light with $\omega_1, \omega_2 \gtrsim E_g/\hbar$ (E_g is the energy gap) is theoretically shown in [294]. This way of generating submillimeter radiation is not convenient, because its realization requires the presence of a quasi-energy spectrum of the "charge carrier + bichromatic electromagnetic field" system, which can exist only in the presence of strong optical pumping.

In this section, in the frame of a two-band model, we consider a mechanism for generating coherent electromagnetic radiation in the microwave and far-IR ranges at resonance transitions between the states of the relative electron–hole movement in Wannier–Mott excitons [277]. The energy spectrum of the excitons is determined by the electrostatic interaction of electrons and holes (Coulomb and exchange contributions), accounting for spin–orbital coupling in the conduction and valence bands, as well as the influence of the crystal field on excitons (the initial splitting of degenerate exciton bands in the crystal field).[6] Since all these values essentially depend on the chemical composition of the crystals, their spatial symmetry, and the constants of the crystal field, there is a great choice of exciton bands for the realization of laser generation in the submillimeter and far-IR regions of the spectrum. Laser generation at transitions between the exciton bands has the characteristic features influenced by the possibility to create an inversion of the exciton band populations by selective coupling of electron–hole pairs into excitons of one (upper) band. It turns out that only a small concentration of excitons in the crystal is necessary to achieve the autoexcitation threshold of this type of laser.

The Hamiltonian of the exciton–photon system for the case of one photon mode and two exciton bands in "the rotating field" approximation (ignoring the antiresonance terms) has the form

$$H = \hbar\omega C^+ C + \sum_{\mathbf{K}} \varepsilon_n(\mathbf{K}) A_{n\mathbf{K}}^+ A_{n\mathbf{K}} + \sum_{\mathbf{K}} \varepsilon_{n'}(\mathbf{K}) A_{n'\mathbf{K}}^+ A_{n'\mathbf{K}}$$

$$+ \sum_{\mathbf{K}} \left[g_{n'n}(0) C A_{n'\mathbf{K}}^+ A_{n\mathbf{K}} + \text{H.c.} \right], \tag{2.27}$$

where C^+ and C are the creation and destruction operators of the photon of frequency ω; $A_{n\mathbf{K}}^+$ and $A_{n\mathbf{K}}$ are the creation and destruction operators of the exciton

---

[6]In semiconductors containing heavy atoms, one must also consider the relativistic corrections that can bring, for example, in lead halogenides, an essential reconstruction of the band structure [295].

with wave vector $\mathbf{K}$ in the band with index n; $\varepsilon_n(\mathbf{K})$ is the exciton energy in the band n; $g_{n'n}(0)$ is the constant of the exciton–photon coupling (Fourier transform of the interaction energy of the excitons with the photons at transitions between exciton bands). Here we do not demonstrate the dependence of operators C^+ and C on the wave vector and polarization of the photon, since in the microwave and IR ranges the dependence of the matrix elements of the transition on the wave vector of the photon may be ignored, and the exciton states should be selected beforehand with features of symmetry which correspond to the dipole-allowed transitions of the given polarization of photons.

The Hamiltonian (2.27) does not contain terms describing the interaction of excitons and photons with a dissipative subsystem and a pumping subsystem to create the inverted population of exciton bands. We shall consider these subsystems phenomenologically.

The equations of motion for operators of photon fields, exciton polarization and the operators of the population difference of the states $|n\mathbf{K}\rangle$ and $|n'\mathbf{K}\rangle$ of different bands have the form:

$$\frac{dC}{dt} = -(i\omega + \varkappa)C - \frac{i}{\hbar}g_{n'n}^*(0)\sum_{\mathbf{K}} S_-^{n'n}(\mathbf{K}),$$

$$\frac{d}{dt}S_-^{n'n}(\mathbf{K}) = -\left[i\omega_{n'n}(\mathbf{K}) + \Gamma_{n'n}\right]S_-^{n'n}(\mathbf{K}) + 2\frac{i}{\hbar}g_{n'n}(0)CS_z^{n'n}(\mathbf{K}),$$

$$\frac{d}{dt}S_z^{n'n}(\mathbf{K}) = \frac{1}{T_1}\left[\frac{1}{2}d_{n'n}^{(0)}(\mathbf{K}) - S_z^{n'n}(\mathbf{K})\right] \tag{2.28}$$

$$+ \frac{i}{\hbar}\left[g_{n'n}^*(0)C^+S_-^{n'n}(\mathbf{K}) - g_{n'n}(0)CS_+^{n'n}(\mathbf{K})\right],$$

$$\frac{d}{dt}S_+^{n'n}(\mathbf{K}) = \left(\frac{d}{dt}S_-^{n'n}(\mathbf{K})\right)^+; \quad \frac{dC^+}{dt} = \left(\frac{dC}{dt}\right)^+,$$

where

$$S_z^{n'n}(\mathbf{K}) = \frac{1}{2}\left(A_{n'\mathbf{K}}^+ A_{n'\mathbf{K}} - A_{n\mathbf{K}}^+ A_{n\mathbf{K}}\right), \quad S_-^{n'n}(\mathbf{K}) = A_{n\mathbf{K}}^+ A_{n'\mathbf{K}},$$

$$S_+^{n'n}(\mathbf{K}) = A_{n'\mathbf{K}}^+ A_{n\mathbf{K}}; \quad \hbar\omega_{n'n}(\mathbf{K}) = \varepsilon_{n'}(\mathbf{K}) - \varepsilon_n(\mathbf{K}).$$

The phenomenological constants are introduced in (2.28) by analogy with the two-level theory of lasers [296, 297]: $\varkappa$ is the constant of photon damping in the exciton–photon interaction; $\Gamma_{n'n} = \frac{1}{2}(\Gamma_n + \Gamma_{n'})$, $\hbar\Gamma_n$, and $\hbar\Gamma_{n'}$ are terms for "the washing away" of bands n and n' due to the interaction of excitons with the dissipative subsystem (the washing away of the exciton band is considered independent of $\mathbf{K}$, or more precisely, as a measure of this washing away we select the value of the constant of exciton damping at interaction with the lattice vibrations in the actual region of the exciton wave vectors, where this interaction is more effective); T_1 is the longitudinal relaxation time of the excitons (the time of establishment of the equilibrium difference of populations of the exciton states $|n\mathbf{K}\rangle$ and $|n'\mathbf{K}\rangle$ due to the longitudinal relaxation of the excitons is not considered to be dependent on $\mathbf{K}$); $d_{n'n}^{(0)}$ is the initial inverted population of exciton bands n and n', which is conditioned by the pumping and all other noncoherent processes in the absence of laser action.

We shall average (2.28) by means of the density matrix of the system and take into account that in the regime of generation the corpuscular nature of the photon field is not practically evidenced:

$$\langle C S_z^{n'n}(\mathbf{K}) \rangle \simeq \langle C \rangle \langle S_z^{n'n}(\mathbf{K}) \rangle,$$
$$\langle C S_\pm^{n'n}(\mathbf{K}) \rangle \simeq \langle C \rangle \langle S_\pm^{n'n}(\mathbf{K}) \rangle. \tag{2.29}$$

Then, representing $\langle C \rangle$ from (2.29) in the form

$$\langle C \rangle = \sqrt{N} e^{-i\Omega t}, \tag{2.30}$$

where N is the number of photons in the mode, after substituting (2.30) in $\frac{d}{dt} S_+^{n'n}(\mathbf{K})$ of (2.28), for the stationary case we obtain

$$\langle S_+^{n'n}(\mathbf{K}) \rangle = -\frac{1}{\hbar} d_{n'n}^{(0)}(\mathbf{K}) g_{n'n}^*(0) \sqrt{N} e^{i\Omega t} \left[\Omega - \omega_{n'n}(\mathbf{K}) - i\Gamma_{n'n} \right]^{-1}. \tag{2.31}$$

Substituting (2.31) in $\frac{d}{dt} S_z^{n'n}(\mathbf{K})$ of (2.28) we have

$$\langle S_z^{n'n}(\mathbf{K}) \rangle = 2 d_{n'n}^{(0)}(\mathbf{K}) \left\{ 1 - 4\Gamma_{n'n} T_1 \left| g_{n'n}(0) \right|^2 N \right.$$
$$\left. \times \frac{1}{\hbar^2} \left[(\Omega - \omega_{n'n}(\mathbf{K}))^2 + \Gamma_{n'n}^2 \right]^{-1} \right\}. \tag{2.32}$$

After substituting (2.32) in $\frac{d}{dt} S_-^{n'n}(\mathbf{K})$ of (2.28) we obtain

$$\langle S_-^{n'n}(\mathbf{K}) \rangle = -\frac{1}{\hbar} d_{n'n}^{(0)}(\mathbf{K}) g_{n'n}(0) \sqrt{N} \left\{ 1 - 4\Gamma_{n'n} T_1 \left| g_{n'n}(0) \right|^2 N \right.$$
$$\left. \times \frac{1}{\hbar^2} \left[(\Omega - \omega_{n'n}(\mathbf{K}))^2 + \Gamma_{n'n}^2 \right]^{-1} \right\} \left[\Omega - \omega_{n'n}(\mathbf{K}) + i\Gamma_{n'n} \right]^{-1} e^{-i\Omega t}. \tag{2.33}$$

From (2.33) we have

$$\frac{d}{dt} \langle C^+ \rangle = (i\omega - \varkappa) \langle C^+ \rangle - \frac{i}{\hbar^2} \left| g_{n'n}(0) \right|^2 N \sum_{\mathbf{K}} \frac{d_{n'n}^{(0)}(\mathbf{K})}{\Omega - \omega_{n'n}(\mathbf{K}) - i\Gamma_{n'n}}$$
$$\times \left\{ 1 - \frac{4}{\hbar^2} \Gamma_{n'n} T_1 \left| g_{n'n}(0) \right|^2 N \left[(\Omega - \omega_{n'n}(\mathbf{K}))^2 + \Gamma_{n'n}^2 \right]^{-1} \right\} e^{i\Omega t}.$$

Again using the substitution (2.30), and after the necessary transformations, we obtain

$$\varkappa + i(\Omega - \omega) = \frac{i}{\hbar^2} |g_{n'n(0)}|^2 \sum_{\mathbf{K}} \frac{d_{n'n}^{(0)}(\mathbf{K})}{\Omega - \omega_{n'n}(\mathbf{K}) - i\Gamma_{n'n}}$$

$$\times \left\{ -1 + \frac{4}{\hbar^2} \Gamma_{n'n} T_1 |g_{n'n}(0)|^2 N \big[(\Omega - \omega_{n'n}(\mathbf{K}))^2 + \Gamma_{n'n}^2 \big]^{-1} \right\}.$$

(2.34)

We choose the real and the imaginary parts in (2.34):

$$2\varkappa = \mathcal{G}_1 - \mathcal{G}_2 N,$$

(2.35)

where

$$\mathcal{G}_1 = \frac{2}{\hbar^2} \Gamma_{n'n} |g_{n'n}(0)|^2 \sum_{\mathbf{K}} \frac{d_{n'n}^{(0)}(\mathbf{K})}{(\Omega - \omega_{n'n}(\mathbf{K}))^2 + \Gamma_{n'n}^2},$$

$$\mathcal{G}_2 = \frac{8}{\hbar^4} \Gamma_{n'n}^2 T_1 |g_{n'n}(0)|^4 \sum_{\mathbf{K}} \frac{d_{n'n}^{(0)}(\mathbf{K})}{[(\Omega - \omega_{n'n}(\mathbf{K}))^2 + \Gamma_{n'n}^2]^2}.$$

(2.36)

Let us multiply both sides of (2.35) by N. Then the first term on the right-hand side will describe the increase of the number of photons due to the induced radiation of the photons by the excitons with the given inverted populations. The sum $\mathcal{G}_2 N$ accounts for the fact that in the process of the laser action the inversion of populations decreases (the effect of "hole burning" in the amplification curve).

From the imaginary part of (2.34) we obtain

$$\Omega - \omega = (\Delta\Omega)_1 + N(\Delta\Omega)_2,$$

(2.37)

where

$$(\Delta\Omega)_1 = -\frac{1}{\hbar^2} |g_{n'n}(0)|^2 \sum_{\mathbf{K}} \frac{d_{n'n}^{(0)}(\mathbf{K})(\Omega - \omega_{n'n}(\mathbf{K}))}{(\Omega - \omega_{n'n}(\mathbf{K}))^2 + \Gamma_{n'n}^2},$$

$$(\Delta\Omega)_2 = \frac{4}{\hbar^4} \Gamma_{n'n} T_1 |g_{n'n}(0)|^4 \sum_{\mathbf{K}} \frac{d_{n'n}^{(0)}(\mathbf{K})(\Omega - \omega_{n'n}(\mathbf{K}))}{[(\Omega - \omega_{n'n}(\mathbf{K}))^2 + \Gamma_{n'n}^2]^2}.$$

(2.38)

The indices 1 and 2 in $\mathcal{G}$ and $\Delta\Omega$ designate, respectively, the absence and presence of saturation.

From (2.35) (taking into account that $\mathcal{G}_2 > 0$) we find the condition of the autoexcitation of the IR laser (the generation threshold)

$$\mathcal{G}_1 \geq 2\varkappa.$$

(2.39)

An analysis of (2.37) and (2.38) shows that the term of the sum $(\Delta\Omega)_1$ describes the displacement of the frequency of generation conditioned by the prolonged influence of the active substance. This shift is directed to the center of the contour of the amplification and is proportional to the intensity of the pumping. The term of the sum $N(\Delta\Omega)_2$, which is proportional to the power of the generation, describes the effect of "the repulsion of downfalls." The shift of the frequency of generation takes place to the side of the center of the radiation line. This result, along with the consequences from (2.35) about the increased number of photons from the induced radiation and the effect of "hole burning" in the amplification curve, is similar to the results for the two-level theory of lasers [296, 297].

As we can see from (2.35)–(2.39), the shift of the frequency of the line of induced radiation and the generation threshold of the IR laser are determined by the value of the inversion of exciton band populations $d_{n'n}^{(0)}(\mathbf{K})$. An evident form $d_{n'n}^{(0)}(\mathbf{K})$ depends on the method of pumping and the function of the distribution of the excitons in the bands. For an equilibrium distribution of excitons in the bands during the optical pumping, we have

$$d_{n'n}^{(0)}(\mathbf{K}) = \frac{8}{V}(\pi\alpha)^{3/2}(N_{n'} - N_n)e^{-\alpha\mathbf{K}^2}, \tag{2.40}$$

where $\alpha = \hbar^2/2m_{\mathrm{ex}}k_0 T$, m_{ex} is the translational effective mass of the exciton, and N_n and $N_{n'}$ are, respectively, the full number of excitons in bands n and n'; V is the volume of the crystal. Here T is the temperature of the crystal at "moderate" levels of optical excitation, when the frequency of exciton–phonon collisions is higher than the frequency of interexciton collisions. If the level of excitation is so high that collisions between excitons take place more often than collisions with phonons, then the exciton dipole–dipole reservoir (EDDR) can be separated. In this case T designates the temperature of the EDDR under the condition that the deviation from equilibrium is not big and a quasi-equilibrium distribution of excitons in the bands is realized.

Intraserial Exciton Transitions   If bands n and n' belong to one exciton series, then for an equilibrium or quasi-equilibrium distribution of excitons in the bands, expressions (2.35) and (2.37) are transformed into the forms

$$\varkappa = F(\Omega)\big[1 + \Phi(\Omega)\big], \tag{2.41}$$

$$\Omega - \omega = F(\Omega)\big[\Phi(\Omega) - 1\big]\big(\Omega - \omega_{n'n}^{(0)}\big)\big[\big(\Omega - \omega_{n'n}^{(0)}\big)^2 + \Gamma_{n'n}^2\big]^{-1}, \tag{2.42}$$

where

$$F(\Omega) = \Gamma_{n'n}(N_{n'} - N_n)\Lambda(\Omega),$$

$$\Phi(\Omega) = 4\Gamma_{n'n}T_1 N\Lambda(\Omega),$$

$$\Lambda(\Omega) = \frac{1}{\hbar^2}|g_{n'n}(0)|^2\big[\big(\Omega - \omega_{n'n}^{(0)}\big)^2 + \Gamma_{n'n}^2\big].$$

Here $\hbar\omega_{n'n}$ is the energetic distance between the "parallel" exciton bands n and n' (the translational exciton masses in both bands are similar).

According to (2.39) and (2.41), during the intraserial exciton transitions the generation threshold of the far-IR laser is determined by the condition

$$
N_{n'} - N_n \geq \frac{\varkappa\hbar^2}{\Gamma_{n'n}} \cdot \frac{(\Omega - \omega_{n'n}^{(0)})^2 + \Gamma_{n'n}^2}{|g_{n'n}(0)|^2}.
\tag{2.43}
$$

As shown from (2.41) and (2.42), due to the intraserial transitions the kinetic energy of the excitons does not influence the conditions of the far-IR laser generation. The threshold of the generation and the shift of the line of the induced radiation depend on the temperature only by means of the phenomenological parameter $\Gamma_{n'n}$.

Interserial Exciton Transitions   In the case of interserial exciton transitions, we have

$$
\omega_{n'n}(\mathbf{K}) = \omega_{n'n}^{(0)} + \frac{\hbar(m_{\mathrm{ex}} - m_{\mathrm{ex}}')}{2m_{\mathrm{ex}}m_{\mathrm{ex}}'}\mathbf{K}^2,
\tag{2.44}
$$

which leads to the additional dependence of the far-IR laser generation on the temperature, conditioned by the difference of the kinetic energy of the excitons in different bands (m_{ex}' is the translational mass of the excitons in the band n'). In this case (2.35)–(2.38) are not calculated precisely, but they may be found approximately. In particular, at low temperatures when the kinetic energy of the exciton satisfies the condition

$$
\frac{\hbar^2 K^2}{2m} < \frac{m_{\mathrm{ex}}'}{m_{\mathrm{ex}} - m_{\mathrm{ex}}'}\hbar(\Omega - \omega_{n'n}^{(0)}),
$$

for the generation threshold we obtain

$$
N_{n'} - N_n \geq \frac{\varkappa}{\Gamma_{n'n}} \cdot \frac{[\hbar(\Omega - \omega_{n'n}^{(0)}) - \frac{m_{\mathrm{ex}} - m_{\mathrm{ex}}'}{m_{\mathrm{ex}}'}k_0 T]^2 + \hbar^2\Gamma_{n'n}^2}{|g_{n'n}(0)|^2}.
$$

The depth of the downfall in the amplification curve conditioned by the downfall of the inverse populations in the process of the laser action as determined by the expression $g_2 N$, also depends upon the temperature.

The additional temperature dependence due to interserial exciton transitions, from (2.37), (2.38), (2.40), and (2.44), is characteristic for the shift of the frequency of generation to the center of the contour of amplification and the effect of "the repulsion of downfalls."

We shall evaluate the effect of the IR-laser radiation by exciton gas. In the sums (2.36) and (2.38) the main contribution is given by the terms of the sum in which $\omega_{n'n}(\mathbf{K}) \sim \Omega$, but $\omega_{n'n}(\mathbf{K}) \sim \omega$; thus we can approximately replace Ω by ω in the sums. Because of the exciton–photon interaction, the shift of $\Omega - \omega_{n'n}^{(0)}$ of the line of radiation near the threshold of the laser generation is less than the contribution to the width of the line from the interaction of the excitons with the dissipative system ($\Omega - \omega_{n'n}^{(0)} < \Gamma_{n'n}$). Thus, after the calculation of the value of constants of the exciton–photon interaction from (2.43), we obtain the following expression for the condition of self-excitation of the IR laser due to interserial exciton transitions:

$$N_{n'} - N_n \geq \left. \frac{\hbar^3 \omega_{\mathbf{Q}} \varepsilon_{\mathbf{Q}} \Gamma_{n'n} \varkappa \mathcal{V}}{2\pi e^2 (E_{n'} - E_n)^2 |\langle n'|\mathbf{e}_{\mathbf{Q}j}\hat{\mathbf{r}}|n\rangle|^2} \right|_{\mathbf{Q}\to 0}, \qquad (2.45)$$

where $\omega_{\mathbf{Q}}$ is the frequency of the photon with wave vector $\mathbf{Q}$ and unit vector of polarization $\mathbf{e}_{\mathbf{Q}j}$, $\varepsilon_{\mathbf{Q}}$ is the dielectric constant on the frequency $\omega_{\mathbf{Q}}$, conditioned by all the excitations of the crystal without the given exciton, and $\hat{\mathbf{r}}$ is the operator of the coordinate that characterises the electron–hole relative movement in the exciton.

For the transition $2P_0 \to 1S$, from (2.45) we find that

$$N_{2P_0} - N_{1S} \geq 0.9 \frac{\hbar^3 \omega_{\mathbf{Q}} \varepsilon_{\mathbf{Q}} \Gamma_{2P_0,1S} \varkappa \mathcal{V}}{\pi (e a_{\mathrm{ex}})^2 (E_{2P_0} - E_{1S})^2}, \qquad (2.46)$$

where a_{ex} is the exciton Bohr radius.

According to (2.46), for the crystal CdS ($\mathcal{V} = 10^{-5}$ cm^3, $\varepsilon_{\mathbf{Q}} = 9.27$, $a_{\mathrm{ex}} = 28.7$ Å [170], $\hbar\omega_{\mathbf{Q}} \simeq E_{2P_0} - E_{1S} = 19$ meV [86], $\varkappa\Gamma_{2P_0,1S} \sim 10^{18}$ s^{-2} [97, 298]), the threshold of the laser generation at the transition $2P_0 \to 1S$ is achieved at $n_{2P_0}^{\mathrm{ex}} - n_{1S}^{\mathrm{ex}} \gtrsim 0.75 \cdot 10^{10}$ cm^{-3}. In the conditions of the inversion of populations at relatively low exciton concentration in the crystal CdS ($n_{\mathrm{ex}} \sim 10^{14}$ cm^{-3}), the laser generation at the transition $2P_0 \to 1S$ begins as soon as the ratio of the exciton concentrations from different bands becomes different from 1 and this difference is not less than 10^{-4}.

2.5 Generation of Coherent Magnons in Magnetic Semiconductors

Coherent collective excitations in crystals have attracted attention because of their possible practical use based on their high frequency and phase stabilities. Akhiezer et al. [50] were the first to indicate the possibility of creating coherent magnons in 1963. Due to the interaction of spin waves with the drifting charge carriers, a binding between the spin and the drifting waves takes place, when their frequencies coincide. This binding and, along with it, the amplification of the spin waves can also occur when the frequencies of the spin and drifting waves are not in resonance [299]. To achieve this, it is sufficient to apply an external constant electric field, which helps to change the drifting speed of the electron. Then, if the drifting speed of the free carriers exceeds the phase speed of the spin wave, the spin wave is amplified. This mechanism of spin wave amplification is similar to the mechanism of acoustic wave amplification in piezo-semiconductors [300]. In the limiting case of the absence of collisions between electrons, the coefficient of absorption of the spin waves is equal to zero, and their absorption or amplification is not possible.

In the classical description of spin wave amplification, the length of the spin wave is considered to be much bigger than the length of the free run of the electron l ($kl < 1$, k is the wave number of the spin wave), and its frequency is less than the frequency of the electron collisions. The amplification of spin waves in a ferromagnetic semiconductor is studied in [301] with electron–magnon interaction in the quantum mechanical limit (case $kl > 1$), and it is shown that spin wave amplification takes place, as in the classical case, if the drifting speed of the charge carriers exceeds the phase speed of the spin waves. Another spin wave amplification method is that of parametric excitation of the spin waves with a given value of the wave vector [52].

We shall consider the method of generation and amplification of the coherent magnons in magnetic semiconductors with their optical excitation in the exciton region of the spectrum. A characteristic peculiarity of magnetic semiconductors is the presence of a strong $s - d(f)$ exchange interaction of the free carriers with localized magnetic moments, which greatly determines the energetic structure of these crystals in the magnetic ordered phase [17]. We note that, in contrast to the "magnetic" excitons of small radius, for whose formation the state of the carriers from the narrow energy bands is important [302], the Wannier–Mott excitons that we consider are formed during the Coulomb coupling of the holes from the valence band with the electrons from the conduction band, the width of which is much greater than the energy of the exchange interaction between free and localized electrons ($W \gg JSM(T)/M(0)$; J is the exchange integral, S is the spin of the localized atom with the unbuilt $d(f)$-shell in the node of the crystal lattice, $M(0)$ is the magnetization of saturation at $T = 0$). If the value of the exchange splitting of the conduction band is much bigger than the splitting of the valence band (e.g., as in $CdCr_2Se_4$ and EuO [303]), then below the phase transition point the exciton spectrum consists of two series: one corresponds to the bound states of the electron from the low spin-polarized conduction subband and the hole of the valence band; the other corresponds to the bound states of the electron from the upper spin-polarized conduction subband and the hole from the valence band.

Strictly speaking, it is necessary to introduce "correct" magnons [304] in the spin wave region for a correct description of the interaction of magnons with the free carriers. Thus, one should consider the contribution of the spins of the electron and hole coupled into an exciton to oscillations of the total angular momentum of the system. With $T \gg T_o$ (T_o is the characteristic temperature depending on the $s - d(f)$ exchange interaction, $d(f) - d(f)$ interionic excange interaction, effective mass of carriers and lattice constant [304]) the spin polaron in ferromagnetic semiconductors with wide band dissociate in the temperature interval $T_o \ll T \ll T_c$ (T_c is the Curie temperature), the exciton–magnon interaction can be considered without taking into account the "correct" magnons.

Exciton–magnon interaction will be considered here in the frame of the approach developed in Sect. 2.4 for the interaction of excitons with low-frequency electromagnetic radiation. Now the difference is that, in the equations of motion, we shall preserve the dependence on the wave vector of the magnon, which could be ignored for photons.

In the "rotating field" approximation, the Hamiltonian of the interaction between excitons and magnons in the model of two exciton bands and one magnon mode has the form [278]

$$H = \hbar\omega_{\mathbf{q}} B_{\mathbf{q}}^+ B_{\mathbf{q}} + \sum_{\mathbf{K}} \varepsilon_\sigma(\mathbf{K}) A_{\mathbf{K},\sigma}^+ A_{\mathbf{K},\sigma}$$

$$+ \sum_{\mathbf{K}} \varepsilon_{\sigma'}(\mathbf{K}+\mathbf{q}) A_{\mathbf{K}+\mathbf{q},\sigma'}^+ A_{\mathbf{K}+\mathbf{q},\sigma'} + \sum_{\mathbf{K}} [G_{\sigma'\sigma}(\mathbf{q}) B_{\mathbf{q}} A_{\mathbf{K}+\mathbf{q},\sigma'}^+ A_{\mathbf{K},\sigma} + \text{H.c.}],$$

$$(2.47)$$

where $B_{\mathbf{q}}^+(B_{\mathbf{q}})$ is the creation (destruction) operator of the magnon with wave vector $\mathbf{q}$ and frequency $\omega_{\mathbf{q}}$ (index $\mathbf{q}$ also includes polarization of the magnon); $A_{\mathbf{K},\sigma}^+(A_{\mathbf{K},\sigma})$ is the creation (destruction) operator of the exciton with energy $\varepsilon_\sigma(\mathbf{K})$ (σ is the index of the exciton band); $A_{\mathbf{K}+\mathbf{q},\sigma'}^+$ and $A_{\mathbf{K}+\mathbf{q},\sigma'}$ are similar operators for the excitons from the band with index σ'; $G_{\sigma'\sigma}(\mathbf{q})$ is the constant of exciton–magnon coupling. Indices σ and σ' distinguish the ground states of excitons of two different series, resulting from the binding into excitons of a hole and electron from the spin-polarized subband with spin projections $\uparrow$ and $\downarrow$, respectively.

In the Hamiltonian (2.47) we omitted terms of the type $B_{\mathbf{q}}^+ A_{\mathbf{K}+\mathbf{q},\sigma}^+ A_{\mathbf{K},\sigma}$ and $B_{\mathbf{q},\sigma} A_{\mathbf{K},\sigma}^+ A_{\mathbf{K}+\mathbf{q},\sigma}$, which are responsible for the processes of intraband scattering of excitons on magnons. The reason is that in the frequency range where the line of interband absorption of magnons by excitons has a maximum, the contribution from the processes of intraband scattering of excitons on magnons can be ignored. The Hamiltonian (2.47) also does not contain terms describing the interaction of excitons and magnons with the dissipative subsystems and the pumping subsystem that create the inversion of populations of exciton bands. As in Sect. 2.4, the pumping and the dissipative subsystems will be considered phenomenologically.

The equations of motion for the operators of the magnon field, exciton polarization, and the operators of the population difference of the states $|\sigma'; \mathbf{K}+\mathbf{q}\rangle$ and $|\sigma; \mathbf{K}\rangle$ of different bands have the form

$$\dot{B}_{\mathbf{q}} = -(\mathrm{i}\omega_{\mathbf{q}} + \varkappa) B_{\mathbf{q}} - \frac{\mathrm{i}}{\hbar} G_{\sigma'\sigma}^*(\mathbf{q}) \sum_{\mathbf{K}} S^-(\sigma'\sigma|\mathbf{K},\mathbf{q}),$$

$$\dot{S}^-(\sigma'\sigma|\mathbf{K},\mathbf{q}) = -[\mathrm{i}\omega_{\sigma'\sigma}(\mathbf{K},\mathbf{q}) + \gamma_{\sigma'\sigma}] S^-(\sigma'\sigma|\mathbf{K},\mathbf{q})$$

$$+ 2\frac{\mathrm{i}}{\hbar} G_{\sigma'\sigma}(\mathbf{q}) B_{\mathbf{q}} S_z(\sigma'\sigma|\mathbf{K},\mathbf{q}),$$

$$\dot{S}_z(\sigma'\sigma|\mathbf{K},\mathbf{q}) = \frac{1}{T_1} \left[\frac{1}{2} d_{\sigma'\sigma}^{(0)}(\mathbf{K},\mathbf{q}) - S_z(\sigma'\sigma|\mathbf{K},\mathbf{q}) \right] \qquad (2.48)$$

$$+ \frac{\mathrm{i}}{\hbar} [G_{\sigma'\sigma}(\mathbf{q}) B_{\mathbf{q}} S^-(\sigma'\sigma|\mathbf{K},\mathbf{q}) - G_{\sigma'\sigma}(\mathbf{q}) B_{\mathbf{q}} S^+(\sigma'\sigma|\mathbf{K},\mathbf{q})],$$

$$\dot{S}^+(\sigma'\sigma|\mathbf{K},\mathbf{q}) = (\dot{S}^-(\sigma'\sigma|\mathbf{K},\mathbf{q}))^+,$$

$$\dot{B}_{\mathbf{q}}^+ = (\dot{B}_{\mathbf{q}})^+,$$

where the effective spin operators are introduced:

$$S_z(\sigma'\sigma|\mathbf{K}, \mathbf{q}) = \frac{1}{2}(A^+_{\mathbf{K}+\mathbf{q},\sigma'}A_{\mathbf{K}+\mathbf{q},\sigma'} - A^+_{\mathbf{K},\sigma}A_{\mathbf{K},\sigma}),$$

$$S^+(\sigma'\sigma|\mathbf{K}, \mathbf{q}) = A^+_{\mathbf{K}+\mathbf{q},\sigma'}A_{\mathbf{K},\sigma},$$

$$S^-(\sigma'\sigma|\mathbf{K}, \mathbf{q}) = A^+_{\mathbf{K},\sigma}A_{\mathbf{K}+\mathbf{q},\sigma'}$$

and the following designations are used:

$$\hbar\omega_{\sigma'\sigma}(\mathbf{K}, \mathbf{q}) = \varepsilon_{\sigma'}(\mathbf{K}+\mathbf{q}) - \varepsilon_\sigma(\mathbf{K}) = JSM(T)/M(0) + \frac{\hbar^2 q^2}{2m_{\text{ex}}} + \frac{\hbar^2 \mathbf{Kq}}{m_{\text{ex}}},$$

$\varkappa$ is the constant of magnon damping at the exciton–magnon interaction; $\gamma_{\sigma'\sigma} = \frac{1}{2}(\gamma_{\sigma'} + \gamma_\sigma)$; $\hbar\gamma_\sigma$ and $\hbar\gamma_{\sigma'}$ are the magnitudes of the washing away of the bands σ and σ'; T_1 is the time of the longitudinal relaxation of the excitons; $d^{(0)}_{\sigma'\sigma}(\mathbf{K}, \mathbf{q})$ is the initial inverse population of the exciton bands.

In approaching the macrocompleted magnon mode ($\langle B_\mathbf{q}\rangle = \sqrt{N_\mathbf{q}}e^{-i\Omega t}$, $N_\mathbf{q}$ is the number of magnons in the mode), the solution of the system of nonlinear equations (2.48) by the method described in Sect. 2.4 leads to expressions for the constant $\varkappa$ of magnon damping,

$$2\varkappa = \mathcal{G}_1 - \mathcal{G}_2 N_\mathbf{q}, \tag{2.49}$$

and for the shift of the frequency of generation Ω relative to the frequency $\omega_\mathbf{q}$ of the magnon in the absence of interaction with excitons,

$$\Omega - \omega_\mathbf{q} = (\Delta\Omega)_1 + N_\mathbf{q}(\Delta\Omega)_2, \tag{2.50}$$

which coincide with formulas (2.35) and (2.37), if we replace N, $g_{n'n}(0)$, $\Gamma_{n'n}$, $d^{(0)}_{n'n}(\mathbf{K})$, and $\omega_{n'n}(\mathbf{K})$ by $N_\mathbf{q}$, $G_{\sigma'\sigma}(\mathbf{q})$, $\gamma_{\sigma'\sigma}$, $d^{(0)}_{\sigma'\sigma}(\mathbf{K}, \mathbf{q})$, and $\omega_{\sigma'\sigma}(\mathbf{K}, \mathbf{q})$. From these expressions it is seen that the condition of the self-excitation of the quantum generator of the magnons ($\mathcal{G}_1 > 2\varkappa$) and the shift of the frequency of generation are determined by the function $d^{(0)}_{\sigma'\sigma}(\mathbf{K}, \mathbf{q})$, whose evident form depends on the method of pumping and the function of the distribution of excitons in the bands. For an equilibrium distribution of excitons we have

$$d^{(0)}_{\sigma'\sigma}(\mathbf{K}, \mathbf{q}) = \frac{8}{V}(\pi\alpha)^{3/2}e^{-\alpha K^2}\left(N_{\sigma'}e^{-\alpha(q^2+2\mathbf{Kq})} - N_\sigma\right), \tag{2.51}$$

where $N_{\sigma'}(N_\sigma)$ is the full number of excitons in the band $\sigma'(\sigma)$.

After replacing in (2.49) and (2.50) Ω by $\omega_{\mathbf{q}}$ (on the same basis as in the case of IR laser) and using formula (2.51), we obtain

$$\mathcal{G}_1 = 2\left(\frac{\alpha}{\pi}\right)^{1/2}\hbar^{-2}\gamma_{\sigma'\sigma}\left|G_{\sigma'\sigma}(\mathbf{q})\right|^2\left(\frac{m_{\mathrm{ex}}}{\hbar q}\right)^2(\tilde{N}_{\sigma'}I_2 - N_{\sigma}I_1),$$

$$\mathcal{G}_2 = 8\left(\frac{\alpha}{\pi}\right)^{1/2}\hbar^{-4}\gamma_{\sigma'\sigma}^2 T_1\left|G_{\sigma'\sigma}(\mathbf{q})\right|^4\left(\frac{m_{\mathrm{ex}}}{\hbar q}\right)^4(\tilde{N}_{\sigma'}I_4 - N_{\sigma}I_3),$$

$$(\Delta\Omega)_1 = -\left(\frac{\alpha}{\pi}\right)^{1/2}\hbar^{-2}\left|G_{\sigma'\sigma}(\mathbf{q})\right|^2\frac{m_{\mathrm{ex}}}{\hbar q}\big[C_{\sigma'\sigma}(\tilde{N}_{\sigma}I_2 - N_{\sigma}I_1) \qquad (2.52)$$

$$- (\tilde{N}_{\sigma'}I_6 - N_{\sigma}I_5)\big],$$

$$(\Delta\Omega)_2 = \left(\frac{\alpha}{\pi}\right)^{1/2}\hbar^{-4}\left|G_{\sigma'\sigma}(\mathbf{q})\right|^4\gamma_{\sigma'\sigma}T_1\left(\frac{m_{\mathrm{ex}}}{\hbar q}\right)^3$$

$$\times\big[C_{\sigma'\sigma}(\tilde{N}_{\sigma}I_4 - N_{\sigma}I_3) - (\tilde{N}_{\sigma'}I_8 - N_{\sigma}I_7)\big],$$

where

$$I_l = \int_{-\infty}^{\infty}\frac{K_z^{\mu}\exp\{-\alpha(K_z^2 + 2\beta q K_z)\}}{[a_{\sigma'\sigma}^2 + (C_{\sigma'\sigma} + K_z)^2]^{\nu}}\,dK_z. \qquad (2.53)$$

For $l = 1, 2, \ldots, 8$, the triples of numbers μ, β, ν have, respectively, the values 001, 011, 002, 012, 101, 111, 102, 112.

In (2.52) and (2.53) we use the designations

$$a_{\sigma'\sigma} = \frac{m_{\mathrm{ex}}}{\hbar q}\gamma_{\sigma'\sigma}, \qquad \tilde{N}_{\sigma'} = N_{\sigma'}e^{-\alpha q^2},$$

$$C_{\sigma'\sigma} = \frac{q}{2} - \frac{m_{\mathrm{ex}}v}{\hbar} + \frac{m_{\mathrm{ex}}}{\hbar^2 q}JSM(T)/M(0), \qquad (2.54)$$

where K_z is the projection of the exciton wave vector on the direction $\mathbf{q}$.

The integral of type $I_1(\mu = \beta = 0; \nu = 1)$ from (2.53) is sometimes found in other problems of microwave spectroscopy in research on the forms of resonance lines, in determining a function of Voigt's form [12]. In Chap. 3 the connection of the integrals I_l with the probability function during the consideration of the exciton–phonon interaction will be shown, for which the dependence on the wave vector of the phonon is as essential as the dependence of the exciton–magnon interaction on the wave vector of the magnon discussed in this section. Here we note that, as a consequence of the large conduction band splitting due to the interaction of free charge carriers with the localized magnetic moments (for example, for the europium chalcogenides $JS \sim 0.5$ eV [303]), the distance among the lowest exciton bands σ and σ', belonging to different series, exceeds the kinetic energy of the excitons which move with the speed of the spin wave:

$$JSM(T)/M(0) > \frac{m_{\mathrm{ex}}v_{\mathrm{s}}^2}{2}.$$

Therefore, during the intraserial transitions with the absorption (emission) of magnons there are the excitons from the lowest band having wave vectors that are closed in a sphere of radius $K_c = C_{\sigma'\sigma}$, where $C_{\sigma'\sigma}$ is determined by (2.54). Note that these excitons do not take part in the exciton–magnon interaction due to the law of energy and impulse conservation. This hampers the observation of the absorption of the magnons by the excitons because of the necessity to create excitons in the lowest band with large values of the wave vector $K > K_c$. However, the stimulated radiation of coherent magnons is possible during the optical creation of excitons in the state with $\mathbf{K} = \mathbf{Q}$ ($\mathbf{Q}$ is the wave vector of the photon) of the upper band. Under conditions of an inverted exciton population of exciton bands, coherent magnons with wave vectors $\mathbf{q} \geq \mathbf{K}_c - \mathbf{Q}$ are emitted. If the width of the lowest band does not exceed the distance between bands σ and σ', then if the condition $aK_c \leq \pi$ is satisfied, coherent magnons with large wave vectors are generated, their length being the order of the distance from the center to the frontier of the Brillouin zone.

The authors of the work [305] displayed the resonance behavior of magnetic circular dichroism and the Faraday effect in the magnetic ordered phase of the single crystal $CdCr_2Se_4$ in the exciton region of the spectrum (during the excitation of photons with the energy $E = 1$–1.5 eV exceeding the energy of the "red-shifted" edge of the absorption). In samples with a high content of indium impurity (~ 4 weights %) the resonance peaks related to the right- and left-circular polarized light were not observed. The Faraday effect was also suppressed sharply in these samples. Thus we expect that, for the experimental observation of the effect of generation of coherent magnons at optical excitation of magnetic semiconductors, it is necessary to have sufficiently pure crystals.

Features of the stoichiometry of crystals may be indicated using the small EPR linewidth in the region of the wide minimum on the temperature dependence $\Delta H(T)$ in the paramagnetic phase. With $T = T_{\min}$ ($T_{\min}$ is the temperature at which the function $\Delta H(T)$ achieves the minimum) the lowest value $\Delta H(T_{\min})$ corresponds to the lowest deviation from stoichiometry [20, 306]. In [20] it was shown that the experimentally observed temperature dependence of the EPR linewidth of monocrystals $CdCr_2S_4$, synthesized by the method of gas-transport reactions, is well approximated at $T > T_{\min}$ by the equation

$$\Delta H(T) = \Delta H(T_{\min}) + A \exp(-\theta/T)$$

with the parameters $\theta = 1210$ K and $A = 1300$ Oe. A noncontradictory explanation of the exponential dependence of $\Delta H(T)$ on the indicated values θ and A was given on the basis of the account of Raman processes with participation of the localized phonons in the conditions when $\theta > T_D$ (T_D is the Debye temperature; for $CdCr_2S_4$ $T_D = 415$ K [307]), leading to reorientation of the spin of the Cr^{3+} ion. Local vibrations are performed near the locations of rapidly relaxing paramagnetic impurity centers (the role of the latter is played by the ions of the two-valent chromium, which appear because of the deviation of the composition of $CdCr_2S_4$ from stoichiometry with the formation of a sulfur deficit [308]). That is why the value of ΔH in the domain of the wide minimum (near T_{min}) indicates the degree of nonstoichiometry of a magnetic semiconductor.

2.6 Exciton Paraelectric Resonance

Paraelectric resonance (PER) as a method of investigation of the impurity centers in solids with proper electric moments was proposed in [41]. The first experiments in this field (PER of dipolar molecules) were described in [309]. Substitutional impurity ions, shifted from their equilibrium positions in the lattice nodes ("noncentral" ions), can be investigated by PER method. The movement of these centers in a crystal is limited by the domain of their location and involves tunneling through the potential barriers. The first information on the observation of PER of noncentral ions was presented in [310]. An analysis of the results of the PER investigation of noncentral ions is given by Deigen and Glinchuk [42].

With PER selective absorption of the energy of an alternative electric field by a system of electric dipoles in the external constant electric field takes place. The directed deformation of the crystal [311] is applied along with the external constant electric field to remove the degeneration of the energy levels of the paraelectric system. The energy of the alternative field is transmitted by the resonance to the system of the electric dipoles, which leads to violation of its thermodynamic equilibrium. The latter is restored by the transmission of an abundance of absorbed energy to the lattice by dipole–lattice relaxation.

In some cases paraelectric systems provide perspective for practical applications in quantum electronics and low-temperature techniques [311]. For this reason, and because the optical electric dipole transitions between some of the states of paraelectric systems are forbidden by the selection rules (which hinders the study of these states by the usual optical methods), expanding the number of objects that can be investigated by the PER method is of interest. To this range of objects we can attribute Wannier–Mott excitons in semiconductors, regardless of the fact that, in contrast to paraelectric centers, the excitons possess translational symmetry. Exciton paraelectric resonance (ExPR) was studied theoretically in [170, 279–282].

We shall consider the degenerate exciton levels that are split in a constant electric field $\mathbf{E}_0$, where the magnitude of the splitting $\Delta(\mathbf{E}_0)$ is in the microwave range. The transitions between split components are excited by the electric component of the microwave field (ExPR). The experimental conditions for the ExPR observations are more favorable in crystals without a center of inversion, in which $\Delta(\mathbf{E}_0) \sim E_0$ may be expected. For this the exciton level must be orbitally degenerated, and its states must be characterized by different values of the proper or induced electric dipole moment. The exciton states that are transformed according to the twofold irreducible representations of the point groups of symmetry D_3 and C_{3v} or to the threefold representations of the symmetry groups T_d and O satisfy these conditions. Doubly degenerate exciton levels are not split by the electric field, if the corresponding exciton states are transformed according to the twofold irreducible representations of the symmetry point groups C_3, S_4, S_6, and T. This is a consequence of time-reversal symmetry.

Linear Stark splitting should show, in particular, the energy levels corresponding to exciton states formed from electron states of the lowest conduction band Γ_6 and hole states of the upper valence band Γ_5, which is split by the spin–orbital interaction into subbands Γ_7 and Γ_8 in the crystal field of symmetry T_d. It relates, for example, to the dipole-active absorption line of orthoexciton $1S(\Gamma_5)$, formed by an electron from the Γ_6 band and a hole from the subband Γ_7 and to the lines as $1S(\Gamma_4)$ and $1S(\Gamma_5)$, corresponding to the exciton formed by an electron from the Γ_6 band and a hole from the Γ_8 subband in the crystals CuCl, CuBr, and CuI [312]. The linear Stark effect has been observed in the electroabsorption on excitons in CuCl [313].

Other objects in which it is possible to observe the linear Stark effect on excitons are semiconductors of type $A_2 B_6$ and $A_3 B_5$ with the structure of zinc blende: ZnS, ZnSe, ZnTe, CdTe and GaP, GaAs, InP.

The absence of a center of inversion does not guarantee the linear splitting of the exciton level by the electric field. For example, in crystals with C_{6v} symmetry, the Γ_5 and Γ_6 states [314] at comparatively small electric fields are split quadratically on the field. This results from general theoretical-group reasons, and it is confirmed on the basis of experiments for CdS by Thomas and Hopfield [315].

The exciton level splittings necessary for observation of ExPR may also be achieved in crystals that possess a center of inversion. In this case in second-order perturbation theory $\Delta(\mathbf{E}_0) \sim E_0^2$. We shall select for consideration ExPR, in the presence of a center of inversion, for triply degenerate quadrupole-active level $1S(\Gamma_5^+)$ of the yellow exciton series of crystal Cu_2O. Level $1S(\Gamma_5^+)$ is probably suitable for the ExPR observation, because it is deep and allows us to apply high electric fields without essential self-ionization of the exciton state. This level is sufficiently narrow and isolated from other levels. For the calculation of $\Delta(\mathbf{E}_0)$ we shall use the wave functions of first-order perturbation theory, taking into consideration the mixing under the action of the electric field of the state Γ_5^+ with the excited states of the type Γ_2^-, Γ_3^-, Γ_4^- and Γ_5^-. At $\mathbf{E}_0 \parallel \langle 1\bar{1}0 \rangle$ the corrections of the second order to the energy of the components of the split level $1S(\Gamma_5^+)$, where it mixes in the constant electric field only with the states $n = 2$ of the yellow exciton series [316], are

$$\varepsilon_{A_1} = E_0^2 \left(\frac{D_1^2}{\delta_1} + \frac{D_2^2}{\delta_2} + \frac{D_2'^2}{\delta_2'} \right),$$

$$\varepsilon_{A_2} = E_0^2 \left(\frac{D_2^2}{\delta_2} + \frac{D_2'^2}{\delta_2'} + \frac{3}{2} \cdot \frac{|D_4|^2}{\delta_4} \right), \tag{2.55}$$

$$\varepsilon_{B_2} = E_0^2 \left(\frac{D_1^2}{\delta_1} + \frac{D_3^2}{\delta_3} + \frac{1}{2} \cdot \frac{|D_4|^2}{\delta_4} \right),$$

where $\delta_i = \varepsilon^0_{\Gamma_5^+(n=1)} - \varepsilon^0_{\Gamma_j^-(n=2)}$, i equal to 1, 2, 3, 4 respectively for j, which is equal to 4, 5, 2, 3; A_1, A_2 and B_2 are irreducible representations of the symmetry point group C_{2v} [127]. The terms D_2^2/δ_2 and $D_2'^2/\delta_2'$ reflect the fact that for $n = 2$ there are two different states of type Γ_5^-.

Matrix elements D_i of the electric dipole moment of transition between states Γ_5^+ and $\Gamma_j^- = \Gamma_4^-, \Gamma_5^-, \Gamma_2^-, \Gamma_3^-$ have the form

$$D_i = e\langle\Psi_{\Gamma_5^+,j_1}^0 \mid \sum_k r_{kj_2} \mid \Psi_{\Gamma_j^-,j_3}^0\rangle, \tag{2.56}$$

where r_{kj_2} is the j_2-projection of the radius vector of the kth electron; j_1, j_2 and j_3 are selected as $D_i \neq 0$. For example, for $\Gamma_j^- = \Gamma_4^-$, $j_1 = 1$, $j_2 = 2$, $j_3 = 3$.

In order to account for mixing of state Γ_5^+ with the odd excited states of all terms of the exciton series, it is sufficient to replace $D_i^2\delta_i^{-1}$ in (2.55) by $\sum_{fn} D_{i,n}^2\delta_{i,n,f}^{-1}$, where f is the number of representations Γ_j for the given n.

The burning of the exciton absorption in the dipole approximation to the level Γ_5^+ in the electric field is affected by the mixing of state Γ_5^+ only with states of type Γ_4^-. In contrast, in the splitting of the level Γ_5^+ in the electric field, the mixing with other states contributes, and these states are transformed according to odd representations. As can be seen from (2.55), it is not possible to achieve a complete splitting without taking into account the mixing with the states of type Γ_2^-, Γ_3^-, and Γ_5^-. Since Stark splittings depend on δ_i^{-1} linearly, the line $1S(\Gamma_5^+)$ of the yellow exciton series of the Cu_2O crystal in the electric field must show, along with the splitting, a shift to the longer wavelengths of the spectrum. The latter is in accordance with the experiments of Nikitine et al. [317]. However, the change of the energy gap in the electric field also leads to a shift of exciton levels.

Estimations according to the formulas (2.55) in a one-electron approximation on the simple exciton wave functions taking into account orbital degeneration of the valence band with $r_{ex(n=1)} = 10$ Å, $r_{ex(n=2)} = 30$–40 Å, $\delta_i = -0.12$ eV, and $E_0 = 160$ kV/cm lead to $\Delta(E_0) \approx (1$–$3) \cdot 10^{-4}$ eV, which corresponds to the value $\Delta(E_0)$, estimated in the experiments on the burning and amplification of the absorption on level Γ_5^+ in the electric field [316]. Stark splitting of the line $1S(\Gamma_5^+)$ in the Cu_2O crystal seems to be of the order of the width of exciton level $1S(\Gamma_5^+)$ at a temperature of 4.2 K in the absence of the electric field ($\Delta\nu_{1/2} = 2.3 \cdot 10^{-4}$ eV [317]) and may be measured by the method of modulational spectroscopy, applying synchronous detection [318].

We shall determine the amplitudes of the paraelectric transitions between the components of level Γ_5^+ with the assumption that the main contribution to the width of the Stark sublevels is caused by exciton–phonon collisions and the width of separate Stark components is not bigger than the width of the exciton line in the absence of an electric field. Matrix elements of the transitions between the components Γ and Γ' of level Γ_5^+ having been split in the electric field for different polarizations $[hkl]$ of the incident microwave radiation are determined by the expression

$$P_{\Gamma\Gamma'}^{[hkl]} = e\langle\Psi_\Gamma \mid \sum_i \mathbf{r}_i \mid \Psi_{\Gamma'}\rangle\mathbf{E}_1^{[hkl]},$$

where Ψ_Γ and $\Psi_{\Gamma'}$ are the wave functions of the exciton on sublevels Γ and Γ' in first-order perturbation theory (taking into account the mixing with the levels Γ_2^-, Γ_3^-, Γ_4^-, and Γ_5^- for $n \geq 2$), and E_1 is the amplitude of the electric component of the microwave.

For some symmetry directions in the crystal we obtain

$$\mathbf{E}_0 \parallel [1\bar{1}0] \quad P_{A_1 B_2}^{[001]} = \frac{E_1}{E_0}(\varepsilon_{B_2} - \varepsilon_{A_2}),$$

$$\mathbf{E}_0 \parallel [001] \quad P_{A_2 B_2}^{[110]} = \frac{E_1}{E_0}(\varepsilon_{A_2} + \varepsilon_{B_2} - 2\varepsilon_{A_1}),$$

$$P_{\Delta_2' \Delta_5}^{[1\bar{1}0]} = P_{\Delta_2' \Delta_5}^{[110]} = P_{A_1 B_2}^{[001]}, \tag{2.57}$$

$$\mathbf{E}_0 \parallel [111] \quad P_{\Lambda_1 \Lambda_3}^{[1\bar{1}0]} = P_{\Lambda_1 \Lambda_3}^{[11\bar{2}]} = \frac{E_1}{3E_0}(3\varepsilon_{B_2} + \varepsilon_{A_2} - 4\varepsilon_{A_1})$$

where Δ_2', Δ_5, and Λ_j are the irreducible representations of the symmetry groups C_{4v} and C_{3v} [319]. The designations A_1, A_2, B_2, B_1 correspond to E_1, E_2, E_3, E_4 from [319]. The other matrix elements for the directions $\mathbf{E}_0$ indicated in (2.57) are equal to 0. Therefore, in the adopted approximation $P_{\Gamma \Gamma'} = 0$ at $\mathbf{E}_1 \parallel \mathbf{E}_0$ for all directions of the electric field $\mathbf{E}_0$ and for the ExPR that we consider, the condition of the perpendicularity of the field, consisting in the fact that the component $\mathbf{E}_1$ is perpendicular to $\mathbf{E}_0$, is performed.

The absorption of the energy of the alternative electric field may be used for detecting and investigating the optically inactive levels. One such level is $1S(\Gamma_2^+)$ of the yellow exciton series of the Cu_2O crystal. The nonzero matrix elements of the transitions between the components of the levels Γ_5^+ and Γ_2^+ are

$$P_{A_1 B_1}^{[110]} = -P_{A_2 B_1}^{[001]} = P_{\Delta_2 \Delta_5}^{[1\bar{1}0]} = P_{\Lambda_2 \Lambda_3}^{[1\bar{1}0]} = -P_{\Lambda_2 \Lambda_3}^{[11\bar{2}]} = e^2 E_0 E_1 D_2 D_5 \left(\frac{1}{\delta_5} - \frac{1}{\delta_2}\right), \tag{2.58}$$

where $\delta_5 = \varepsilon_{\Gamma_2^+(n=1)}^0 - \varepsilon_{\Gamma_5^-(n=2)}^0$, D_5 is the matrix element of the electric dipole moment of the transition, similar to D_i from (2.56), between states Γ_2^+ and Γ_5^-, and B_1, Δ_2, and Λ_2 are the irreducible representations according to which state Γ_2^+ is transformed with the decrease of the symmetry of the crystal up to C_{2v}, C_{4v}, and C_{3v} that takes place, respectively, at $\mathbf{E}_0 \parallel [1\bar{1}0]$, $\mathbf{E}_0 \parallel [001]$, and $\mathbf{E}_0 \parallel [111]$ [319]. It is seen from (2.58) that the condition of the perpendicularity of the fields is performed for the transitions between states Γ_2^+ and Γ_5^-.

The expressions (2.57) and (2.58) for $P_{\Gamma \Gamma'}^{[hkl]}$ are obtained without the specificity of the type of exciton wave functions. The latter enter only in the matrix elements D_i. In accordance with the formulas (2.55) the matrix elements D_i can be determined from the experimental data for the splittings and the shifts of exciton lines in the electric field. If this is possible, then one can select optimal conditions for the ExPR observation on the excitons, immediately resulting from the experimental data. The selection of optimal conditions for the ExPR observation can also be performed if the data on the polarizability of the excitons are known. The large diamagnetism [320] of the excitons and their large polarizability [321] cause high oscillator strengths for the paraelectric transitions. The values D_i can also be determined from the experiments on burning and amplification of exciton lines of light absorption in an electric field, or estimated by means of direct theoretical calculations in the one-electron approximation [281, 282, 316].

We shall estimate the power absorbed by the unit volume of the crystal at the ExPR transition from level A_1 to level B_2 by the formula

$$W_{A_1 B_2} = \frac{\pi}{2\hbar}(N_{A_1} - N_{B_2})\omega_{A_1 B_2}\left(\frac{E_1}{E_0}\right)^2 (\varepsilon_{B_2} - \varepsilon_{A_1})^2 g(\omega),$$

where $\omega_{A_1 B_2}$ is the frequency of transition, N_{A_1} and N_{B_2} are the populations of levels A_1 and B_2, and $g(\omega)$ is the function of the form of the ExPR line.

In the high temperature approximation ($\varepsilon_{B_2} - \varepsilon_{A_1} \ll k_0 T$) at $\varepsilon_{B_2} - \varepsilon_{A_1} = 2$ cm^{-1}, $T = 77$ K, $E_0 = 160$ kV/cm, concentration of excitons $n_{ex} = 10^{15}$ cm^{-3}, and with the supposition that the ExPR linewidth is approximately equal to the resonance frequency, we obtain $W_{A_1 B_2} \approx 5 \cdot 10^{-8}$ W/cm^3. Let us remark that the fluctuations of the power of a microwave generator are less in comparison with the estimated value of the power absorbed by a unit of the volume of the substance with ExPR transitions on excitons. For example, for a klystron of 3 cm spectral range, the level of noise in the 2.5 MHz band, shifted with respect to the fundamental frequency of the klystron on the value of 30 MHz, is 10^{-11}–10^{-12} W and sharply decreases with an increase of the intermediate frequency from 30 MHz to 90 MHz [12]. Therefore, the experimental detection of ExPR in semiconductors is feasible.

In relatively pure crystals, in which the concentration of impurities is $n_{imp} \lesssim 10^{16}$ cm^{-3}, the width of the exciton levels is determined mainly by exciton–phonon and exciton–exciton collisions. In this case, the time of the free run of the exciton at low temperatures is $\tau_{path} \sim 10^{-10}$–$10^{-9}$ s [171]. Consequently, ExPR may be observed on the frequencies $\omega > 10^9$–10^{10} s^{-1}.

The form of the ExPR line is also determined by exciton–phonon and exciton–exciton interaction. These interactions are as effective as dipole–lattice and dipole–dipole interaction for localized electric dipoles (for OH$^-$ dipoles in the KCl crystal the electric dipole–lattice relaxation time at $T = 11$ K is equal to 10^{-11} s) [309], but the electric dipole–dipole relaxation time at 2.4 K and $n_{OH^-} = 3 \cdot 10^{16}$ cm^{-3} is 10^{-10} s [311]. Similarly as with the local centers we could introduce for the excitons the concepts of electric dipole–lattice and dipole–dipole interaction indicated above. However, an important peculiarity of excitons distinguishing them from the local centers is the possibility of intraband exciton–phonon scattering.

Absorption of the energy of the electric field by excitons can also occur and then, when the processes of the paraelectric relaxation are not effective. This possibility is influenced by the presence of the difference of populations of exciton levels under conditions of optical pumping. The numbers of filling excitons on the split electric field sublevels are determined by pumping and depend both on the processes of relaxation between sublevels and the lifetime of the excitons on these sublevels relative to the processes of annihilation.

2.7 Isotopic Shift of Exciton Paraelectric Resonance in Cu_2O Crystals

The PER on excitons considered in the preceding section may be both intraserial (by which ExPR transitions take place between the states belonging to one exciton series) and interserial. An interesting peculiarity of interserial ExPR is the possibility for it to manifest isotopic effects. Qualitatively it may be explained on the basis of the existing experimental data on the isotopic shift of the exciton lines in the optical spectrum of Cu_2O crystals. According to [322–324], with isotopic substitution of the isotope ^{16}O by ^{18}O in the crystal Cu_2O a shift of all the lines of the exciton absorption takes place. Each of all four exciton series are shifted as a whole without a significant change of the Rydberg constant (the yellow and green series are shifted to shorter wavelengths of the spectrum, the dark blue and light blue series toward longer wavelengths of the spectrum). In the substituted crystal the energy of the longitudinal optical phonon of symmetry Γ_4^-, caused by the shift of the oxygen ions, decreases by 25 cm^{-1}. Changes of the lattice constant are not detected.

As isotopic substitution leads to a shift of the exciton series as a whole, it will not be displayed by intraserial ExPR, but it may be detected in the spectra of interserial ExPR, if the shift of the ExPR line is large enough for the measurement. For excitons of the yellow and green series of Cu_2O crystals, the isotopic shift of the interserial ExPR lines, as will be shown below, occurs in the short-wavelength region of the spectrum and has a magnitude ~ 2.3 cm^{-1} that may be easily measured.

The basis for the theoretical calculation of the considered effect of the isotopic shift of the interserial ExPR lines is the method of canonical transformation, which helps us to obtain renormalized energies of the electrons and holes due to the interaction with phonons before and after the isotopic substitution.

The Hamiltonian of the system of electrons and holes interacting with the phonons has the form

$$H = H_0 + H_1 + H_2, \tag{2.59}$$

where

$$H_0 = \sum_{kl} E_{\Gamma_6^+ l}(\mathbf{k}) a^+_{\mathbf{k}\Gamma_6^+ l} a_{\mathbf{k}\Gamma_6^+ l} + \sum_{km} E_{\Gamma_8^- m}(\mathbf{k}) a^+_{\mathbf{k}\Gamma_8^- m} a_{\mathbf{k}\Gamma_8^- m}$$

$$+ \sum_{kn} E_{\Gamma_7^{+*} n}(\mathbf{k}) B^+_{\mathbf{k}\Gamma_7^{+*} n} B_{\mathbf{k}\Gamma_7^{+*} n} + \sum_{ks} E_{\Gamma_8^{+*} s}(\mathbf{k}) B^+_{\mathbf{k}\Gamma_8^{+*} s} B_{\mathbf{k}\Gamma_8^{+*} s}$$

$$+ \sum_{\mathbf{q}} \hbar\omega_0 C^+_{\mathbf{q}} C_{\mathbf{q}}, \tag{2.60}$$

$$H_1 = \frac{i}{\sqrt{N}} \sum_{\mathbf{k}\mathbf{q}jl} \{\theta_e(\mathbf{k},\mathbf{q};j,l) a^+_{\mathbf{k}+\mathbf{q},\Gamma_6^+ j} a_{\mathbf{k}\Gamma_6^- l}$$

$$+ \theta_e^*(\mathbf{k},-\mathbf{q};j,l) a^+_{\mathbf{k}\Gamma_8^- l} a_{\mathbf{k}-\mathbf{q},\Gamma_6^+ j} - \theta_h(\mathbf{k},\mathbf{q};j,l) B^+_{-\mathbf{k}\Gamma_8^{+*} l} B_{-\mathbf{k}-\mathbf{q},\Gamma_7^{+*} j}$$

$$- \theta_h^*(\mathbf{k},-\mathbf{q};j,l) B^+_{-\mathbf{k}+\mathbf{q},\Gamma_7^{+*} j} B_{-\mathbf{k}\Gamma_8^{+*} l}\} (C_{\mathbf{q}} - C^+_{-\mathbf{q}}),$$

$$H_2 = \frac{i}{\sqrt{N}} \sum_{\mathbf{k}\mathbf{q}j} \vartheta(\mathbf{q}) \left(a^+_{\mathbf{k}+\mathbf{q},\Gamma_6^+ j} a_{\mathbf{k}\Gamma_6^+ j} + a^+_{\mathbf{k}+\mathbf{q},\Gamma_8^- j} a_{\mathbf{k}\Gamma_8^- j} \right.$$

$$\left. - B^+_{-\mathbf{k}\Gamma_7^{+*} j} B_{-\mathbf{k}-\mathbf{q},\Gamma_7^{+*} j} - B^+_{-\mathbf{k}\Gamma_8^{+*} j} B_{-\mathbf{k}-\mathbf{q},\Gamma_8^{+*} j} \right) (C_{\mathbf{q}} - C^+_{-\mathbf{q}}). \qquad (2.61)$$

In (2.60)–(2.61) $C_{\mathbf{q}}^+$ is the creation operator of the longitudinal optical phonon with wave vector $\mathbf{q}$, $a^+_{\mathbf{k}\Gamma\gamma}$ and $B^+_{\mathbf{k}\Gamma*\gamma}$ are, respectively, the creation operators of the electron and hole with wave vector $\mathbf{k}$ in the band characterized by the irreducible representation Γ; $E_{\Gamma\gamma}(\mathbf{k})$ and $E_{\Gamma*\gamma}(\mathbf{k})$ are the energy of the electron and hole in the corresponding bands, γ is the line of representation, and N is the number of atoms in the lattice. By $\vartheta(\mathbf{q})$ and $\theta_{e(h)}(\mathbf{k},\mathbf{q};j,l)$ we designate the Fourier transform of the energy of the electron (hole)–phonon interaction for the cases of intraband and interband scattering, respectively. The function $\vartheta(\mathbf{q})$ has the form (only the Fröhlich interaction is taken into consideration)

$$\vartheta(\mathbf{q}) = \left(2\pi e^2 \hbar\omega_0 / \varepsilon^* V_0 q^2\right)^{1/2} \langle \Gamma\gamma, \mathbf{k}_1 | e^{i\mathbf{q}\mathbf{r}} | \Gamma\gamma, \mathbf{k}_2 \rangle,$$

where $1/\varepsilon^* = 1/\varepsilon_\infty - 1/\varepsilon_0$, ω_0 is the frequency of the longitudinal optical phonon, and V_0 is the volume of the elementary cell. The form of the functions $\theta_e(\mathbf{k},\mathbf{q};j,l)$ and $\theta_h(\mathbf{k},\mathbf{q};j,l)$ must be determined.

Electron (hole)–phonon interaction is eliminated by means of the Fröhlich unitary transformation of the Hamiltonian (2.59)

$$\tilde{H} = e^{-S} H e^S.$$

The operator S satisfies the condition

$$[H_0, S] + H_1 = 0.$$

After averaging on the phonon subsystem we obtain the effective Hamiltonian of the electrons and holes with the renormalized energy. At $T = 0$ the expressions for the energy of electrons in the bands Γ_6^+, Γ_8^- and holes in the bands Γ_7^+ and Γ_8^+, renormalized by intraband and interband scattering on phonons, can be presented in the form

$$\tilde{E}_{\Gamma_\nu}(\mathbf{k}) = E_{\Gamma_\nu}(\mathbf{k}) + \Delta_{\Gamma_\nu}^{\text{intra}} + \Delta_{\Gamma_\nu}^{\text{inter}}, \qquad (2.62)$$

where $\Gamma = \Gamma_6^+, \Gamma_8^-, \Gamma_7^+$, and Γ_8^+; $\Delta_{\Gamma_\nu}^{\text{intra}}$ and $\Delta_{\Gamma_\nu}^{\text{inter}}$ are energy corrections influenced, respectively, by intra- and interband scattering of electrons and holes on the phonons.

The energy correction $\Delta_{\Gamma_\nu}^{\text{intra}}$ from (2.62) is determined by the expression

$$\Delta_{\Gamma_\nu}^{\text{intra}} = -\frac{1}{N} \sum_q \frac{|\vartheta(\mathbf{q})|^2}{E_{\Gamma_\nu}(\mathbf{k}+\mathbf{q}) - E_{\Gamma_\nu}(\mathbf{k}) + \hbar\omega_0},$$

which in the approximation [325–327]

$$\frac{\hbar^2(\mathbf{kq})}{m_{\Gamma_\nu}(\hbar\omega_0 + \hbar^2 q^2/2m_{\Gamma_\nu})} \ll 1$$

is transformed to

$$\Delta_{\Gamma_\nu}^{\text{intra}} = -\frac{1}{2}\frac{e^2}{\varepsilon^*} U_{\Gamma_\nu}. \tag{2.63}$$

Here $U_{\Gamma_\nu} = \sqrt{2m_{\Gamma_\nu}\omega_0/\hbar}$ is the inverse polaron radius and m_{Γ_ν} is the mass of the band electron (hole).

Corrections influenced by interband scattering have the form

$$\Delta_{\Gamma_6^+,j}^{\text{inter}} = -\frac{1}{N} \sum_q \frac{|\theta_e(\mathbf{k},\mathbf{q};j,m)|^2}{E_{\Gamma_8^-}(\mathbf{k}+\mathbf{q}) - E_{\Gamma_6^+}(\mathbf{k}) + \hbar\omega_0},$$

$$\Delta_{\Gamma_8^-,j}^{\text{inter}} = \frac{1}{N} \sum_{qm} \frac{|\theta_e(\mathbf{k},\mathbf{q};j,m)|^2}{E_{\Gamma_8^-}(\mathbf{k}) - E_{\Gamma_6^+}(\mathbf{k}+\mathbf{q}) - \hbar\omega_0},$$

$$\Delta_{\Gamma_7^{+*}j}^{\text{inter}} = \frac{1}{N} \sum_{qm} \frac{|\theta_h(\mathbf{k},\mathbf{q};j,m)|^2}{E_{\Gamma_8^+}(\mathbf{k}-\mathbf{q}) - E_{\Gamma_7^+}(\mathbf{k}) + \hbar\omega_0},$$

$$\Delta_{\Gamma_8^{+*}j}^{\text{inter}} = -\frac{1}{N} \sum_{qm} \frac{|\theta_h(\mathbf{k},\mathbf{q};j,m)|^2}{E_{\Gamma_8^+}(\mathbf{k}) - E_{\Gamma_7^+}(\mathbf{k}-\mathbf{q}) - \hbar\omega_0},$$

$$\tag{2.64}$$

where

$$\theta_{e(h)}(\mathbf{k},\mathbf{q};j,m) = \left(\frac{2\pi e^2 \hbar\omega_0}{V_0\varepsilon^* q^2}\right)^{1/2} \mathcal{I}_{e(h)}. \tag{2.65}$$

Here

$$\mathcal{I}_e = \int \Psi_{\Gamma_6^+ j,\mathbf{k}_1}^*(\mathbf{r}) e^{i\mathbf{qr}} \Psi_{\Gamma_8^- m,\mathbf{k}_2}(\mathbf{r}) \, d\mathbf{r}, \tag{2.66}$$

$$\mathcal{I}_h = \int \Psi_{\Gamma_7^+ j,\mathbf{k}_1}^*(\mathbf{r}) e^{i\mathbf{qr}} \Psi_{\Gamma_8^+ m,\mathbf{k}_2}(\mathbf{r}) \, d\mathbf{r}. \tag{2.67}$$

In formulas (2.66) and (2.67) $\Psi_{\Gamma m,\mathbf{k}}(\mathbf{r})$ is the Bloch wave function of the electron (hole) in the conduction (valence) band which may be expressed through Wannier functions localized in the nodes of the lattice:

$$\Psi_{\Gamma\mathbf{k}}(\mathbf{r}) = \frac{1}{\sqrt{N}} \sum_{\mathbf{g}} e^{i\mathbf{k}\mathbf{g}} W_\Gamma(\mathbf{r} - \mathbf{g}). \tag{2.68}$$

From formulas (2.66) and (2.68) we have

$$\mathcal{I}_e = \delta_{\mathbf{k}_1 - \mathbf{k}_2, \mathbf{q}} \sum_{\mathbf{f}} e^{-i\mathbf{f}(\mathbf{k}_2 + \mathbf{q})} \int W^*_{\Gamma_6^+ j}(\mathbf{r} - \mathbf{f}) e^{i\mathbf{q}\mathbf{r}} W_{\Gamma_8^- m}(\mathbf{r}) \, d\mathbf{r}$$

$$\approx \delta_{\mathbf{k}_1 - \mathbf{k}_2, \mathbf{q}} \int W^*_{\Gamma_6^+ j}(\mathbf{r}) e^{i\mathbf{q}\mathbf{r}} W_{\Gamma_8^- m}(\mathbf{r}) \, d\mathbf{r}. \tag{2.69}$$

Since the conduction bands Γ_6^+ and Γ_8^- are formed mainly from $4s$ and $4p$ states of the copper atom, to estimate the integral (2.69) we can use as functions $W_{\Gamma_6^+}$ and $W_{\Gamma_8^-}$ the atomic functions Ψ_{4s} and Ψ_{4p}. If by this we preserve in $\mathcal{I}_e$ (that depends on $\mathbf{q}$) only the term which gives the biggest contribution in summation (integration) over $\mathbf{q}$ into (2.64), then from (2.65) we obtain the following expression for $\theta_e(q)$:

$$\theta_e(q) \cong i \left(\frac{2\pi e^2 \hbar \omega_0}{\varepsilon^* V_0} \right)^{1/2} \frac{\sqrt{5} a_s}{[1 + (a_s q/2)^2]^3}, \tag{2.70}$$

in which the radius of the copper atom $4 a_B/Z_a$ is replaced by the radius a_s of the atom in the crystal.

In order to obtain the expression for $\theta_h(\mathbf{k}, \mathbf{q}; j, m)$ we use the wave functions of the valence electron in the bands Γ_7^+ and Γ_8^+ [328]:

$$\Psi_{\Gamma_7^+, 1/2} = -\frac{i\mathcal{A}}{\sqrt{3}} \frac{R_{32}(r)}{r^2} \big[XY|\alpha\rangle + (YZ + iXZ)|\beta\rangle \big],$$

$$\Psi_{\Gamma_7^+, -1/2} = -\frac{i\mathcal{A}}{\sqrt{3}} \frac{R_{32}(r)}{r^2} \big[-XY|\beta\rangle + (YZ - iXZ)|\alpha\rangle \big],$$

$$\Psi_{\Gamma_8^+, 3/2} = \frac{i\mathcal{A}}{\sqrt{6}} \frac{R_{32}(r)}{r^2} \big[2XY|\beta\rangle + (YZ - iXZ)|\alpha\rangle \big],$$

$$\Psi_{\Gamma_8^+, 1/2} = -\frac{i\mathcal{A}}{\sqrt{2}} \frac{R_{32}(r)}{r^2} (YZ - iXZ)|\beta\rangle, \tag{2.71}$$

$$\Psi_{\Gamma_8^+, -1/2} = \frac{i\mathcal{A}}{\sqrt{2}} \frac{R_{32}(r)}{r^2} (YZ + iXZ)|\alpha\rangle,$$

$$\Psi_{\Gamma_8^+, -3/2} = \frac{i\mathcal{A}}{\sqrt{6}} \frac{R_{32}(r)}{r^2} \big[2XY|\alpha\rangle - (YZ + iXZ)|\beta\rangle \big],$$

where $|\alpha\rangle$ and $|\beta\rangle$ are the spin functions, $R_{32}(r)$ is the radial function of the $3d$ state of the copper ion, and $\mathcal{A}$ is the normalization constant.

Using (2.65), (2.67), and (2.71), we obtain

$$\theta_h(q) \cong i \left(\frac{2\pi e^2 \hbar \omega_0}{\varepsilon^* V_0} \right)^{1/2} \frac{q a_i^2}{[1 + (a_i q/2)^2]^5} \cdot \frac{2\sqrt{2}}{3}. \tag{2.72}$$

As in the previous case, we preserved in (2.72) only the term which will give the biggest contribution in integration over $\mathbf{q}$. The radius of the copper ion $3a_B/Z_i$ was replaced by the radius a_i of the ion in the crystal.

The addition to the electron energy in the conduction band Γ_6^+ due to electron–phonon scattering with transition of the electron from band $\Gamma_6^+(\Gamma_8^-)$ to band $\Gamma_8^-(\Gamma_6^+)$ according to (2.64) and (2.70) is determined by the expression [325–327]

$$\Delta_{\Gamma_6^+}^{inter} \approx -\frac{35 e^2}{64 \varepsilon^* a_s} \cdot \frac{\hbar \omega_0}{E_B - E_Y + \hbar \omega_0 + \hbar^2 q_{1 max}^2 / 2 m_{\Gamma_8^-}}, \tag{2.73}$$

where E_B and E_Y are the energy gaps for the blue and yellow series, respectively. The value $q_{1 max} = \frac{2}{\sqrt{5}} a_s^{-1}$ is found from the condition of the maximum of function

$$f_1(q) = q^2 / [1 + (q a_s/2)^2]^6.$$

For the corrections to the energies of holes in bands Γ_7^+ and Γ_8^+ influenced by interband scattering $\Gamma_7^+ \rightleftarrows \Gamma_8^+$ of holes on phonons, on the basis of formulas (2.64) and (2.72) we get [325–327]

$$\Delta_{\Gamma_7^{+*}}^{inter} \approx \frac{0.031 e^2}{\varepsilon^* a_i} \cdot \frac{\hbar \omega_0}{E_G - E_Y + \hbar \omega_0 + \hbar^2 q_{2 max}^2 / 2 m_{\Gamma_8^+}},$$
$$\Delta_{\Gamma_8^{+*}}^{inter} \approx -\frac{0.031 e^2}{\varepsilon^* a_i} \cdot \frac{\hbar \omega_0}{E_G - E_Y - \hbar \omega_0 - \hbar^2 q_{2 max}^2 / 2 m_{\Gamma_7^+}}. \tag{2.74}$$

Here E_G is the energy gap for the green series, and $q_{2 max} = a_i^{-1}$ is determined from the condition of the maximum of the function

$$f_2(q) = q^4 / [1 + (q a_i/2)^2]^{10}. \tag{2.75}$$

The numerical estimations of the values of the shifts of bands Γ_6^+, Γ_8^-, Γ_7^+, and Γ_8^+ and the corresponding shifts of exciton levels at inter- and intraband scattering of electrons and holes on phonons were determined according to formulas (2.63), (2.73), and (2.74) at the following values of the parameters [325–330]: $E_Y = 17523$ cm^{-1}, $E_G = 18588$ cm^{-1}, $E_B = 21220$ cm^{-1}, $m_{\Gamma_6^+} = 0.99 m_0$, $m_{\Gamma_8^-} = 0.26 m_0$, $m_{\Gamma_7^+} = 0.69 m_0$, $m_{\Gamma_8^+} = 1.776 m_0$, $\varepsilon_0 = 7.5$, $\varepsilon_\infty = 6.25$, $\hbar \omega_0 = 660$ cm^{-1}, $a_s = 0.8d$, and $a_i = 0.4d$ ($d = 4.263$ Å is the constant of the lattice). The results are given in Tables 2.1 and 2.2.

Table 2.1 The energy shift of bands Γ_6^+, Γ_7^+, and Γ_8^+ at intra- and interband scattering of electrons and holes on the longitudinal optical phonons of symmetry Γ_4^- before and after isotopic substitution $^{16}O \rightarrow {}^{18}O$ in Cu$_2$O crystals

	$\Delta_{\Gamma_6^+}^{intra}$ (cm^{-1})	$\Delta_{\Gamma_6^+}^{inter}$ (cm^{-1})	$\Delta_{\Gamma_7^{+*}}^{intra}$ (cm^{-1})	$\Delta_{\Gamma_7^{+*}}^{inter}$ (cm^{-1})	$\Delta_{\Gamma_8^{+*}}^{intra}$ (cm^{-1})	$\Delta_{\Gamma_8^{+*}}^{inter}$ (cm^{-1})
Before isotopic substitution	−227.28	−26.57	−190.8	4.91	−306.11	2.53
After isotopic substitution	−222.93	−25.62	−187.1	4.735	−300.17	2.43
Isotopic shift	4.35	1	3.7	−0.18	5.94	−0.10

Table 2.2 The change of energy gaps E_Y and E_G for the yellow and green exciton series of the crystal Cu$_2$O and energies of the interserial transitions Y $\rightleftarrows$ G at intra- and interserial scattering of electrons and holes on the longitudinal optical phonon Γ_4^- before and after isotopic substitution $^{16}O \rightarrow {}^{18}O$

	ΔE_Y (cm^{-1})	ΔE_G (cm^{-1})	G $\rightleftarrows$ Y (cm^{-1})
Before isotopic substitution	−67.96	49.73	−
After isotopic substitution	−66.185	−49.19	−
Isotopic shift	1.775	−0.54	2.315

From Table 2.2 it is seen that under conditions of independence of the exciton Rydberg constant on the isotopic substitution that was observed experimentally in the crystal Cu$_2$O [322, 323], upon substitution of ^{16}O by ^{18}O, all resonance lines are shifted by the same value to shorter wavelengths, due to interserial exciton transitions of the yellow and green series. Analogical isotopic shifts are also characteristic for other intraserial exciton transitions in Cu$_2$O (yellow $\rightleftarrows$ blue, blue $\rightleftarrows$ green series).

The fine structure of the exciton absorption spectrum in Cu$_2$O crystals is experimentally well studied up to $n = 10$ [329, 331]. The large number of exciton states manifested in this case allows us to carry out interserial resonances and use them to investigate the isotopic effects. In particular, at electric dipole transitions between the exciton levels $1SG(^4P_{3/2})$ and $2PY(^2P_{3/2})$ of the green and yellow series (absorption of long-wavelength IR radiation), the distance between which is $\hbar\omega_{1SG \rightarrow 2PY} = E_{2PY}(^2P_{3/2}) - E_{1SG}(^4P_{3/2}) = 76.63$ cm^{-1} [329], the relative isotopic shift has the value $\Delta\omega_{1SG \rightarrow 2PY}/\omega_{1SG \rightarrow 2PY} = 1.72 \cdot 10^{-2}$. The relative isotopic shift becomes even bigger with a decrease in the frequency of transition. Thus for the interserial paraelectric exciton transitions $5SG \rightleftarrows 5PY$ (the distance between the exciton levels in this case is 1.61 cm^{-1}) the relative isotopic shift is equal to $\Delta\omega_{5SG \rightarrow 5PY}/\omega_{5SG \rightarrow 5PY} = 0.82$. For this transition, as is seen, the shifts of the resonance frequency at isotopic substitution are on an order comparable with the frequency of the transition, e.g., the isotopic shift in this case is giant.

Calculation of the interaction of electrons and holes with optical phonons by means of the deformation potential may slightly change the values of the isotopic shift. For intraband scattering processes the ratio of the square modules of the electron (hole)–phonon coupling constants by deformation and polarization interactions is determined by the expression [332]

$$\frac{|\vartheta_{e(h)}^{def}(\mathbf{q})|^2}{|\vartheta_{e(h)}^{pol}(\mathbf{q})|^2} = \frac{T_{\mathbf{q}}}{\hbar\omega_0}, \quad T_{\mathbf{q}} = \frac{\hbar^2\mathbf{q}^2}{2m_{e(h)}}. \tag{2.76}$$

Thus, at low values of the quasi-impulses and large phonon energies, the polarization interaction is determining. According to (2.76) in Cu_2O crystals the interaction of the electrons in band Γ_6^+ and the holes in bands Γ_7^+ and Γ_8^+ with the longitudinal optical phonon of symmetry Γ_4^- by deformation potential becomes actual only with values of the wave vectors of the phonons equal, respectively, to $1.46 \cdot 10^7$ cm^{-1}, $1.22 \cdot 10^7$ cm^{-1}, and $1.95 \cdot 10^7$ cm^{-1}.

In concluding this section, we remark that on the basis of measurements of the isotopic shifts of interserial exciton resonances, from (2.64)–(2.75), the radii a_s and a_i of the atom and the copper ion, respectively, in Cu_2O crystals can be determined.

Chapter 3
Exciton Acoustic Resonance

Under the excitation of ultra- and hypersonic oscillations in semiconductors, these oscillations interact with the natural oscillations of the lattice, various impurity centers and defects, free carriers, and all other elementary excitations that exist in crystals.

The influence of impurity centers or point defects on ultrasonic attenuation has been considered in many papers, e.g., [333–336]. The presence of point defects affects the absorption of ultrasound in two ways: (1) impurities directly scatter the ultrasonic wave (scattering of low-frequency acoustic waves on point defects has a Rayleigh nature), and (2) the impurities scatter phonons and thus change their contribution to the absorption of ultrasound. The direct effect of the point charges on the ultrasound absorption is negligibly small. The second effect is significant if the thermal phonons are scattered on the impurities more frequently than on each other, for which large concentrations of impurities are required [333]. The dislocations can influence the attenuation of the elastic waves. However, the dislocations in the hypersonic range of frequences, for which there are actual acoustic effects on excitons, appear to be significantly lower than those in the ultrasonic range [337].

Before we consider the acoustic absorption caused by excitons and biexcitons, the beginning of this chapter briefly describes the mechanisms of ultrasound absorption by lattice vibrations and conduction electrons. These processes inevitably accompany the exciton absorption of ultra- and hypersound and should be considered when distinguishing the exciton's contribution from the total ultra- and hypersound absorption.

The coefficient of the lattice absorption of ultrasound has different temperature and frequency dependencies for different mechanisms of phonon–phonon interaction. This allows us to see the acoustic absorption by excitons on the background of the lattice absorbtion. The acoustic effects on excitons at sufficiently low exciton concentrations are one to three orders of magnitude higher than the lattice absorption of hypersound.

The contribution of free charge carriers to the acoustic absorption can also be separated from the hypersonic absorption by excitons, which, in turn, is different for the processes of intra- and interband scattering of excitons on phonons. This

I. Geru, D. Suter, *Resonance Effects of Excitons and Electrons*,
Lecture Notes in Physics 869, DOI 10.1007/978-3-642-35807-4_3,
© Springer-Verlag Berlin Heidelberg 2013

difference is due to an incoincidence between the resonance frequencies of acoustic absorption lines at intra- and interband transitions of excitons (intraband scattering is most effective for values of phonon wave vectors of the order of the inverse exciton radius, while the position of the maximum of interband scattering is determined by the distance between the exciton bands), as well as by different dependencies of the absorption coefficient on the phonon hypersonic wave vector for the intra- and interband transitions.

The attenuation of polarized hypersound on excitons in anisotropic crystals is strongly anisotropic. It increases with the increasing difference of the effective masses of charge carriers bound into excitons, which becomes significant at high frequencies (in the hypersonic range). This causes large values of the absorption co-efficient of the hypersonic waves by the excitons—giant values ($\sim 10^2$–10^3 dB/cm) are acquired at low exciton concentrations of $\sim 10^{14}$ cm^{-3} for direct-gap semiconductors for polarized hypersound in $A_2 B_6$ piezo-semiconductors.

The incoincidence between intra- and interband acoustic absorption lines is further used in considering the instability of the system of excitons and resonant hypersonic phonons and the construction of a two-band theory of phonon maser (or, more exactly, faser) generation by excitons. This theory has a development similar to that from Chap. 2 for the theory of generation of coherent magnons in magnetic semiconductors with a redefinition of the physical sense of its constituent quantities. An important consequence of the theory of phonon maser generation by excitons is the low threshold of self-excitation of this generator, which is attained at middle concentrations of excitons in the crystals.

The effects of inter-exciton interaction are manifested at high concentrations of excitons in crystals. If attractive forces are displayed between the excitons, and biexcitons are formed, then the absorption of hypersound by biexcitons is more efficient than the absorption of hypersound by excitons in the same range of frequencies.

The main results of this chapter are published in [170, 338–341].

3.1 The Effect of Lattice Vibrations and Free Carriers on Ultrasonic Attenuation in Crystals

There are two methods for considering acoustic absorption by lattice vibrations depending on the lifetime τ of the thermal phonons in comparison with the period of ultrasound wave vibrations. In the quantum mechanical method, when $\omega \tau \gg 1$ ($\omega = v_s |\mathbf{q}|$ is the ultrasound frequency), the absorption is considered as a scattering of ultrasonic phonons on thermal phonons, with a subsequent summation over all phonons. The energy and quasi-momentum of these phonons are well defined, and they satisfy the conservation laws. If the lifetime of thermal phonons is much less than the period of the ultrasonic wave, $\omega \tau \ll 1$, then the uncertainty in the energy and in the quasi-momentum of ultrasonic phonons become greater than the values of energy and quasi-momentum of these phonons. In this case, the thermal phonons are scattered so fast that the field of the ultrasound deformations does not

have time to change substantially, and the acoustic absorption is due to the deviation of the thermal phonons distribution from the equilibrated one in the presence of
ultrasonic waves. The shift of the distribution to an equilibrated one is accompanied
by a change in entropy due to energy loss of the ultrasonic wave.

When the condition $\omega\tau \gg 1$ applies, the phase velocity of heat waves is limited
because of dispersion and the longitudinal ultrasonic waves cannot interact with
most of the thermal waves, since the speed of the thermal phonons is less than the
speed of the longitudinal ultrasonic waves. The transverse ultrasonic waves have a
lower speed and can interact with the longitudinal thermal waves.

Herring drew attention to the fact that in anisotropic crystals there are lines crossing or touching the various phonon branches $\omega(\mathbf{q})$ in the space of wave vectors
$\mathbf{q}$ [342]. The presence of these lines of intersection means that processes such as
$l_0 + l_1 \rightarrow l_2$ (l_0 is a longitudinal ultrasonic phonon, l_1 and l_2 are longitudinal thermal
phonons), which are forbidden in the first approximation in an isotropic medium,
are allowed in anisotropic crystals [343]. However, this mechanism provides a very
small value of ultrasonic attenuation and leads to frequency and temperature dependencies that diverge from the experimental data. In the works [344, 345] the influence of the finite relaxation time of the thermal phonons on the possible phonon–
phonon interactions is considered. It is shown that, for frequencies of ultrasound
and temperatures where $\omega\tau \ll (\theta/T)^2$ (θ is the Debye temperature) but the condition $\omega\tau \gg 1$ is still satisfied, the laws of conservation do not impose restrictions on
the processes of the collinear interaction of ultrasonic waves with high-frequency
thermal phonons. Herewith the attenuation of both longitudinal and transverse ultrasonic waves is proportional to ωT^4. With decreasing temperature the relaxation
time of the thermal phonons increases and the inequality $\omega\tau \ll (\theta/T)^2$ ceases to
apply.

In the case $\omega\tau \gg (\theta/T)^2$ the uncertainty in the thermal phonon energy becomes
less than the imbalance of the energy in the process $l_0 + l_1 \rightarrow l_2$. Therefore, this
process is forbidden, and the attenuation of the ultrasonic waves sharply decreases
(it does not depend on the frequency and is proportional to T^7). In that case, the
attenuation of longitudinal ultrasonic phonons in processes of $l_0 + t_1 \rightarrow t_2$ type
(t_1 and t_2 are transversal thermal phonons) also does not depend on the frequency,
but it is proportional to T^9.

The temperature dependencies of the lattice ultrasound absorption coefficient
$\Gamma \sim T^7$ and $\Gamma \sim T^9$ along with the lack of dependence of ultrasonic attenuation
on its frequency have been observed experimentally at low temperatures [346–348].

Ultrasonic interactions with the free charge carriers in semiconductors possessing an inversion center are described using the deformation potential. In a series of
papers [349–351] the coefficient of the ultrasound absorption using the deformation
potential was calculated, and it was shown that its value is strongly dependent on the
carrier energy spectrum. However, most important for practical applications are the
studies of the absorption of ultrasound in piezo-semiconductors, in which the ultrasonic wave is accompanied by a wave of the longitudinal electric field (piezoelectric
field). Gurevich [300] and Hutson and White [352, 353] developed a macroscopic
theory of the interaction of ultrasonic waves with mobile charge carriers in the case

where $q l_e \ll 1$ (l_e is the average length of the free run of the charge carriers), in which the ultrasonic wave is regarded as a classic one, creating a perturbation in the distribution of charge carriers.

Gurevich and Kagan [354] have examined the absorption of ultrasound in piezoelectric semiconductors for the case where $q l_e \gg 1$. Using the equations of motion of the piezoelectric conductive medium,

$$\rho \ddot{U}_i = \lambda_{iklm} \frac{\partial^2 U_l}{\partial x_k \, \partial x_m} - \beta_{l,ik} \frac{\partial^2 \varphi}{\partial x_k \, \partial x_l},$$

$$\varepsilon_{ik} \frac{\partial^2 \varphi}{\partial x_i \, \partial x_k} - 4\pi \beta_{i,kl} \frac{\partial^2 U_k}{\partial x_i \, \partial x_l} = -4\pi e n',$$
(3.1)

the authors found in the linear approximation on φ an expression for the ultrasound absorption coefficient, which for Boltzmann statistics at a quadratic and isotropic conduction electron spectrum and satisfying the inequalities $\hbar q \ll p_0$, $\omega \ll q p_0 / m_e$, $\hbar \omega \ll p_0^2 / m_e$ (p_0 is the average thermal momentum of the electron, m_e is the electron mass) takes the form

$$\Gamma_e = \frac{8\sqrt{2} \pi^{5/2} e^2 \sqrt{m_e} n_0}{\rho v_s q \varepsilon^2 (k_0 T)^{3/2}} \left[\frac{\beta_{l,mi} \xi_m q_l q_i}{q^2 + \varkappa_{\mathcal{D}}^2} \right]^2.$$
(3.2)

Here ρ is the density of the crystal, $\mathbf{U}$ is the vector of the shift in the ultrasonic wave, λ_{iklm} is the tensor of the elastic moduli, $\beta_{i,kl}$ is the piezoelectric tensor, ε_{ik} is the tensor of the dielectric permeability of the crystal, φ is the piezoelectric potential, n' is the excess (compared with ionic) concentration of the conduction electron, n_0 is the equilibrium concentration of the conduction electrons, e is the electron charge, ξ and v_s are the unit polarization vector and group velocity for the given oscillation, $\varkappa_{\mathcal{D}}$ is the inverse Debye–Hückel shielding radius, and k_0 is the Boltzmann constant.

At $q \ll \varkappa_{\mathcal{D}}$ conduction electrons almost completely shield the electric fields that occur during the propagation of acoustic waves in the piezoelectric semiconductor and the speed of ultrasound is obtained as if the piezoelectric properties of the medium were absent. The mechanism of ultrasonic attenuation in this case has the same nature as the mechanism of attenuation of plasma waves. In the opposite extreme case ($q \gg \varkappa_{\mathcal{D}}$) the shielding is not important, and the expression for the ultrasonic velocity takes the same form as in the piezoelectric dielectrics.

The effect of the capture of free carriers of the charge in the traps, which can be in the crystals, was investigated in [355, 356]. When the period of the ultrasonic wave is much smaller than the relaxation time τ_c of the charge carriers with respect to the capture during trapping ($\omega \tau_c \gg 1$), the effect of the latter on the absorption of ultrasound can be neglected, since in this case the capture on the traps does not have time to occur. Experimentally, this situation occurs namely in $A_2 B_6$ piezoelectric semiconductors in the hypersonic frequency range. According to [356], in the crystal CdS, $\tau_c \sim 10^{-10}$–10^{-9} s, and for hypersonic frequencies $\omega \sim 10^{11}$ rad/s the condition $\omega \tau_c \gg 1$ is well satisfied. Therefore, the decrease in the number of free carriers due to their capture in the traps is negligibly small, and the influence of this effect on the absorption coefficient of the hypersound by free carriers can be ignored. It turns out that the acoustic effects on excitons are most significant in the hypersonic range of frequencies (and not in the ultrasonic one). Thus, when considering these effects, the slight decrease in the number of excitons due to the capture in the trap of a small number of free electrons and holes before they are coupled into the excitons cannot be taken into account.

3.2 Resonant Absorption of Hypersound at Intraband Exciton Scattering on Phonons

The absorption of acoustic waves by an exciton gas during the processes of intraband scattering of excitons on phonons is most effective when the phonon wave vectors are comparable with the inverse exciton radius ($q \sim a_{ex}^{-1}$). Therefore, in direct-band semiconductors, where the effective Bohr radii of excitons do not exceed a few tens of angstroms, acoustic bands represent the resonance curves, and the coefficient of intraband absorption of acoustic waves by excitons reaches a maximum in the hypersonic frequency range ($\sim 10^{11}$–10^{12} s^{-1}). The interaction of hypersound with the excitons can be so strong that even a greatly reduced interaction due to shielding by free charges can lead in some cases to large acoustic effects.

The absorption of sound due to intraband scattering of excitons on phonons (the interaction of excitons with longitudinal piezoelectric vibrations), for the case when the exciton free run length is much larger than the hypersonic wavelength ($q l_{ex} \gg 1$), was first considered in [357]. On the basis of the Hamiltonian interaction of an exciton with a longitudinal hypersonic wave, these authors obtained an expression for the absorption coefficient of the hypersound Γ_{ex} for a nondegenerate exciton gas (the Maxwell–Boltzmann statistics), which at $m_{ex} v_s^2 \ll k_0 T$, $\hbar \omega \ll (2 m_{ex} v_s^2 k_0 T)^{1/2}$, $q a_{ex} \ll 1$, and $k_0 T < Ry^{ex}$ (Ry^{ex} is the exciton Rydberg constant) for a "nearly isotropic" crystal in a zeroth anisotropy approximation $\varepsilon_{ij} = \varepsilon \delta_{ij}, \varepsilon = \sqrt{\varepsilon_\parallel \varepsilon_\perp}$ is given in the form

$$\Gamma_{ex} \approx \frac{8.68(2\pi)^{5/2} e^2 m_{ex}^{1/2} n_{ex}}{\rho \omega \varepsilon^2 (k_0 T)^{3/2}} \left[\frac{\beta_{i,jl} q_i q_l \xi_j}{q^2 + \varkappa_D^2} \right]^2, \tag{3.3}$$

where ω is the hypersonic frequency, m_{ex} is the translational mass of the exciton, and n_{ex} is the exciton concentration.

If we replace m_{ex}, n_{ex}, and 8.68, respectively, by m_e, n_0, and $8\sqrt{2}$, formula (3.3) is transformed into expression (3.2), obtained in [354] for the coefficient of ultrasound absorption in semiconductors in the process of scattering of the electrons by phonons. As seen from (3.3) and (3.2), in the zeroth anisotropy approximation, the contribution to the total hypersound absorption coefficient due to the excitons, cannot be separated from the contribution to the acoustic absorption due to the free charge carriers, as the temperature, frequency, and concentration dependencies for both mechanisms are identical.[1]

The numerical evaluation on the basis of formula (3.3) for A-type excitons in a CdS crystal when the exciton concentration $n_{ex} = 10^{14}$ cm^{-3}, the concentration of shielding carriers $n_{ex} \sim 10^{13}$ cm^{-3}, the frequency of the hypersound $\omega = 10^{10}$ rad/s, and $T = 4$ K, is $\Gamma_{ex} \simeq 1.2$ dB/cm [357]. It follows from this estimation that the experimental observation of hypersound absorption by excitons is possible in principle. However, as noted above, if we simultaneously satisfy all inequalities that are necessary to derive (3.3), in a "nearly isotropic" crystal it is practically difficult to separate the contributions from free charge carriers and excitons in the total absorption of the hypersound.

We note, first of all, that the estimation given in [357] corresponds to the attenuation of hypersound, not at the maximum curve of the acoustic absorption, but at lower hypersonic frequencies. Secondly, because of the strong anisotropy of the matrix element of exciton–phonon intraband interaction in a uniaxial piezosemiconductor (see formulas (3.10a), (3.10b) and (3.11) below), the transition to higher frequencies, at which an essential difference between the effective masses of the electron and hole forming a free exciton exists, is accompanied by an increase in the hypersound attenuation coefficient. This coefficient in the region of maximum of the acoustic absorption reaches values of order 10^2–10^3 dB/cm. Under the same conditions, but far from resonance and without taking into consideration the differences of the effective masses of the carriers, the coefficient is only ~ 1 dB/cm or less.

The Hamiltonian interaction of an exciton with a longitudinal acoustic wave in a piezoelectric semiconductor[2] for $ql_{ex} \gg 1$ has the form

$$H = V_{\mathbf{q}} B_{\mathbf{q}} \big(C_1 e^{i g_1 \mathbf{q} \cdot \mathbf{r}} - C_2 e^{-i g_2 \mathbf{q} \cdot \mathbf{r}} \big) e^{i \mathbf{q} \cdot \mathbf{R}} + \text{H.c.}, \qquad (3.4)$$

where

$$V_{\mathbf{q}} = -\left(\frac{\hbar}{2 \rho L^3 \omega} \right)^{1/2} \frac{4 \pi e \beta_{i,jl} q_i q_l \xi_j}{\varepsilon_{ij} q_i q_j + 4 \pi e^2 \mathcal{K}_{\mathbf{q}}^0(\omega)}. \qquad (3.5)$$

Here, $B_{\mathbf{q}}$ is the annihilation operator of a phonon with wave vector $\mathbf{q}$ (index $\mathbf{q}$ also includes the phonon polarization); $\mathbf{R}$, and $\mathbf{r}$ are the radius vector of the exciton center of the inertia and the electron–hole radius vector, $\mathbf{r} = \mathbf{r}_e - \mathbf{r}_h$; g_1 and g_2 are the ratios of the effective masses of the hole and electron to the exciton effective mass m_{ex}; C_1 and C_2 are dimensionless constants, taking into account the deformation of the electron shells during the atomic vibrations, for the electron and the hole, respectively [358]; L^3 is the volume of the main region of the crystal. The function

---

[1]It is possible to evidentiate the acoustic absorption on the excitons on the background of the acoustic absorption on the free charge carriers in this case only, if the concentration of the excitons is much higher than the concentration of the free carriers of the current.

[2]The interaction of excitons with hypersonic phonons through the deformation potential will be considered in the next section.

$\mathcal{K}_q^0(\omega)$ takes into account the shielding effect of the electron–phonon interaction by free charges [359]. In formula (3.5) summation over the double repeated indices is intended.

On the basis of the Hamiltonian (3.4), in the next section we calculate the absorption coefficient of the hypersound in the exciton system. For convenience in comparing the contributions to the total absorption of the hypersound by elastic and inelastic scattering of excitons on the phonons, these processes will be considered in the next section in a unified way. The absorption coefficients of the hypersound at intra- and interband scattering of excitons on phonons are calculated, taking into account the broadening of the exciton bands due to the finite lifetime of the exciton states.

3.3 Resonant Absorption of Hypersound During Transitions Between Exciton Subbands

In uniaxial anisotropic semiconductors the Stark splittings of the degenerate exciton bands in a crystal field in order of magnitude are comparable with the energy of the hypersonic phonons [170, 338, 360]. It is therefore of interest to investigate the mechanism of the absorption of hypersound at transitions between exciton subbands split in a crystal field. In a number of cases, as will be shown below, the intensity of the acoustic absorption bands caused by the transitions between exciton bands is greater than the intensity of the acoustic absorption lines associated with the intraband transitions in the same range of frequencies.

We shall take into account the finite lifetime of the exciton states by means of a formal transition to the complex frequencies in the exciton wave function:

$$\omega_{\lambda\lambda'}(\mathbf{K}, \mathbf{K}') \to \omega_{\lambda\lambda'}(\mathbf{K}, \mathbf{K}') - \frac{1}{2}i(\gamma_\lambda + \gamma_{\lambda'}),$$

$$\hbar\omega_{\lambda\lambda'}(\mathbf{K}, \mathbf{K}') \equiv \varepsilon_{\lambda'}(\mathbf{K}') - \varepsilon_\lambda(\mathbf{K}).$$

The broadenings $\hbar\gamma_\lambda$ and $\hbar\gamma_{\lambda'}$ of the subbands λ and λ' are considered to be independent of $\mathbf{K}$. As a measure for such broadenings, attenuation constants of excitons in subbands λ and λ' due to thermal phonons in the frequency range where they are maximal are chosen.

The absorption coefficient of the hypersound at intra- and interband scattering of excitons on the hypersonic phonons, taking into account the finite lifetime of exciton states and under the condition $ql_{\text{ex}} \gg 1$ (the quantum mechanical approach), is determined by the expression [338]

$$\Gamma_{\lambda\lambda'}(\mathbf{q}) = \frac{2}{\hbar^2}\left(\frac{\alpha}{\pi}\right)^{1/2}\left|U_{\lambda\lambda'}^{\text{piezo}}(\mathbf{q})\right|^2\left(\frac{m_{\text{ex}}}{\hbar q}\right)^2\frac{\gamma_{\lambda\lambda'}}{v_{\text{s}}}\Lambda_{\lambda\lambda'}(\mathbf{q}), \tag{3.6}$$

where

$$\Lambda_{\lambda\lambda'}(\mathbf{q}) = \int_{-\infty}^{\infty} \frac{1}{a_{\lambda\lambda'}^2 + (b_{\lambda\lambda'} + K_z)^2}$$

$$\times \left[N_\lambda e^{-\alpha K_z^2} - N_{\lambda'} e^{-\alpha(K_z+q)^2} \right] dK_z,$$

$$U_{\lambda\lambda'}^{\text{piezo}}(\mathbf{q}) = V_{\mathbf{q}} \langle \lambda | (C_1 e^{i g_1 \mathbf{q}\cdot\mathbf{r}} - C_2 e^{-i g_2 \mathbf{q}\cdot\mathbf{r}}) | \lambda' \rangle,$$

$$\gamma_{\lambda\lambda'} = \frac{1}{2}(\gamma_\lambda + \gamma_{\lambda'}), \qquad a_{\lambda\lambda'} = \frac{m}{\hbar q} \gamma_{\lambda\lambda'}, \qquad \alpha = \frac{\hbar^2}{2 m_{\text{ex}} k_0 T},$$ (3.7)

$$b_{\lambda\lambda'} = \frac{q}{2} + \frac{m}{\hbar^2 q} \Delta_{\lambda\lambda'} - \frac{m_{\text{ex}} v_{\text{s}}}{\hbar},$$

$$\Delta_{\lambda\lambda'} = \varepsilon_{\lambda'}^0 - \varepsilon_\lambda^0, \qquad \varepsilon_{\lambda(\lambda')}^0 \equiv \varepsilon_{\lambda(\lambda')}(\mathbf{K})|_{\mathbf{K}=0}.$$

Here K_z is the projection of the exciton wave vector in the direction $\mathbf{q}$ and $N_\lambda (N_{\lambda'})$ is the total number of excitons in the band $\lambda(\lambda')$. The presence of the integral (3.7) in the expression for $\Gamma_{\lambda\lambda'}(\mathbf{q})$ from (3.6) containing the projection of the exciton wave vector K_z, is connected with the consideration in the Hamiltonian (3.4) of only one hypersonic mode.

The function $\Lambda_{\lambda\lambda'}(\mathbf{q})$ from (3.7) contains integrals I_l, determined by (2.53), with $l = 1$ and 2, in which $C_{\sigma\sigma'}$ is replaced by $b_{\lambda\lambda'}$. These integrals are calculated by means of the Fourier transform. As a result, we get

$$\Lambda_{\lambda\lambda'}(\mathbf{q}) = N_\lambda I_1 - \tilde{N}_{\lambda'} I_2, \qquad \tilde{N}_{\lambda'} = e^{-\alpha q^2},$$

where

$$I_1 = \frac{\pi}{2 a_{\lambda\lambda'}} \left\{ e^{2 i \alpha a_{\lambda\lambda'} b_{\lambda\lambda'}} \left[1 - \Phi\left(\sqrt{\alpha}(a_{\lambda\lambda'} + i b_{\lambda\lambda'}) \right) \right] \right.$$

$$\left. + e^{-2 i \alpha a_{\lambda\lambda'} b_{\lambda\lambda'}} \left[1 - \Phi\left(\sqrt{\alpha}(a_{\lambda\lambda'} - i b_{\lambda\lambda'}) \right) \right] \right\} e^{\alpha(a_{\lambda\lambda'}^2 - b_{\lambda\lambda'}^2)}.$$ (3.8)

In (3.8) $\Phi(z)$ is the probability function. The integral I_2 is obtained from I_1 by replacing the $b_{\lambda\lambda'}$ by $b_{\lambda\lambda'} - q$ and multiplying the result by $e^{\alpha q^2}$. When satisfying the inequality $a_{\lambda\lambda'} \ll b_{\lambda\lambda'}$, in the argument of the probability function and in the exponents from (3.8), we can ignore the terms containing $a_{\lambda\lambda'}$. Then for $\Lambda_{\lambda\lambda'}(q)$ we obtain

$$\Lambda_{\lambda\lambda'}(\mathbf{q}) \approx \frac{\pi}{a_{\lambda\lambda'}} \left\{ N_\lambda e^{-\alpha b_{\lambda\lambda'}^2} - N_{\lambda'} e^{-\alpha(b_{\lambda\lambda'}-q)^2} \right\}.$$ (3.9)

The approximation (3.9) holds in a hypersonic frequency range of $\omega \sim 2\pi \cdot 10^{11}$ rad/s at $\gamma_{\lambda\lambda'} < \omega$. In this case expression (3.6) for $\Gamma_{\lambda\lambda'}(\mathbf{q})$ is identical with the corresponding expression for the absorption coefficient of the hypersound at $\gamma_{\lambda\lambda'} = 0$.

Thus, although for the exciton bands broadened due to interaction with thermal phonons, the possibility of replacing the Lorentzian shape by the δ-function in the formula for the density of states in the exciton band while calculating the probability of exciton–phonon transitions is not obvious, such a substitution may be made only in the case of $\gamma_{\lambda\lambda'} < \omega$. Therefore, we will further consider the processes of absorption of hypersound at intra- and interband transitions for the case $\gamma_{\lambda\lambda'} = 0$.

In uniaxial A_2B_6 piezo-semiconductors in which the exciton band with $n = 2$ is split in a crystal field of axial symmetry into subbands $2P_0$, $2S$, and $2P_{\pm 1}$, the hypersound absorption coefficients during scattering of excitons on phonons in the lowest subband with $n = 2$ (subband $2P_0$) and upon transitions between subbands $2P_0$ and $2S$ in the approximation of $\gamma_{\lambda\lambda'} < \omega$ are given by

$$\Gamma_{2P_0,2P_0}(\mathbf{q}) = A_{2P_0,2P_0}\left[\frac{1 - \sigma g_1^2 a_{ex}^2 q^2}{(1 + g_1^2 a_{ex}^2 q^2)^4}C_1 - \frac{1 - \sigma g_2^2 a_{ex}^2 q^2}{(1 + g_2^2 a_{ex}^2 q^2)^4}C_2\right]^2, \quad (3.10a)$$

$$\Gamma_{2P_0,2S}(\mathbf{q}) = 9A_{2P_0,2S}a_{ex}^2 q^2 \cos^2\theta\left[\frac{1 - g_1^2 a_{ex}^2 q^2}{(1 + g_1^2 a_{ex}^2 q^2)^4}g_1 C_1\right.$$
$$\left. + \frac{1 - g_2^2 a_{ex}^2 q^2}{(1 + g_2^2 a_{ex}^2 q^2)^4}g_2 C_2\right]^2, \quad (3.10b)$$

where

$$A_{\lambda\lambda'} = \frac{2m_{ex}}{\hbar^3\omega}\sqrt{\pi\alpha}|V_{\mathbf{q}}|^2\left[N_\lambda e^{-\alpha b_{\lambda\lambda'}^2} - N_{\lambda'}e^{-\alpha(b_{\lambda\lambda'}-q)^2}\right],$$
$$\sigma = 1 + 2(3\cos^2\theta - 1). \quad (3.11)$$

Here θ is the angle between the C_6 symmetry axis of the crystal and the propagation direction of the hypersound.

Expressions (3.6) and (3.10a), (3.10b) for the absorption coefficients of hypersound by excitons are valid when using inequality $ql_{ex} \gg 1$ along with inequality $\omega\tau_{ex\text{-}ph} \gg 1$, where $\tau_{ex\text{-}ph}$ is the exciton relaxation time at interaction with acoustic phonons. At low frequencies, when these inequalities are violated, the interaction of hypersound with excitons can be studied in analogy to the interaction of an acoustic wave of low [300, 352, 353] and intermediate [361, 362] frequency with current carriers.

As seen from (3.10a), (3.10b), the fact that the effective masses of the electron and hole are different leads to an increase in the intraband absorption of the hypersound. In particular, at $C_1 = C_2$ and $g_1 = g_2$, intraband absorption of the hypersound is equal to zero, whereas by taking into account differences in the effective mass of electrons and holes ($g_1 \neq g_2$) in accordance with (3.10a), (3.10b) we have $\Gamma_{2P_0,2P_0}(\mathbf{q}) \neq 0$.

The processes of interband absorption of hypersound essentially depend on the ratio between the kinetic energy of an exciton that has moved with hypersonic speed and the distance $\Delta_{\lambda\lambda'}$ between exciton bands in the center of the Brillouin zone. If $\Delta_{\lambda\lambda'} > m_{ex}v_s^2/2$, then excitons from the lower band, the wave vectors of which are enclosed in a sphere of radius $K_c = b_{\lambda\lambda'}$ (the value $b_{\lambda\lambda'}$ is defined by the formula (3.7)), do not participate in the processes of interband absorption of hyper-

Fig. 3.1 The dependence of
the absorption coefficient
$\Gamma_{\lambda\lambda'}$ of longitudinal
hypersound on phonon wave
vector for $(2P_0)$
exciton–phonon scattering
and at interband transitions
$2P_0 \rightarrow 2S$ in CdS and ZnO
crystals at $\Delta n_{ex} =$
$n_{2P_0} - n_{2S} = 10^{14}$ cm^{-3},
$T = 4.2$ K, $C_1 = 0.9$, $C_2 = 1$,
$\varkappa_{\mathcal{D}} = 7.5 \cdot 10^4$ cm^{-1}
(concentration of screening
carriers $n_e \sim 10^{13}$ cm^{-3}):
$1, 2, 3$—$\Gamma^{CdS}_{2P_0,2P_0}$, $\Gamma^{CdS}_{2P_0,2S}$,
$\Gamma^{ZnO}_{2P_0,2P_0}$ ($\Gamma_{\lambda,\lambda'} = \Gamma \cdot$
10^2 dB/cm). 4—$\Gamma^{ZnO}_{2P_0,2S} =$
$\Gamma \cdot 2 \cdot 10^2$ dB/cm

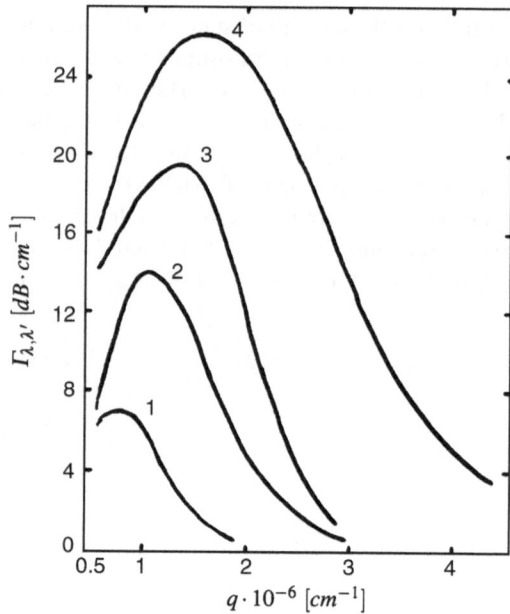

sound. This restriction follows from the energy and quasi-momentum conservation
laws in the exciton–phonon interaction. In this case, during the interband scattering
of excitons on phonons, only the excitons with large quasi-momenta ($K > K_c$) contribute, as a result of which the lines of interband absorption of hypersound have
large widths. In the case of an inverse inequality ($\Delta_{\lambda\lambda'} < m_{ex} v_s^2/2$), the restriction
on the exciton wave vectors is absent, and in the processes of interband absorption
of the hypersound the entire exciton band makes a contribution.

Figure 3.1 shows the dependence of the absorption coefficient of longitudinal hypersound on the phonon quasi-momentum (formula (3.10b)) for the exci-
ton transitions $2P_0 \rightarrow 2S$ (curves 2 and 4) for CdS ($\Delta_{2P_0,2S} > m_{ex} v_s^2/2$) and ZnO
($\Delta_{2P_0,2S} < m_{ex} v_s^2/2$) crystals, respectively, at $\Delta n_{ex} = n_{2P_0} - n_{2S} = 10^{14}$ cm^{-3} and
$T = 4.2$ K. The numerical calculations were performed for the case of hypersound
propagation along the axis of symmetry C_6 with the following crystal parameters:

CdS

$$m_e^{\parallel} = 0.153 m_0, \quad m_h^{\parallel} = 5 m_0, \quad m_e^{\perp} = 0.171 m_0, \quad m_h^{\perp} = 0.7 m_0, \quad \varepsilon = 9.27,$$

$$\text{Ry}^{ex} = 0.028 \text{ eV [363]}, \quad v_s = 4.37 \cdot 10^5 \text{ cm/s}, \quad \rho = 4.82 \text{ g/cm}^3,$$

$$\varkappa_{\mathcal{D}} = 7.5 \cdot 10^4 \text{ cm}^{-1} \quad \text{(concentration of shielding carriers } 10^{13} \text{ cm}^{-3}),$$

$$\Delta_{2P_0,2S} = 0.4 \text{ meV}, \quad \beta_{3,11} = -0.732,$$

$$\beta_{3,33} = 1.32, \quad \beta_{1,13} = -0.63 \cdot 10^5 \text{ CGSE charge units/cm}^2 \text{ [364, 365]},$$

$$C_1 = 0.9, \quad C_2 = 1.0.$$

ZnO

$$m_e = 0.38m_0, \quad m_h = 1.8m_0, \quad \varepsilon = 8.47, \quad Ry^{ex} = 0.059 \text{ eV } [363],$$

$$v_s = 5.5 \cdot 10^5 \text{ cm/s},$$

$$\rho = 5.6 \text{ g/cm}^3, \quad \varkappa_{\mathcal{D}} = 7.5 \cdot 10^4 \text{ cm}^{-1}, \quad \Delta_{2P_0,2S} = 0.1 \text{ meV}; \quad \beta_{3,11} = -0.48,$$

$$\beta_{3.33} = 3.3,$$

$$\beta_{1.13} = -0.93 \cdot 10^5 \text{ CGSE charge units/cm}^2 \text{ [364, 365]}, \quad C_1 = 0.9, \quad C_2 = 1.0.$$

To compare the effects of intra- and interband absorption of hypersound by excitons, Fig. 3.1 also shows the absorption of the longitudinal hypersound, calculated by formula (3.10a), which is due to the scattering of excitons inside the subband $2P_0$ in crystals of CdS and ZnO (curves 1 and 3), respectively.

The long-wave curves of the absorption coefficient of hypersound in Fig. 3.1 are cut at the values of phonon wave vectors which still have the inequalities $ql_{ex} \gg 1$ and $\omega\tau_{ex\text{-}ph} \gg 1$ required for quantum mechanical consideration of exciton–phonon interaction. With a further decrease in the frequency of hypersound ω, both of these inequalities are violated. The evaluation of $\tau_{ex\text{-}ph}$ according to [358] for the interaction of excitons with hypersound via the piezoelectric potential gives $\tau_{ex\text{-}ph}^{(CdS)} \approx 5 \cdot 10^{-10}$ s and $\tau_{ex\text{-}ph}^{(ZnO)} \approx 2 \cdot 10^{-10}$ s at $T = 4.2$ K. Since the value of $v_s/v_{ex} \simeq 0.38$ for CdS and $\simeq 0.53$ for ZnO (v_{ex} is the average thermal velocity of excitons), in both cases the criterion of applicability of quantum mechanical consideration of the hypersound interaction with excitons is determined by the inequality $\omega\tau_{ex\text{-}ph} \gg 1$, which is applied when $\omega \gtrsim (5–10) \cdot 10^{10}$ rad/s.

In the numerical calculations of $\Gamma_{\lambda\lambda'}(\mathbf{q})$ the interaction of hypersound with excitons via the deformation potential was not considered. According to [366, 367], a comparative measure of the effectiveness of the interaction of electrons (holes) with phonons through piezoelectric or the deformation potentials (in the one-dimensional case) is determined by the ratio

$$\eta = \frac{E_{piezo}^2}{E_{def}^2} = \left(\frac{\beta e v_s}{C_d \varepsilon}\right)^2 \frac{1}{\omega^2}, \tag{3.12}$$

where $E_{piezo} = \frac{\beta}{\varepsilon}S$ and $E_{def} = \frac{q}{e}C_d S$ are the electric field intensities at the same deformation of the crystal lattice S; C_d is the deformation potential constant. An estimate η for the case $\mathbf{q} \parallel C_6$ ($\beta \equiv \beta_{3,33}$) under the assumption that the deformation potential constants for the valence and conduction bands coincide by the order of magnitude $C_d \simeq (C_{d,\Gamma_9})_{max} \lesssim 4$ eV (CdS) and $C_d \simeq (C_{d,\Gamma_7}^V)_{max} \lesssim 3$ eV (ZnO) [367], gives

$$\eta_{CdS} \simeq 0.61 \cdot 10^{26}\omega^{-2}, \quad \eta_{ZnO} \simeq 3.3 \cdot 10^{26}\omega^{-2}.$$

Since the operators of the interaction of excitons with phonons via piezoelectric and deformation potentials have the same dependence on the coordinates of the

Fig. 3.2 Temperature
dependence of the coefficient
of interband absorption of
hypersound $\Gamma_{2P_0,2S}(T)$ in the
crystals ZnO and CdS at
frequencies of
$\omega = 7.5 \cdot 10^{11}$ rad/s and
$\omega = 4.37 \cdot 10^{11}$ rad/s,
respectively.
1—$\Gamma_{2P_0,2S}^{\mathrm{ZnO}} = \Gamma \cdot 10^3$ dB/cm,
2—$\Gamma_{2P_0,2S}^{\mathrm{CdS}} = \Gamma \cdot 10^2$ dB/cm

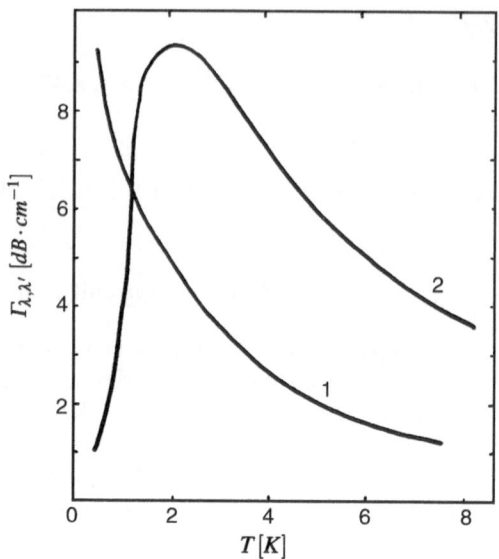

charge carriers, then in the approximation of $m_e \approx m_h$ and $a_{ex}q < 2$ the value η_{ex}
is determined by (3.12), if in the latter we make the replacements $C_d \to |C_d^c - C_d^v|$
and $\beta \to \beta|C_1 - C_2|$. Thus, at the same deformation S in the CdS and ZnO crystals
at frequencies $\omega \sim (10^{11}–10^{12})$ rad/s the contribution of piezoelectric potential in
the exciton–phonon interaction is greater than the contribution of the deformation
potential, which can be ignored.

An interesting feature of the interband absorption of hypersound is the temperature dependence

$$\Gamma_{\lambda\lambda'}(T) \sim \frac{1}{\sqrt{T}} \exp\left\{-\frac{1}{k_0 T} \cdot \frac{m_{ex}}{2\hbar^2 q}\left(\Delta_{\lambda\lambda'} + \frac{\hbar^2 q^2}{2m_{ex}} - \hbar\omega\right)\right\}. \qquad (3.13)$$

From formula (3.13) it can be seen that the temperature dependence of the func-
tion $\Gamma_{\lambda\lambda'}(T)$ is determined by the ratio between $\Delta_{\lambda\lambda'}$ and $m v_{ex}^2/2$, the kinetic energy
of an exciton moving at the speed of hypersound. In the case $m v_{ex}^2/2 > \Delta_{\lambda\lambda'}$ with
decreasing temperature the acoustic band becomes narrower, and in the limit $T \to 0$
it transforms into the δ-shaped peak. Otherwise ($m v_{ex}^2/2 < \Delta_{\lambda\lambda'}$) the absorption
band also becomes narrower. However, at $T \to 0$ interband acoustical absorption
on excitons is absent ($\Gamma_{\lambda\lambda'} = 0$), since at $k_0 T < \hbar^2 K_c^2/2m_{ex}$ the excitons from the
lower band do not interact with hypersound.

The temperature dependencies of the interband absorption of hypersound
$\Gamma_{2P_0,2S}^{(\mathrm{ZnO})}(T)$ and $\Gamma_{2P_0,2S}^{(\mathrm{CdS})}(T)$ are given in Fig. 3.2 at fixed frequencies $\omega_1 = 7.5 \cdot$
10^{11} rad/s and $\omega_2 = 4.37 \cdot 10^{11}$ rad/s, respectively (curves 1, 2). The different be-
havior of curves 1 and 2 with temperature is caused by the fact that, for exciton
bands $2P_0$ and $2S$, in the CdS crystal the inequality $\Delta_{2P_0,2S} > m_{ex} v_s^2/2$ is applied,
whereas in ZnO the reverse inequality occurs.

The sharp decrease in acoustic absorption on excitons in CdS at $T < T_{\mathrm{max}}$ (T_{max} is the temperature at which $\Gamma_{2P_0,2S}(T)$ attains its maximum) occurs due to a sharp decrease in the number of excitons of band $2P_0$ that take part in the exciton–phonon transitions. With decreasing temperature, the share of excitons with wave vectors that lie outside the sphere of radius $K_{\mathrm{c}} = b_{2P_0,2S}$ decreases, and the relative share of excitons with wave vectors inside the sphere of radius K_{c} increases; the latter do not contribute to the absorption process. The shape of curve 1 shows that, as the temperature decreases, the acoustic absorption band not only narrows, but also shifts to a longer wavelength.

The contribution of the thermal phonons to the total absorption of hypersound is estimated on the basis of experimental data with the calculation of the dependence $\Gamma_{\mathrm{lat}} \sim \omega T^n$ ($n \geq 4$). For the ZnO crystal it has been found experimentally that, in the microwave region, $\Gamma_{\mathrm{lat}} \sim \omega^{0.8} T^7$ (at $\omega/2\pi = 8.8$ GHz and $T = 40$ K $\Gamma_{\mathrm{lat}} = 1$ dB/cm) [348], from which it follows that at $\omega \sim 10^{11}$ rad/s and $T = 4.2$ K we obtain $\Gamma_{\mathrm{lat}} \ll 1$ dB/cm. We have used a background of such small lattice absorption coefficients of the hypersound, that the acoustic effects on excitons at moderate optical excitation levels of the crystals ($\Gamma_{\mathrm{ex}} \sim 10^2$–$10^3$ dB/cm) are giant.

3.4   Induced Instability in a System of Excitons and Strictly Resonant Hypersonic Phonons

In the typical piezoelectric semiconductors CdS and ZnO, the acoustic lines of the hypersound interband absorption by excitons are more intensive in comparison with the intraband absorption lines (Fig. 3.1). The renormalized frequency of the external hypersonic wave $\Omega_{\mathbf{q}}$ due to exciton–phonon interaction, taking into account only the interband transitions in the model of two exciton bands in the rotating field approximation, satisfies the equation [339]

$$\varkappa + \mathrm{i}(\Omega_{\mathbf{q}} - \omega_{\mathbf{q}}) = \frac{\mathrm{i}}{\hbar^2}|U_{\lambda\lambda'}(\mathbf{q})|^2 \sum_{\mathbf{K}} \frac{d_{\lambda\lambda'}^{(0)}(\mathbf{K}, \mathbf{q})}{\Omega_{\mathbf{q}} - \omega_{\lambda\lambda'}(\mathbf{K}, \mathbf{q}) - \mathrm{i}\gamma_{\lambda\lambda'}}$$
$$\times \left[4\frac{\gamma_{\lambda\lambda'} T_1}{\hbar^2}|U_{\lambda\lambda'}(\mathbf{q})|^2 N_{\mathbf{q}} \frac{1}{(\Omega_{\mathbf{q}} - \omega_{\lambda\lambda'}(\mathbf{K}, \mathbf{q}))^2 + \gamma_{\lambda\lambda'}^2} - 1\right],$$

(3.14)

which appears in the same form that was used in deriving (2.34). Here, $\omega(\mathbf{q})$ is the frequency of the phonons of the hypersonic mode before the interaction with excitons, $\varkappa$ is a constant characterizing the attenuation in the initial hypersonic mode ($2\varkappa = Q\omega_{\mathbf{q}}$, where Q is the quality factor of the unloaded acoustic resonator), T_1 is the longitudinal relaxation time of the excitons, $N_{\mathbf{q}}$ is the total number of phonons in the hypersonic mode, $d_{\lambda\lambda'}^0(\mathbf{K}, \mathbf{q})$ is the initial populations inversion of the exciton bands λ and λ', $\hbar\omega_{\lambda\lambda'}(\mathbf{K}, \mathbf{q}) = \varepsilon(\mathbf{K} + \mathbf{q}) - \varepsilon(\mathbf{K})$; $U_{\lambda\lambda'}(\mathbf{q})$ is the constant of the

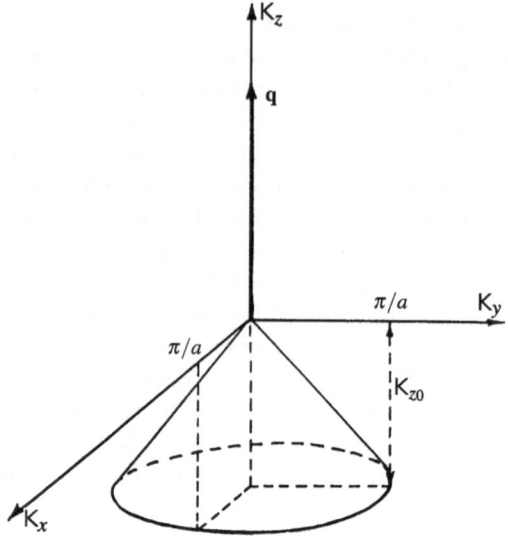

Fig. 3.3 The region of **K**-space for allowed values of the exciton wave vectors (inside the cone) at resonance interaction of excitons with hypersonic phonons with wave vector **q**

exciton–phonon coupling, which for the interaction of excitons with lattice vibrations via the piezoelectric potential is equal to $U_{\lambda\lambda'}^{\text{piezo}}(\mathbf{q})$ from (3.7). The interaction via the deformation potential is determined by

$$U_{\lambda\lambda'}^{\text{def}}(\mathbf{q}) = iq\sqrt{\frac{\hbar}{2\rho\omega_{\mathbf{q}}L^3}}\langle\lambda|\left(C_{\mathrm{d}}^{(1)}e^{ig_1\mathbf{q}\mathbf{r}} - C_{\mathrm{d}}^{(2)}e^{-ig_2\mathbf{q}\mathbf{r}}\right)|\lambda'\rangle,$$

where $C_{\mathrm{d}}^{(1)}$ and $C_{\mathrm{d}}^{(2)}$ are the deformation potential constants characterizing the interaction of the electrons and holes with the lattice vibrations, L^3 is the volume of the crystal.

According to Sect. 3.3, in the case $\gamma_{\lambda\lambda'} < \omega_{\mathbf{q}}$ the expression for the coefficient of interband absorption of hypersound, obtained by taking into account the unsteadiness of exciton states, coincides with the $\Gamma_{\lambda\lambda'}(\mathbf{q})$ at $\gamma_{\lambda\lambda'} = 0$. Thus, in the approximation of $\gamma_{\lambda\lambda'} = 0$, (3.14) can be solved assuming $\gamma_{\lambda\lambda'} = 0$.

For a fixed value and direction of phonon wave vectors of the hypersonic mode, we have

$$\omega_{\lambda\lambda'}(\mathbf{K}, \mathbf{q}) = \omega_{\mathbf{q}}; \quad \mathbf{K}\mathbf{q} = \text{const.} \tag{3.15}$$

In a coordinate system, the z axis of which coincides with the direction of the propagation of the hypersound, and the zth component of the wave vectors of the excitons interacting with the phonons, satisfies the equation

$$K_z = \frac{m_{\text{ex}}}{\hbar q}\left(\omega_{\mathbf{q}} - \frac{\Delta_{\lambda\lambda'}}{\hbar}\right) - \frac{1}{2}q \equiv K_{z0} = \text{const.} \tag{3.16}$$

Figure 3.3 indicates the orientations of the exciton wave vector that are allowed by the laws of conservation of energy and momentum at $\gamma_{\lambda\lambda'} = 0$. One can see that, for a fixed direction of propagation of the hypersound, in the processes of exciton–phonon interaction only those excitons are involved for which the projection of the wave vectors in the direction $\mathbf{q}$ is equal to K_z^0, while K_x and K_y are changed within the entire first Brillouin zone.

Equation (3.14), accounting for (3.15) and (3.16), has simple solutions, which under conditions of inverse population of the exciton bands can lead to instability in the exciton–phonon system. In the case of "integral" pumping of the exciton bands, the function $d_{\lambda\lambda'}^{(0)}(\mathbf{K}, \mathbf{q})$ at the equilibrium distribution of excitons on quasi-momenta has the form

$$d_{\lambda\lambda'}^{(0)}(\mathbf{K}, \mathbf{q}) = \frac{8}{L^3}(\pi\alpha)^{3/2} e^{-\alpha K^2}\left(N_\lambda e^{\frac{\Delta_{\lambda\lambda'} - \hbar\omega_\mathbf{q}}{k_0 T}} - N_{\lambda'}\right). \tag{3.17}$$

After integrating equation (3.14) for all exciton wave vectors that are enclosed inside the cone, as shown in Fig. 3.3, and taking into account (3.17), we obtain [339][3]

$$\varkappa + i(\Omega_\mathbf{q} - \omega_\mathbf{q}) = \frac{2}{L}\sqrt{\frac{\alpha}{\pi}} \cdot \frac{1}{\hbar^2}|U_{\lambda\lambda'}(\mathbf{q})|^2 e^{-\alpha K_{z0}^2}$$
$$\times \left(N_\lambda e^{\frac{\Delta_{\lambda\lambda'} - \hbar\omega_\mathbf{q}}{k_0 T}} - N_{\lambda'}\right) \cdot \frac{1}{\gamma_{\lambda\lambda'} + i(\Omega_\mathbf{q} - \omega_\mathbf{q})}$$
$$\times \left[1 - 4\frac{\gamma_{\lambda\lambda'} T_1}{\hbar^2}|U_{\lambda\lambda'}(\mathbf{q})|^2 N_\mathbf{q} \cdot \frac{1}{\gamma_{\lambda\lambda'}^2 + (\Omega_\mathbf{q} - \omega_\mathbf{q})^2}\right]. \tag{3.18}$$

The real part of this equation determines the total number of hypersonic phonons $N_\mathbf{q}$ in the system depending on the parameters of the crystal and the pumping of the exciton bands:

$$\varkappa = \frac{2}{L}\sqrt{\frac{\alpha}{\pi}} \cdot \frac{\gamma_{\lambda\lambda'}}{\hbar^2}|U_{\lambda\lambda'}(\mathbf{q})|^2 e^{-\alpha K_{z0}^2} \frac{N_\lambda e^{\frac{\Delta_{\lambda\lambda'} - \hbar\omega_\mathbf{q}}{k_0 T}} - N_{\lambda'}}{\gamma_{\lambda\lambda'}^2 + (\Omega_\mathbf{q} - \omega_\mathbf{q})^2}$$
$$\times \left[1 - 4\frac{\gamma_{\lambda\lambda'} T_1}{\hbar^2}|U_{\lambda\lambda'}(\mathbf{q})|^2 N_\mathbf{q} \cdot \frac{1}{\gamma_{\lambda\lambda'}^2 + (\Omega_\mathbf{q} - \omega_\mathbf{q})^2}\right]. \tag{3.19}$$

Equation (3.19) allows us to establish the existence of absolute instability in the system of excitons and resonant hypersonic phonons [339, 368]. According to (3.19) the generation threshold of the hypersonic vibrations (on the threshold of generation

---

[3]Here the integration domain of K_x and K_y is expanded, as usual, outside the reduced Brillouin zone to $\pm\infty$.

$N_{\mathbf{q}} = 0$, above the threshold, $N_{\mathbf{q}} > 0$) is defined from the condition

$$
N_\lambda \exp\left\{ \frac{\Delta_{\lambda\lambda'} - \hbar\omega_{\mathbf{q}}}{k_0 T} \right\} - N_{\lambda'} \geq \hbar^2 \frac{\varkappa}{\gamma_{\lambda\lambda'}} \cdot \frac{L}{2} \sqrt{\frac{\pi}{\alpha}}
$$
$$
\times \frac{1}{|U_{\lambda\lambda'}(\mathbf{q})|^2} [\gamma_{\lambda\lambda'}^2 + (\Omega_{\mathbf{q}} - \omega_{\mathbf{q}})^2] e^{\alpha K_{z0}^2}.
$$

By attempting to determine from (3.19) the condition of the existence of convective instability, we come across the same difficulties (connected with the negative values of the phonon filling numbers) that exist in the quasi-classical theory of lasers and are connected with the negative photon filling numbers [296, 297]. The coefficient of hypersound amplification by excitons can be found by using perturbation theory (see the next section).

The imaginary part of (3.18), under the condition that the frequency of the hypersonic mode $\omega_{\mathbf{q}}$ is renormed due to the exciton–phonon interaction ($\Omega_{\mathbf{q}} - \omega_{\mathbf{q}} \neq 0$), leads to the condition

$$
\varkappa = -\gamma_{\lambda\lambda'}, \tag{3.20}
$$

where $\varkappa$ is determined by expression (3.19).

Equations (3.19) and (3.20) have the following solutions:

$$
\Omega_{\mathbf{q}} = \omega_{\mathbf{q}} \pm \sqrt{\frac{A}{2\varkappa}\left(1 \pm \sqrt{1 - 4\varkappa B/A}\right) - \gamma_{\lambda\lambda'}^2}, \tag{3.21a}
$$

$$
\Omega_{\mathbf{q}} = \omega_{\mathbf{q}} \pm \sqrt{-\frac{A}{2\gamma_{\lambda\lambda'}}\left(1 \pm \sqrt{1 + 4\gamma_{\lambda\lambda'} B/A}\right) - \gamma_{\lambda\lambda'}^2}, \tag{3.21b}
$$

where

$$
A = \frac{1}{\hbar^2} \frac{2}{L} \sqrt{\frac{\alpha}{\pi}} \gamma_{\lambda\lambda'} |U_{\lambda\lambda'}(\mathbf{q})|^2 e^{-\alpha K_{z0}^2} \left(N_\lambda e^{\frac{\Delta_{\lambda\lambda'} - \hbar\omega_{\mathbf{q}}}{k_0 T}} - N_{\lambda'}\right),
$$
$$
B = 4\frac{\gamma_{\lambda\lambda'} T_1}{\hbar^2} N_{\mathbf{q}} |U_{\lambda\lambda'}(\mathbf{q})|^2. \tag{3.22}
$$

Expression (3.21b) is justified only when the condition $\Omega_{\mathbf{q}} \neq \omega_{\mathbf{q}}$ is applied.

During interaction of the hypersonic mode with the excitons, its attenuation takes place or the exciton–phonon system becomes unstable. Depending on the signs of the magnitudes of A and $\varkappa$ in the absolute values of the radical expressions from (3.21a) and (3.21b), amplification and generation of the hypersonic vibrations are possible. In particular, according to (3.21a), in the absence of saturation and in the absence of a populations inversion of the exciton bands ($A < 0$) at $\varkappa > 0$, we have

$$
\Omega_{\mathbf{q}} = \omega_{\mathbf{q}} - i\gamma_{\lambda\lambda'}',
$$

where

$$\gamma'_{\lambda\lambda'} = \gamma_{\lambda\lambda'} + \sqrt{\frac{|A|}{2\varkappa}\left(1 + 4\varkappa\frac{B}{A}\right)}.$$

Under conditions of population inversion of the exciton bands ($A > 0$), from (3.21b) we find that

$$\Omega_{\mathbf{q}}^{(1,2)} = v_s q \pm \sqrt{c^2(\mathbf{q})q^2 - \gamma_{\lambda\lambda'}^2}, \tag{3.23}$$

where

$$c(\mathbf{q}) = \frac{1}{\hbar q}\left(\frac{\alpha}{\pi L^2}\right)^{1/4} |U_{\lambda\lambda'}(\mathbf{q})| e^{-\frac{1}{2}\alpha K_{z0}^2}$$

$$\times \left\{\sqrt{1 + 8\gamma_{\lambda\lambda'} T_1 \sqrt{\frac{\pi}{\alpha}} \cdot \frac{N_{\mathbf{q}}}{N_\lambda e^{\frac{\Delta_{\lambda\lambda'} - \hbar\omega_{\mathbf{q}}}{k_0 T}} - N_{\lambda'}} e^{\alpha K_{z0}^2} - 1}\right\}^{1/2}.$$

Here v_s is the hypersound velocity, and $c(\mathbf{q})$ is some characteristic velocity of the waves being a "mixture" of excitons in bands λ and λ' with an inverse difference of populations and hypersonic phonons (exciton–elastic waves). In the absence of exciton–phonon interaction, $c(\mathbf{q}) = 0$.

Expression (3.23) for the dispersion law of the system of interacting excitons and strictly resonant hypersonic phonons is completely similar in form to the solution of the dispersion equation for the wave kinematics problem for a mechanical model in which the vibrational mechanical system with attenuation is translated as a whole along some directions in space, while the oscillators are in a state of unstable equilibrium [369]. If $c^2(\mathbf{q})q^2 < \gamma_{\lambda\lambda'}^2$, the exciton–phonon system becomes unstable under conditions of populations inversion of the exciton bands.

3.5 Phonon Maser on the Exciton Transitions

The interaction of hypersound with excitons in semiconductors can have important practical applications in the creation of quantum amplifiers and in phonon generation. A realization of the stimulated radiation of the coherent phonons by excitons in a frequency range of 10^{11}–10^{12} GHz will allow us to extend the number of existing methods of hypersound generation. These methods could include: phonon generation in thermal pulses [370], phonon generation in the electron recombination process that occurs as a consequence of the ionization of small donors in an electric field [371], the excitation of hypersound by means of microwave cavities [39], hypersound generation at the resonance frequency during mixing on nonlinear media of two crossing laser fascicles in a crystal [39], excitation of acoustic waves of high frequency (up to a frequency of the order of 10^{13} GHz) by means of the piezoelectric effect [372], hypersound generation in the regime of stimulated phonon induction [373], and others.

The effect of hypersound amplification by excitons was considered in [357, 374], in which the main emphasis was on the drifting mechanism of the amplification that takes place when the drifting speed of the excitons exceeds the phase speed of the acoustic waves—in complete analogy to the case of acoustic wave amplification by drifting charge carriers [300]. The nonequilibrium distribution of excitons on quasi-momenta in the band, necessary in this case for the appearance of drifting excitons, can be challenged by a carrying away of excitons by charged particles in the external electric field [375]. It is also possible to amplify and generate hypersound in the processes of nonelastic scattering of excitons on the hypersonic vibrations with "de-excitation" of the inner state of the exciton (e.g., the transitions $2S \to 1S$ [374] and $2S \to 2P_0$ [170, 338]).

The results of the semiclassical approach to a phonon maser on exciton transitions for a model with two exciton bands and one phonon mode [170, 338] are given below. In the framework of this model, the description of the processes of emission by the excitons of coherent phonons, magnons and photons leads to similar results in many aspects. Therefore, while developing the theory of phonon masers on excitons, we use the results obtained in Sects. 2.4 and 2.5 for the stimulated radiation by excitons of coherent microwaves and IR radiation in anisotropic semiconductors and for the generation of coherent magnons in magnetic semiconductors.

The Hamiltonian of the exciton–phonon system for the case of one hypersonic mode and two exciton bands in the rotating field approximation has the form

$$H = \hbar\omega_{\mathbf{q}} B_{\mathbf{q}}^+ B_{\mathbf{q}} + \sum \varepsilon_\lambda(\mathbf{k}) A_{\mathbf{k},\lambda}^+ A_{\mathbf{k},\lambda}$$
$$+ \sum_{\mathbf{k}} \varepsilon_{\lambda'}(\mathbf{k}+\mathbf{q}) A_{\mathbf{k}+\mathbf{q},\lambda'}^+ A_{\mathbf{k}+\mathbf{q},\lambda'}$$
$$+ \sum_{\mathbf{k}} [U_{\lambda\lambda'}(\mathbf{q}) B_{\mathbf{q}} A_{\mathbf{k}+\mathbf{q},\lambda'}^+ A_{\mathbf{k},\lambda} + \text{H.c.}], \qquad (3.24)$$

where $B_{\mathbf{q}}^+$ and $B_{\mathbf{q}}$ are the phonon creation and destruction operators of frequency $\omega_{\mathbf{q}}$, $A_{\varkappa,\nu}^+$ and $A_{\varkappa,\nu}(\varkappa = \mathbf{k}, \mathbf{k}+\mathbf{q}; \nu = \lambda, \lambda')$ are the exciton creation and destruction operators with wave vector $\varkappa$ in the band ν, and $U_{\lambda'\lambda}(\mathbf{q})$ is the constant of the exciton–phonon coupling. In the Hamiltonian (3.24) terms of the type $B_{\mathbf{q}} A_{\mathbf{k},\nu}^+ A_{\mathbf{k}+\mathbf{q},\nu}$ and $B_{\mathbf{q}}^+ A_{\mathbf{k}+\mathbf{q},\nu}^+ A_{\mathbf{k},\nu}$ are omitted, being responsible for the processes of intraband scattering of excitons on phonons. This is true in the case when, in some spectral range, intraband scattering of the hypersonic phonons on excitons is less effective than interband absorption of hypersound by excitons, and in this range of frequencies we can ignore the processes of intraband scattering. This situation takes place, for example, for the transitions $2P_0 \rightleftarrows 2S$ in the crystals CdS and ZnO at frequencies $\omega_{\mathbf{q}} \simeq 5.68 \cdot 10^{11}$ rad/s and $\omega_{\mathbf{q}} \simeq 1.1 \cdot 10^{12}$ rad/s, respectively (Fig. 3.1). The Hamiltonian (3.24) also does not contain terms describing the interaction of the excitons and the hypersonic field with dissipative subsystems and the pumping subsystem. These subsystems, like those in Sects. 2.4 and 2.5, are considered phenomenologically.

The equations of motion for the operators of the hypersonic field, exciton polarization, and difference of the operators of the number of excitons in the states $|\lambda', \mathbf{k} + \mathbf{q}\rangle$ and $|\lambda, \mathbf{k}\rangle$ coincide in form with (2.48). The system obtained from five nonlinear operator equations is averaged by using the density matrix of the exciton–phonon system, and we shall consider that, in the regime of generation, the corpuscular nature of the hypersonic field is not displayed. The latter allows us to carry out unhooking of the type (2.29). In the approximation of a macrofilled hypersound mode ($\langle B_{\mathbf{q}}\rangle = \sqrt{N_{\mathbf{q}}}e^{-i\Omega t}$, $N_{\mathbf{q}}$ is the number of phonons in the mode) the solution of the averaged system of equations is found by the method described in Sect. 2.4. As a result, we obtain the following expressions for the attenuation coefficient of the hypersound,

$$\varkappa = \frac{1}{2}(\mathcal{G}_1 - \mathcal{G}_2 N_{\mathbf{q}}), \tag{3.25}$$

and for the shift of the frequency of generation Ω relative to the frequency of the hypersonic phonons $\omega_{\mathbf{q}}$ in the absence of interaction with the excitons,

$$\Omega - \omega_{\mathbf{q}} = (\Delta\Omega)_1 + N_{\mathbf{q}}(\Delta\Omega)_2, \tag{3.26}$$

which coincide with formulas (2.35) and (2.37), if we substitute N, $g_{n'n}(0)$, $\Gamma_{n'n}$, $\omega_{n'n}(\mathbf{K})$, and $d^0_{nn'}(\mathbf{K})$ by $N_{\mathbf{q}}$, $U_{\lambda'\lambda}(\mathbf{q})$, $\gamma_{\lambda'\lambda}$, $\omega_{\lambda'\lambda}(\mathbf{K}, \mathbf{q})$, and $d^{(0)}_{\lambda'\lambda}(\mathbf{K}, \mathbf{q})$, respectively.

If, under conditions of the pumping action (for example, during the resonant excitation by polarized light of the upper exciton band) the excitons remain thermal, then the function $d^{(0)}_{\lambda\lambda'}(\mathbf{K}, \mathbf{q})$ is determined by formula (2.51). After the substitution in the expressions (3.25) and (3.26) of Ω by $\omega_{\mathbf{q}}$ (on the same basis as in the case of radiation by excitons of the coherent photons and magnons), for $\mathcal{G}_i$ and $(\Delta\Omega)_i$ from (3.25) and (3.26) for an equilibrium distribution of excitons we obtain formulas (2.52), in which the coupling constant $G_{\sigma'\sigma}(\mathbf{q})$ is substituted by $U_{\lambda'\lambda}(\mathbf{q})$, and instead of $C_{\sigma'\sigma}$ we take the expression $b_{\lambda'\lambda}$ from (3.7). Therefore, all the qualitative conclusions with respect to the effects of "hole burning" in the amplification curve, "repulsion of the downfalls," and the shift of the generation frequency due to the tightening influence of the active substance remain the same as in the case of the exciton quantum generators of magnons, microwaves, and IR photons that we have considered above.

The existence of the equilibrium distribution of excitons in the bands at high levels of optical pumping of crystals is confirmed experimentally in [149, 376]. The authors of [376] showed, on the basis of measurements of the exciton–phonon absorption spectrum in the crystal CdS, that under conditions of high exciton concentration ($n_{\mathrm{ex}} \sim 10^{15}$ cm^{-3}, $T = 77$ K) created by two-photon optical excitation, they are described by the equilibrium function of the distribution. At lower temperatures and higher levels of pumping ($n_{\mathrm{ex}} = 1.6 \cdot 10^{17}$ cm^{-3}, $T = 10$ K) the excitons become hot with $T_{\mathrm{ex}} = 28$ K, and their equilibrium distribution in the band is transformed to a quasi-equilibrium one [149]. We note that in [376] the excitons were excited near the bottom of the exciton band. In the presence of excitons in states with large $\mathbf{K}$, one observes the luminescence of unequilibrated excitons [377].

It is convenient to express the integrals I_l (determined by the expression (2.53) in which $C_{\sigma'\sigma}$ is substituted by $b_{\lambda'\lambda}$) by a probability function for finding the threshold of generation and the frequency shift of the phonon maser. This connection for I_1 is given by formula (3.8). The other integrals are connected with I_1 by the relationships

$$
\begin{aligned}
I_3 &= -\frac{1}{2a_{\lambda'\lambda}} \cdot \frac{\partial I_1}{\partial a_{\lambda'\lambda}}, \\[6pt]
I_5 &= -\frac{1}{2\alpha} \cdot \frac{\partial I_1}{\partial b_{\lambda'\lambda}}, \\[6pt]
I_7 &= -\frac{1}{4\alpha a_{\lambda'\lambda}} \cdot \frac{\partial^2 I_1}{\partial a_{\lambda'\lambda}\,\partial b_{\lambda'\lambda}}, \\[6pt]
I_{2p} &= e^{\alpha q^2}\mathcal{I}_{2p-1} \quad (p = 1,\dots,4),
\end{aligned}
\tag{3.27}
$$

where $\mathcal{I}_l$ designates the expression obtained from I_l upon substitution of $b_{\lambda'\lambda}$ by $b_{\lambda'\lambda} - q$.

From the condition $\mathcal{G}_1 \geq 2\varkappa$ and with formulas (2.53), (3.8), and (3.27) we shall determine the inverse population of exciton bands necessary to begin hypersound generation. For the line of radiation $2S \to 2P_0$ in the crystal CdS at $T = 4.2$ K, hypersound generation at the frequency $\omega_g = 5.68 \cdot 10^{11}$ rad/s begins at $n_{ex}^{2S} - n_{ex}^{2P_0} = 5 \cdot 10^{13}$ cm^{-3}. In the crystal ZnO the beginning of hypersound generation corresponds to a larger inverse population of the exciton bands (at the frequency $\omega_g = 1.1 \cdot 10^{12}$ rad/s at $n_{ex}^{2S} - n_{ex}^{2P_0} = 6 \cdot 10^{14}$ cm^{-3}). In the process of fulfilling these estimations it was considered that $\varkappa = \frac{1}{2}(\omega_q/Q) \sim 10^6$ s^{-1}, because at hypersonic frequencies the quality factor of the nonloaded cavity is $Q \sim 10^5$–10^6 [12].

The frequencies for which the threshold of self-excitation of the phonon maser was estimated correspond to the situation when the intraband exciton–phonon effects are small in comparison with the interband ones. For the selected frequencies the interband absorption of hypersound in the crystals CdS and ZnO is, respectively, 4 and 5.5 times stronger than the intraband absorption. In the CdS crystal hypersound generation begins at a smaller difference of the population of exciton bands in comparison with the corresponding difference in the crystal ZnO. This is connected with the difference of the parameters of the crystals. In the first case $mv_s^2/2 < \Delta_{2S,2P_0}$ and the excitons from the lower band $2P_0$, the wave vectors of which are enclosed in the interior of a sphere with radius $b_{2S,2P_0}$, do not participate in the processes of exciton–phonon interaction.

From formulas (3.8) and (3.27) and the condition $\mathcal{G}_1 \geq 2\varkappa$, it follows that in the process of satisfying the inequality $(a_{\lambda'\lambda}/b_{\lambda'\lambda}) \ll 1$, the generation threshold of the phonon maser does not depend on $\gamma_{\lambda'\lambda}$, because in this approximation I_1 and $I_2 \sim \gamma_{\lambda'\lambda}^{-1}$. The lack of dependence of the generation threshold on $\gamma_{\lambda'\lambda}$ is connected with the fact that in the process of phonon emission the excitons in the final state have a large value of quasi-momentum.

The frequency shift $\Omega - \omega_q$ near the generation threshold, as the calculation shows, is much less than the width of the acoustic band of radiation. The small value of this shift is due to the relatively small difference of populations of the exciton bands and the large width of the line of interband absorption of hypersound.

Thus, the generation threshold of hypersound by excitons is achieved at comparatively small excitation levels in the crystals, and this generation is accompanied by hypersound amplification. The calculation according to perturbation theory accounting for the broadening of exciton bands leads to the next expression for the coefficient of hypersound amplification by excitons:

$$\Gamma^a_{\lambda'\lambda}(\mathbf{q}) = v_s^{-1}\mathcal{G}_1.$$

In the crystal CdS at $n_{ex}^{2S} - n_{ex}^{2P_0} = 3 \cdot 10^{14}$ cm^{-1} the coefficient of hypersound amplification at the frequency $\omega = \omega_g$ has the value $\Gamma^a_{2S,2P_0} = 6.8 \cdot 10^2$ dB/cm.

3.6   Resonance Absorption of Hypersound by Biexcitons

In Chap. 1 the effects of the optical evidence for exciton molecules (biexcitons) in semiconductors were considered. These molecules are formed at high levels of excitations in crystals in the case when the attractive forces between excitons prevail over the repulsive forces between them. The existence of biexcitons in the crystals, apart from the optical evidence, must be present in all other processes in which they participate. Coulomb binding of free electrons and holes into excitons leads to acoustic absorption depending on the type of function of relative electron–hole motion in the exciton, which may be evidentiated on the background of absorption acoustic bands of free carriers. By analogy, excitons coupled into biexcitons must lead to an additional contribution to the total acoustic absorption of hypersound.

We shall consider elastic scattering of biexcitons on hypersonic phonons in semiconductors in the case where the length of the free run of the biexciton is much longer than the length of the hypersonic wave ($ql_B \gg 1$) [170, 341]. We also assume that the energy of hypersonic phonons is less than the biexciton binding energy. This assumption is usually fulfilled even for extreme hypersonic phonon frequencies.

The Hamiltonian of the biexciton–phonon interaction in the case of one hypersonic mode is represented in the form [341]

$$H_{\text{biex-ph}} = \theta_{\mathbf{q}} B_{\mathbf{q}} e^{i\mathbf{q}\mathbf{R}} + \text{H.c.},$$

where

$$\theta_{\mathbf{q}} = V_{\mathbf{q}} \sum_j \mathcal{B}^j_{\mathbf{q}} \tag{3.28}$$

is the constant of the biexciton–phonon interaction, $B_{\mathbf{q}}$ is the phonon destruction operator with wave vector $\mathbf{q}$, and $\mathbf{R}$ is the coordinate of the biexciton center of mass. In (3.28) $V_{\mathbf{q}}$ characterizes the effectivity of the interaction of the charge carriers forming the biexciton with the hypersonic wave. For interaction by means of the piezopotential, $V_{\mathbf{q}}$ is found using formula (3.5), and $\mathcal{B}_{\mathbf{q}}^{j}$ is determined by the expression

$$
\mathcal{B}_{\mathbf{q}}^{j} = \left\{ C_1^{j} \exp\left[i \frac{m_{\mathrm{h},j}}{m_j} \mathbf{q} \mathbf{r}_j \right] \right.
$$
$$
\left. - C_2^{j} \exp\left[-i \frac{m_{\mathrm{e},j}}{m_j} \mathbf{q} \mathbf{r}_j \right] \right\} \exp\left[i\varphi(j) \frac{1}{M} \mathbf{q} \mathbf{r} \right], \tag{3.29}
$$

where $m_{\mathrm{e},j}$, $m_{\mathrm{h},j}$, m_j, and M are the effective masses of the electron, hole, exciton, and biexciton; $\mathbf{r}$ and $\mathbf{r}_j$ are the coordinates of the relative movement of excitons in the biexciton and the relative electron–hole movement in the jth exciton; C_1^{j} and C_2^{j} are dimensionless constants of the order of unity, accounting for the deformations of the electron shells during the atomic vibrations for the electron and the hole of the jth exciton (in the general case $C_1^{j} \neq C_2^{j}$). In formula (3.29) j takes only two values; thus $\varphi(1) = m_2$ and $\varphi(2) = -m_1$.

After the standard calculations of the matrix elements of biexciton–phonon interaction and the determination of the balance for the processes with the absorption and emission of phonons using Maxwell–Boltzmann statistics, we find the following expression for the coefficient of the intraband absorption of hypersound by biexcitons:

$$
\Gamma_{\mathrm{biex}}(\mathbf{q}) = \left(\frac{2\pi M}{k_0 T} \right)^{1/2} \frac{n_{\mathrm{biex}}}{\hbar^2 \omega} \Phi(\mathbf{q}, T) \left(1 - e^{-\frac{\hbar\omega}{k_0 T}} \right)
$$
$$
\times \exp\left\{ -\frac{M v_{\mathrm{s}}^2}{2k_0 T} \left(\frac{\hbar}{2M v_{\mathrm{s}}} q - 1 \right)^2 \right\}, \tag{3.30}
$$

where

$$
\Phi(\mathbf{q}, T) = \sum_n P_n(T) \Phi_n(\mathbf{q}),
$$
$$
\Phi_n(\mathbf{q}) = \left| \langle n | \theta_{\mathbf{q}} | n \rangle \right|^2. \tag{3.31}
$$

Here n is the total quantum number determining the internal state of the biexciton; $P_n(T)$ is the probability of the nth state, and n_{biex} is the concentration of biexcitons.

In the case when biexcitons during intraband scattering on the hypersonic phonons are in the ground $1S$ state, the calculation with inclusion of the matrix element $\langle n | \theta_{\mathbf{q}} | n \rangle$ in formula (3.31) leads to the expression

$$
\langle 1S | \theta_{\mathbf{q}} | 1S \rangle = 2^8 V_{\mathbf{q}} \sum_j \frac{1}{[4 + \frac{1}{M^2} \varphi^2(j) a^2 q^2]^2}
$$
$$
\times \left\{ \frac{C_1^{j}}{[4 + (\frac{m_{\mathrm{h},j}}{m_j})^2 a_{\mathrm{ex},j}^2 q^2]^2} - \frac{C_2^{j}}{[4 + (\frac{m_{\mathrm{e},j}}{m_j})^2 a_{\mathrm{ex},j}^2 q^2]^2} \right\}, \tag{3.32}
$$

where a and $a_{ex,j}$ are the effective Bohr radii of the biexciton and jth exciton, respectively.

The calculations on formulas (3.6) and (3.30), taking into account (3.32), show that the coefficient of absorption of the hypersound by biexcitons $\Gamma_{biex}(\mathbf{q})$ has a maximum in the microwave range in the same region of frequencies where the intraband absorption of the hypersound on the excitons is maximal. For a CdS crystal, the ratio of these maximal coefficients of hypersound absorption is approximately equal to 6.

The acoustic manifestation of exciton–exciton interaction has already been observed experimentally. In the germanium crystal an absorption of ultrasound by electron–hole drops was found [379], forecast theoretically by Keldysh and Tikhodeev [380]. However, the authors of the work [379] did not observe the absorption of the ultrasound, caused by the noncondensed phase of excitons, in either the presence or the absence of electron–hole drops. From the results obtained in Sects. 3.2 and 3.3, this is not surprising, since the intraband absorption of the ultrasonic phonons by free excitons in this experiment took place on the long-wave wing of the acoustic band, the maximum of which is in the hypersonic range of frequencies.

Chapter 4
Double Resonances

In a number of cases it is convenient to investigate the specificity of the discrete structure of an energy spectrum in conditions when two alternative fields of different frequency act over the system simultaneously, one of which is saturating. Here, in contrast to the system with a quasi-energy spectrum which will be considered in Chap. 7, saturating is considered the external variable field with an intensity whose magnitude is sufficient to smooth the difference of populations of the corresponding states, but not sufficient to transform the energy spectrum to a quasi-energy one.

The experimental technique of double resonances possesses an evident advantage in comparison with the usual resonance that consists in simultaneous increasing of the sensitivity and spectral resolution of the method. If the saturating field has a high frequency and the registration is performed at a low frequency, then the sensibility of the method is determined by the difference of populations which corresponds to the frequency of saturation, and the spectral resolution is determined by the interaction leading to the low-frequency splitting of the energy levels. A typical example of the realization of these possibilities is the method of electron–nuclear double resonance (ENDOR) [53].

This chapter investigates the fine structure of ENDOR spectra [170, 381, 382] on the basis of the Pikus–Luttinger method of invariants [43, 44, 383, 384] and gives the theory of the new electron–nuclear double magnetoacoustic resonance (acoustical ENDOR) [59, 170, 385]. It provides a theoretical examination of the excitons captured by the nitrogen isoelectron acceptors in GaP crystal with the formation of localized biexcitons for which the existence of a new resonance is shown—hole–nuclear double resonance (HONDOR) [170, 386, 387]. Radio–optical double resonance (RODOR) is also considered on free excitons in the crystal CuCl [388].

The use of the Pikus–Luttinger method of invariants in radio spectroscopy allowed one to determine the important role of symmetry coordinates, transformed over the odd representations of the point group of symmetry, into acoustical ENDOR and in the processes of one-phonon spin–lattice relaxation, to give evidence for the effect of the acoustic nonequivalence of nuclei that passing into one another during inversion, in the case of spin system excitation by polarized phonons. This

I. Geru, D. Suter, *Resonance Effects of Excitons and Electrons*,
Lecture Notes in Physics 869, DOI 10.1007/978-3-642-35807-4_4,
© Springer-Verlag Berlin Heidelberg 2013

method allowed one to show that in a localized biexciton in the crystal GaP:N the electron spins are compensated, and the exchange and magnetic dipole–dipole interactions are performed only between the holes, which corresponds completely to the experiment [271, 272].

In the methodological aspect the convenient form of the time-reversal operator in the radio–spectroscopic applications of the method of invariants for a system with arbitrary angular momenta is found, and the geometric interpretation of the Kramers theorem is carried out. The connection between point groups of magnetic symmetry and the time-reversal transformation for systems with integer and half-integer angular momentum is shown. The convenience of applying the two-Bose operator representation of angular momentum together with the standard technique of spin operators is also shown. On the basis of unitarity properties of two-boson spin operators of a special type that we have introduced, the invariance of the operators of angular momentum projections in the two-Bose operator representation relative to the orthogonal addition and orthogonal reduction of the basis of the angular momentum space is demonstrated.

The main results of this chapter are published in [59, 71–73, 170, 381, 382, 385–393].

4.1 Pikus–Luttinger Method of Invariants and Its Applications to ENDOR and Acoustical ENDOR

The method of invariants used by Luttinger to investigate the cyclotron resonance of the holes in Ge [43] and generalized by Pikus for any problems of solid state physics [44, 383, 384] allows us to determine the general type of the Hamiltonian $\mathcal{H}(\mathcal{K})$, if the law of transformation of the basic functions and component $\mathcal{K}_i$ is known (the symbol $\mathcal{K}$ designates the vector and tensor values on which the Hamiltonian depends). We have used this method to investigate hyperfine and spin–phonon interactions [59, 170, 381, 382, 389].

To describe a hyperfine interaction by the method of invariants, we shall present the spin system in the form of two subsystems (electron and nuclear) characterized by the effective spin of the paramagnetic center S and the spin of the nucleus I. The value S determines $(2S+1)^2$ of the linearly independent matrices A_l of dimension $(2S+1) \times (2S+1)$, which under action of the operators of symmetry G of the rotation group are transformed into one another according to the representation T_A:

$$D_S(G^{-1}) A_l D_S(G) = \sum_{l'=1}^{(2S+1)^2} G_{l'l}^{T_A} A_{l'}.$$

Here $D_S(G)$ is the matrix of the irreducible representation D_S of the rotation group. The matrices B_m $(m = 1, 2, \ldots, (2I+1)^2)$ are transformed in an analogical way, determined by the value of the nuclear spin.

In the case of superhyperfine interaction of the system of $(2I + 1)^2$ linearly independent matrices, it is necessary to have a correspondence for every nucleus with which the superhyperfine interaction is considered. Then the matrices $B(m|\mathbf{R}_k)$, referring to corresponding nodes $\mathbf{R}_k$ ($k = 1, 2, \ldots, p$) of the crystal lattice, are transformed according to the law

$$D_I(G^{-1})B(m|\mathbf{R}_k)D_I(G) = \sum_{m'=1}^{(2I+1)^2} \sum_{k'=1}^{p} G_{m'm}^{T_B} G_{k'k}^{\Pi} B(m'|\mathbf{R}_{k'}),$$

where T_B and Π are representations to transform the matrices B_m and radius vectors $\mathbf{R}_k$, which determine the positions of the nuclei. The dimension of the representation Π is equal to the number of nuclei with which the superhyperfine interaction of the effective spin S takes place. An evident assumption is made here that the motions in the coordinate and spin spaces are independent, and the representation T, according to which the matrices $B(m|\mathbf{R}_k)$ are transformed, may be expressed in the form of the direct product of the representations Π and T_B:

$$T = T_B \otimes \Pi.$$

The representations T_A and T_B in turn may be expressed in the form of the direct product of the irreducible representations of the rotation group [44, 170], and thus we have

$$T_A = D_S \otimes D_S^* = \sum_{\lambda=0}^{2S} D_\lambda, \qquad T = \Pi \otimes D_I \otimes D_I^* = \Pi \otimes \sum_{\lambda=0}^{2I} D_\lambda.$$

Thus, T_A, T_B, and T are reducible representations.

It is convenient to present the bases of representations T_A and T_B through spherical functions, which are necessary, when used with the Stevens operator-equivalent method [394], to pass to the irreducible tensor operators of the angular momentum. In [170, 390] it was demonstrated that this transition is easier to perform if, instead of repeated applications of the commutation relations for the angular momentum operators, we use the generalized commutation relations between polynomials in the components of the angular momentum. In Appendix A we have given the irreducible tensor operators in this way: $Y_M^L(J)$ for $J \leq 7/2$. We have used spherical functions with phase multipliers determined in accordance with [126]. The operators $Y_M^L(J)$ for $J \leq 5/2$ were obtained earlier by Koster and Statz [395] using another method. The function operators for $J \leq 5/2$ found by means of Wigner's coefficients are also given in [396].

The irreducible tensor operators in the method of invariants are initial in the construction of the interaction operators of spin systems between themselves and with the external fields. The factual decomposition of reducible representations T_A and T_B into irreducible ones of the symmetry point groups is essentially simplified using Leushin's tables [397].

The basis function operators of irreducible representations of the symmetry point groups $\Gamma_\gamma \in \Pi \otimes D_I \otimes D_I^*$ are found by using a combination of "node" functions, transformed according to irreducible representations $\Gamma_\alpha \in \Pi$ with the function operators that are transformed according to the representations $\Gamma_\beta \in D_I \otimes D_I^*$, applying the coefficients of angular momentum coupling for the point groups. Thus, to account for "node" symmetry, the basis vectors $B(0|\mathbf{R}_k)$ should be considered as the usual functions depending on spatial coordinates. In order to build the invariants, it is necessary to combine the basis function operators of the irreducible representations contained in $D_S \otimes D_S^*$ with the basis function operators of the representations contained in $\Pi \otimes D_I \otimes D_I^*$. The number of invariants is equal to the number of unit representations contained in the direct product $D_S \otimes D_S^* \otimes D_I \otimes D_I^* \otimes \Pi$. The only invariants $\mathcal{H}_i$ that will be included in the Hamiltonian are those that also remain invariant relative to the time-reversal transformation:

$$U_S U_{I1} U_{I2} \cdots U_{Ip} \mathcal{H}_i^* U_{Ip}^+ \cdots U_{I2}^+ U_{I1}^+ U_S^+ = \mathcal{H}_i,$$

where the unitary operator U_S acts only on the electron subsystem, and U_{Ik} ($k = 1, 2, \ldots, p$) only in the spin space of the nucleus situated in the kth node.

Different ways of constructing the unitary operator U_J for spin systems are indicated in [170, 391]. All of them are reduced to the fact that, for a system with angular momentum J, the matrix of the operator U_J coincides with the matrix that performs the transition from the representation D_J^* to D_J, the properties of which are considered in [398]. The matrices of the operator U_J in the basis of the functions that are being transformed according to the irreducible representations of all symmetry point groups are presented in Appendix B.

In Appendix C the geometrical interpretation of the time-reversal operation is given, and on its basis it is found that magnetic symmetry groups under a time-reversal transformation are essentially different (both in quantity and structure) for cases of integer and half-integer angular momentum. It then follows that the known 58 "younger" point groups or magnetic symmetry groups [399, 400] that are isomorphic to the corresponding antisymmetry groups can be related only to systems with integer total angular momentum. For systems with resulting half-integer spin, the magnetic symmetry groups are not isomorphic to the Shubnikov antisymmetry groups, but to the Belov four-color symmetry groups [392]. In this case there exist four younger point groups of four-color symmetry influenced by a time-reversal transformation.

Spin–phonon interaction may be examined by the Pikus–Luttinger method of invariants in the same way as for hyperfine interaction. The difference consists in the necessity to take into account the mechanical representation Ξ, on which the symmetry coordinates of the vibrational system are transformed. In the case of modulation of the hyperfine interaction by lattice vibrations, the number of spin–phonon invariants is determined by the number of unit representations contained in the direct product $D_s \otimes D_s^* \otimes D_I \otimes D_I^* \otimes \Pi \otimes \Delta$, where $\Delta = \Xi$ for one-phonon processes and $\Delta = [\Xi]^n + \{\Xi\}^n$ for n-phonon processes. One must then consider all the mechanical degrees of freedom of the evidentiated "quasi-molecule," and there are no grounds to discard some of them, since the system is not free but is tied to the crystal.

The large number of problems solved by ENDOR spectroscopy (distribution of the electron cloud for local and impurity centers [15], effects of covalency in semiconductors and complex compounds [6], distribution of the spin density of the unpaired electrons in free radicals [401], the processes of spin relaxation [6, 15], hyperfine fields in ferroelectrics [402], and others) determines the actuality of the method of invariants in ENDOR for investigating the hyperfine interaction with the distant nuclei and the effects of high order, to which the high degrees of spin operators correspond.

Let us consider some peculiarities of ENDOR in the fluorite lattice. The ENDOR spectrum of $CaF_2:Eu^{2+}$ on the ^{151}Eu and ^{152}Eu nuclei cannot be adequately interpreted based on Abraham–Pryce's spin Hamiltonian; therefore, it is necessary to attract additional terms of high order, including diagonal components of the operator of octupole interaction [403]. Below, on the basis of the method of invariants, we shall obtain the spin Hamiltonian of superhyperfine interaction of paramagnetic ions, substituting Ca^{2+} in CaF_2, with the nearest nuclei of fluorine $(1, 1'; 2, 2'; 3, 3'; 4, 4')$ situated in the lattice nodes in the directions $[111], [\bar{1}\,\bar{1}\,\bar{1}], [\bar{1}11], [1\bar{1}1], [11\bar{1}], [\bar{1}\,\bar{1}1], [1\bar{1}\,\bar{1}]$, and $[\bar{1}11]$.

Representation Π_8, on which the nodes are transformed, is reduced to the following irreducible representations of symmetry point group O_h:

$$\Pi_8 = \Gamma_1^+ \oplus \Gamma_2^- \oplus \Gamma_4^- \oplus \Gamma_5^+.$$

The basis functions of the even irreducible representations Γ_1^+ and Γ_5^+ have the form:

$$\Gamma_1^+: \quad \frac{1}{\sqrt{8}}(\mathbf{R}_1 + \mathbf{R}_{1'} + \mathbf{R}_2 + \mathbf{R}_{2'} + \mathbf{R}_3 + \mathbf{R}_{3'} + \mathbf{R}_4 + \mathbf{R}_{4'}),$$

$$\Gamma_5^+: \quad \begin{cases} \frac{1}{\sqrt{8}}(\mathbf{R}_1 + \mathbf{R}_{1'} + \mathbf{R}_4 + \mathbf{R}_{4'} - \mathbf{R}_2 - \mathbf{R}_{2'} - \mathbf{R}_3 - \mathbf{R}_{3'}), \\ \frac{1}{\sqrt{8}}(\mathbf{R}_1 + \mathbf{R}_{1'} + \mathbf{R}_2 + \mathbf{R}_{2'} - \mathbf{R}_3 - \mathbf{R}_{3'} - \mathbf{R}_4 - \mathbf{R}_{4'}), \\ \frac{1}{\sqrt{8}}(\mathbf{R}_1 + \mathbf{R}_{1'} + \mathbf{R}_3 + \mathbf{R}_{3'} - \mathbf{R}_2 - \mathbf{R}_{2'} - \mathbf{R}_4 - \mathbf{R}_{4'}). \end{cases} \tag{4.1}$$

Combining the basic functions (4.1) with the basic function operators of the irreducible representations contained in $D_J \otimes D_J^* \otimes D_{1/2} \otimes D_{1/2}^*$ (J is the angular momentum of the paramagnetic ion, $1/2$ is the spin of the nucleus F), we find spin invariants of the hyperfine interaction with precision up to the terms of the fourth degree on the angular momentum operators [170]:

$$\mathcal{H}_F = A_0 \xi_1 \eta_1 + A_1 \sum_p \xi_{4p}^{(1)} \eta_{4p}^{(1)} + A_2 \sum_p \xi_{4p}^{(1)} \eta_{4p}^{(2)}$$

$$+ A_3 \sum_p \xi_{4p}^{(2)} \eta_{4p}^{(1)} + A_4 \sum_p \xi_{4p}^{(2)} \eta_{4p}^{(2)} + A_5 \sum_p \xi_{5p}^{(1)} \eta_{5p}^{(1)}$$

$$+ A_6 \sum_p \xi_{5p}^{(1)} \eta_{5p}^{(3)} + A_7 \sum_p \xi_{5p}^{(2)} \eta_{5p}^{(2)}, \tag{4.2}$$

where A_i are the spin Hamiltonian constants, p has the values x, y, and z, and the expressions for ξ and η are given in Appendix D. In (4.2) only those operators are preserved that are invariant both relative to $G \in O_h$ and to time-reversal transformation.

The spin Hamiltonian (4.2) causes transitions with the selection rules $\Delta M_J = 0$ and $\Delta M_k^F = \pm 1$, where M_k^F is the spin projection of the kth nucleus F in the direction of the magnetic field. For orientation of the magnetic field along the symmetry axis of the fourth order, the value of the ENDOR frequency is determined only by two constants of superhyperfine interaction:

$$\hbar \omega_{M_J, M_k^F; M_J, M_k^F + 1} = \frac{1}{\sqrt{8}} M_J \left\{ A_1 - \frac{1}{4\sqrt{10}} A_2 \left[5M_J^2 - 3J(J+1) + 1 \right] \right\}.$$

The other ENDOR frequencies are not given here. Note only that the nuclei connected to each other by the operation of spatial inversion have similar ENDOR frequencies.

The presence of the additional items in the spin Hamiltonian leads to shifts of hyperfine levels that, in turn, lead to splittings and shifts of the ENDOR lines affecting their thin structure [170, 381]. Let us determine this structure for interaction of a localized electron with nuclei in an octahedral environment [381].

In the presence of a magnetic field, the interaction of the localized electron with a couple of equivalent nuclei situated on the z axis, with precision up to cubic terms on the spin operators, is described by the spin Hamiltonian

$$\mathcal{H} = \mathcal{H}_0 + \mathcal{H}_1,$$

where

$$\mathcal{H}_0 = g\beta \mathbf{H} \mathbf{S} + a(\mathbf{I}_k + \mathbf{I}_{k'})\mathbf{S} + b\left[3(I_{kz} + I_{k'z})S_z - (\mathbf{I}_k + \mathbf{I}_{k'})\mathbf{S} \right]$$
$$+ Q\left[I_{kz}^2 + I_{k'z}^2 - \frac{2}{3}I(I+1) \right] - g_N \beta_N \mathbf{H}(\mathbf{I}_k + \mathbf{I}_{k'}), \tag{4.3}$$

$$\mathcal{H}_1 = a_1\left[(I_{kx}^3 + I_{k'x}^3)S_x + (I_{ky}^3 + I_{k'y}^3)S_y + (I_{kz}^3 + I_{k'z}^3)S_z \right]$$
$$+ a_2\left[2(I_{kz}^3 + I_{k'z}^3)S_z - (I_{ky}^3 + I_{k'y}^3)S_y - (I_{kx}^3 + I_{k'x}^3)S_x \right]$$
$$+ a_3\left[(I_{kz}^2 I_{kx} + I_{k'z}^2 I_{k'x} - I_{kx}I_{ky}^2 - I_{k'x}I_{k'y}^2)S_x \right.$$
$$\left. + (I_{kz}^2 I_{ky} + I_{k'z}^2 I_{k'y} - I_{ky}I_{kx}^2 - I_{k'y}I_{k'x}^2)S_y \right]. \tag{4.4}$$

Here g, g_N and β, β_N are the electron and nuclear g-factor and magneton; x, y, z are the cubic axes of the crystal, $\mathcal{H}_0$ is the usual spin Hamiltonian, and $\mathcal{H}_1$ contains additional terms admitted by the properties of symmetry.

Expressions (4.3) and (4.4) compose the spin Hamiltonian if the direction of the magnetic field $\mathbf{H}$ does not coincide with any of the axes of symmetry. In the opposite case, one must account for the equivalence of more than two nuclei. The spin wave function in zero-order approximation will be presented in the form of a product of the electron spin function and the symmetric (antisymmetric) spin function of a couple of physical equivalent nuclei [170]:

$$\chi_{SI_kI_{k'}}^{(s,a)}\left(M_S, M_{I_k}, M_{I_{k'}}, M'_{I_k}, M'_{I_{k'}}\right)$$
$$= \frac{1}{\sqrt{2}}|SM_S\rangle\left\{|I_kM_{I_k}\rangle\big|I_{k'}M'_{I_{k'}}\big\rangle \pm \big|I_kM'_{I_k}\big\rangle|I_{k'}M_{I_{k'}}\rangle\right\}. \tag{4.5}$$

The indices s and a designate, respectively, symmetric and antisymmetric spin functions.

In first-order perturbation theory, the levels of energy of the spin system are determined by the expression

$$E_{M_S,M_{I_k},M_{I_{k'}}} = g\beta H M_S + a M_S(M_{I_k} + M_{I_{k'}})$$
$$+ b\theta M_S(M_{I_k} + M_{I_{k'}}) + \frac{1}{2}Q\theta\left[M_{I_k}^2 + M_{I_{k'}}^2 - \frac{2}{3}I(I+1)\right]$$
$$- g_N\beta_N H(M_{I_k} + M_{I_{k'}}) + \frac{1}{2}a_1 M_S\left\{(M_{I_k}^3 + M_{I_{k'}}^3)(5\theta_1 - 3)\right.$$
$$+ (M_{I_k} + M_{I_{k'}})[3I(I+1) - 1](1 - \theta_1)\right\}$$
$$+ \frac{1}{2}a_2 M_S\left\{(M_{I_k}^3 + M_{I_{k'}}^3)(5\theta_2 - 3\theta) + (M_{I_k} + M_{I_{k'}})\right.$$
$$\times [3I(I+1) - 1](\theta - \theta_2)\right\} + \frac{1}{2}a_3\theta_3 M_S\left\{(M_{I_k} + M_{I_{k'}})\right.$$
$$\times [3I(I+1) - 1] - 5(M_{I_k}^3 + M_{I_{k'}}^3)\right\},$$

where

$$\theta = 3n^2 - 1, \qquad \theta_1 = l^4 + m^4 + n^4,$$
$$\theta_2 = 3n^4 - \theta_1, \qquad \theta_3 = n^2(n^2 - 1) + 2l^2m^2.$$

Here l, m, n are the leading cosines of the vector of magnetic field intensity $\mathbf{H}$ with respective cubic axes x, y, z.

The frequencies of the spin–nuclear transitions with the selection rules $\Delta M_S = 0$, $\Delta(M_{I_k} + M_{I_{k'}}) = \pm 1$ for the pair of equivalent nuclei situated on the z axis are equal:

$$\begin{aligned} \nu_1^\pm &= \nu_0^\pm \pm \theta Q/\hbar \pm \Delta_1, & \nu_2^\pm &= \nu_0^\pm \pm \Delta_2, \\ \nu_3^\pm &= \nu_0^\pm \mp \theta Q/\hbar \pm \Delta_1, \end{aligned} \tag{4.6}$$

where

$$\Delta_1 = \varepsilon_0 + 13\varepsilon_1, \qquad \Delta_2 = \varepsilon_0 + \varepsilon_1,$$

$$\varepsilon_0 = \frac{1}{4h}\left[3I(I+1)-1\right]\left[a_1(1-\theta_1)+a_2(\theta-\theta_2)+a_3\theta_3\right],$$

$$\varepsilon_1 = \frac{1}{16h}\left[a_1(5\theta_1-3)+a_2(5\theta_2-3\theta)-5a_3\theta_3\right],$$

$$h\nu_0^{\pm} = \frac{a}{2}+\frac{b}{2}\theta \mp g_N\beta_N H.$$

Formulas (4.6) were obtained for the case when $Q > 0$ and $\varepsilon_1 > 0$. The expressions for the frequencies of spin–nuclear transitions for the pairs of nuclei situated on axes x and y are obtained from (4.6) by replacing $lmn \rightarrow mnl$ and $lmn \rightarrow nlm$, respectively.

The availability of additional items (4.4) in the spin Hamiltonian leads, from (4.6), to a violation of the equidistance of the quadrupole triplet components in the ENDOR spectrum. With $n = \pm 1/\sqrt{3}$, without accounting for perturbation $\mathcal{H}_1$, the quadrupole triplet is degenerate in a single line. Due to interaction $\mathcal{H}_1$, this line is split into two components, the distance between which is equal to

$$|\Delta\nu| = \frac{1}{12h}\left\{45\left[(a_1-a_2)(l^4+m^4)-2a_3l^2m^2\right]+2\left[5(a_2+a_3)-11a_1\right]\right\}.$$

Perturbation $\mathcal{H}_1$, as we can see from (4.4) and (4.5), also leads to forbidden transitions in the ENDOR spectrum with selection rules $\Delta(M_{I_k}+M_{I_{k'}})=\pm 2$.

The effects of high order hyperfine interaction must be considered when they are displayed experimentally, as occurs in a number of cases for the invariants of the third and fifth power on the spin operators for the invariants of octupole interaction and others [3, 5, 403].

Let us consider further the spin–phonon invariants in the presence of hyperfine interaction. These invariants describe the processes of nuclear spin–lattice relaxation and the effects of acoustic perturbation of the spin system. A specific interest in the last case is that of transitions between hyperfine levels under the action of acoustic waves under conditions of double resonance. Since the transitions between electron spin levels are caused by the magnetic component of an electromagnetic wave, and those between hyperfine levels are caused by acoustic waves, we shall name this type of resonance electron–nuclear double magnetoacoustic resonance (acoustical ENDOR) [59, 170].

In first-order perturbation theory, isotropic dynamic hyperfine interaction does not induce spin–phonon transitions without change of the electron spin projection. Therefore, as a perturbation we must use the anisotropic part of the dipole–dipole and quadrupole interactions of the dynamic spin Hamiltonian. Although the constant of the anisotropic spin–phonon interaction is less than the constant of isotropic interaction of the spin with phonons, the probability of transition between hyperfine levels is sufficiently greater for registration on acoustical ENDOR spectra [59].

The acoustical ENDOR method is more sensitive than the method of direct registration of the acoustic action on the spin system, for reasons analogical to those by which the ENDOR method is more sensitive than EPR.

As we build the acoustical ENDOR theory, we shall proceed from the Pikus–Luttinger method of invariants; based on this method, it is simpler to understand the peculiarities of this resonance, influenced by the symmetry properties. We shall consider the dynamic hyperfine interaction between a local electron center ($S = 1/2$) or an impurity center with spin S and nuclei with spin I in an octahedral environment. As the basic functions of the vibrational subsystem we shall select the coordinates of symmetry Q_j. In the case of local electron centers Q_j ($j = 1, \ldots, 18$) are transformed according to representation $\mathcal{E}_{18}$, which is expanded into the irreducible representations of the symmetry group O_h:

$$\mathcal{E}_{18} = \Gamma_1^+ \oplus \Gamma_3^+ \oplus \Gamma_4^+ \oplus \Gamma_5^+ \oplus 2\Gamma_4^- \oplus \Gamma_5^-.$$

The spin–phonon invariants are built by combining coordinates of symmetry $Q_1, Q_2, \ldots, Q_{18}$ with the basic function operators of the irreducible representations contained in $D_{1/2} \otimes D_{1/2}^*$, $D_I \otimes D_I^*$ and Π_6, and they are influenced by two types of "coupling":

$$[\mathcal{E}_{18}] \leftrightarrow D_{1/2} \otimes D_{1/2}^* \otimes D_I \otimes D_I^* \otimes [\Pi_6],$$

$$\{\mathcal{E}_{18}\} \leftrightarrow D_{1/2} \otimes D_{1/2}^* \otimes D_I \otimes D_I^* \otimes \{\Pi_6\},$$

where

$$[\mathcal{E}_{18}] = \Gamma_1^+ \oplus \Gamma_3^+ \oplus \Gamma_4^+, \qquad \{\mathcal{E}_{18}\} = 2\Gamma_4^- \oplus \Gamma_5^-,$$

$$[\Pi_6] = \Gamma_1^+ \oplus \Gamma_3^+, \qquad \{\Pi_6\} = \Gamma_4^-.$$

The first type of coupling applies to even representations, and the second to odd representations of the symmetry group O_h characterizing the spin–vibrational system.

Because of the evenness of the spin projection operators relative to the spatial inversion, one usually considers that the spin–phonon interaction operator cannot linearly contain the coordinates of symmetry $Q_j^{(-)}$, which are transformed on the odd representations. However, these coordinates of symmetry can enter the spin Hamiltonian if its nonphonon part is transformed according to the odd representation. It is sufficient if the representation Π contains an odd representation Γ_δ^- such that

$$\Gamma_\delta^- \otimes \Gamma_\gamma^- \in D_S \otimes D_S^* \otimes D_I \otimes D_I^*,$$

where Γ_γ^- is the irreducible representation according to which $Q_j^{(-)}$ is transformed.

With $S = 1/2$, $I = 3/2$ the unit representation is contained 144 times in the direct product of the representations $\mathcal{E}_{18} \otimes D_{1/2} \otimes D_{1/2}^* \otimes D_{3/2} \otimes D_{3/2}^* \otimes \Pi_6$; therefore, there is also the same number of spin–phonon invariants. Many of these invariants are forbidden by time-reversal symmetry, but after rejecting those that are not invariant relative to the time reversal, 71 independent invariants remain. With a precision up to the square terms on the nuclear spin projection operators, the Hamiltonian of spin–phonon interaction has the form

$$\mathcal{H}_{s-ph} = \mathcal{H}_+ + \mathcal{H}_-, \tag{4.7}$$

where the indices $+$ and $-$ correspond to the consideration of the symmetry coordinates that are transformed on the basis of the even and odd representations. The operator $\mathcal{H}_+$ has the invariants

$$
\begin{aligned}
\mathcal{H}_+ = {} & a_{1,\alpha_1} \sum_k \mathbf{I}_k \mathbf{S} Q + a_{2,\alpha_2} \sum_k \big[(3 I_{kz} S_z - \mathbf{I}_k \mathbf{S}) Q_2 \\
& + \sqrt{3}(I_{kx} S_x - I_{ky} S_y) Q_3 \big] + a_{3,\alpha_3} E \sum_k \big\{ 3 \big[I_{kz}^2 - 2I(I+1) \big] Q_2 \\
& + \sqrt{3}\big(I_{kx}^2 - I_{ky}^2 \big) Q_3 \big\} + a_{4,\alpha_4} \sum_k \big[(I_{ky} S_z + I_{kz} S_y) Q_{5x} + \Delta_4 \big] \\
& + a_{5,\alpha_5} E \sum_k \mathbf{P}_k Q_5 + 3 a_{6,\alpha_6} E Q_1 \big[I_{1x}^2 + I_{4x}^2 - 2I(I+1) + \Delta_6 \big] \\
& + a_{7,\alpha_7} E \big\{ 3 \big[I_{3z}^2 + I_{6z}^2 - 2I(I+1) \big] Q_2 + \sqrt{3}\big(I_{1x}^2 + I_{4x}^2 \\
& - I_{3x}^2 - I_{6x}^2 \big) Q_3 + \Delta_7 \big\} + a_{8,\alpha_8} Q_1 \Big[3 I_x^{14} S_x + \Delta_8 - \sum_k \mathbf{I}_k \mathbf{S} \Big] \\
& + a_{9,\alpha_9} \Big[(Q_2 - \sqrt{3} Q_3) \Big(\sum_k I_{kx} - 3 I_x^{14} \Big) S_x + (Q_2 + \sqrt{3} Q_3) \\
& \times \Big(\sum_k I_{ky} - 3 I_y^{25} \Big) S_y + 2 Q_3 \Big(3 I_z^{36} - \sum_k I_{kz} \Big) S_z \Big] \\
& + a_{10,\alpha_9} \big[(\sqrt{3} Q_2 + Q_3) I_{\bar{x}36}^{x25} S_x + (\sqrt{3} Q_2 - Q_3) I_{\bar{y}36}^{y14} S_y \\
& + 2 Q_3 I_{\bar{z}14}^{z25} S_z \big] + a_{11,\alpha_{10}} \Big\{ \Big[\Big(3 I_z^{36} - \sum_k I_{kz} \Big) S_y \\
& + \Big(3 I_y^{25} - \sum_k I_{ky} \Big) S_z \Big] Q_{5x} + \Delta_{11} \Big\} \\
& + a_{12,\alpha_{10}} \big[(I_{\bar{z}25}^{z14} S_y + I_{\bar{y}36}^{y14} S_z) Q_{5x} + \Delta_{12} \big] \\
& + a_{13,\alpha_{11}} E \Big[\Big(3 P_x^{14} - \sum_k P_{kx} \Big) Q_{5x} + \Delta_{13} \Big] \\
& + a_{14,\alpha_{12}} \sum_k \big[(I_{kz} S_y - I_{ky} S_z) Q_{4x} + \Delta_{14} \big]
\end{aligned}
$$

$$+ a_{15,\alpha_{13}} E\left(P^{x25}_{\overline{x}36} Q_{4x} + \Delta_{15}\right)$$

$$+ a_{16,\alpha_{14}} \left\{\left[3\left(I^{36}_z S_y - I^{25}_y S_z\right) + \sum_k (I_{ky} S_z - I_{kz} S_y)\right] Q_{4x} + \Delta_{16}\right\}$$

$$+ a_{17,\alpha_{14}} \left[\left(I^{z14}_{\overline{z}25} S_y + I^{y36}_{\overline{y}14} S_z\right) Q_{4x} + \Delta_{17}\right], \tag{4.8}$$

where the summation is carried out on the nuclei of the first coordination sphere. In formula (4.8) and henceforth[1] the following designations are used:

$$I^{ll'}_\xi = I_{l\xi} \pm I_{l'\xi}, \qquad I^{\xi ll'}_{\eta m m'} = I_{l\xi} + I_{m\eta} \pm (I_{l'\xi} + I_{m'\eta}),$$

$$I^{\xi ll'}_{\overline{\eta} m m'} = I_{l\xi} - I_{m\eta} \pm (I_{l'\xi} - I_{m'\eta}), \qquad P_\xi = \frac{1}{2}(I_\eta I_\zeta + I_\zeta I_\eta),$$

$$P^{ll'}_\xi = P_{l\xi} \pm P_{l'\xi}, \qquad P^{\xi ll'}_{\eta m m'} = P_{l\xi} + P_{m\eta} \pm (P_{l'\xi} + P_{m'\eta}),$$

$$P^{\xi ll'}_{\overline{\eta} m m'} = P_{l\xi} - P_{m\eta} \pm (P_{l'\xi} - P_{m'\eta}).$$

(4.9)

Here ξ is one of the axes x, y, z; η and ζ are two others; E is the unit 2×2 matrix; Δ_i is the expression obtained by cyclic permutation in one direction of indices x, y, z and 14, 25, 36 (nuclei 1, 2, and 3 are situated on the axes x, y, z and are connected by the operation of inversion with nuclei 4, 5, and 6, respectively) in the operator part of the ith Hamiltonian term.[2] The magnitudes $a_{1,\alpha_1}, \ldots, a_{17,\alpha_{14}}$ are constants of the spin–phonon interaction. Through $\alpha_2, \ldots, \alpha_{14}$ we have designated direct products of irreducible representations of the group O_h, generating unit representations Γ_{1g}, according to which the operating part that corresponds to the given coupling constant is transformed. Every direct product contains four representations:

$$\Gamma^+_1 \otimes \Gamma^+_4 \otimes \Gamma^+_4 \otimes \Gamma^+_1, \qquad \Gamma^+_1 \otimes \Gamma^+_4 \otimes \Gamma^+_4 \otimes \Gamma^+_3,$$

$$\Gamma^+_1 \otimes \Gamma^+_3 \otimes \Gamma^+_1 \otimes \Gamma^+_3, \qquad \Gamma^+_1 \otimes \Gamma^+_4 \otimes \Gamma^+_4 \otimes \Gamma^+_5,$$

$$\Gamma^+_1 \otimes \Gamma^+_5 \otimes \Gamma^+_1 \otimes \Gamma^+_5, \qquad \Gamma^+_3 \otimes \Gamma^+_3 \otimes \Gamma^+_1 \otimes \Gamma^+_1,$$

$$\Gamma^+_3 \otimes \Gamma^+_3 \otimes \Gamma^+_1 \otimes \Gamma^+_3, \qquad \Gamma^+_3 \otimes \Gamma^+_4 \otimes \Gamma^+_4 \otimes \Gamma^+_1,$$

$$\Gamma^+_3 \otimes \Gamma^+_4 \otimes \Gamma^+_4 \otimes \Gamma^+_3, \qquad \Gamma^+_3 \otimes \Gamma^+_4 \otimes \Gamma^+_4 \otimes \Gamma^+_5,$$

$$\Gamma^+_3 \otimes \Gamma^+_5 \otimes \Gamma^+_1 \otimes \Gamma^+_5, \qquad \Gamma^+_1 \otimes \Gamma^+_4 \otimes \Gamma^+_4 \otimes \Gamma^+_4,$$

$$\Gamma^+_3 \otimes \Gamma^+_5 \otimes \Gamma^+_1 \otimes \Gamma^+_4, \qquad \Gamma^+_3 \otimes \Gamma^+_4 \otimes \Gamma^+_4 \otimes \Gamma^+_4.$$

---

[1] The subscript in (4.9) refers to $\mathcal{H}_-$ from (4.10).

[2] Δ_7 is obtained during cyclic permutation of indices x, y, z and 14, 25, 36 in the opposite directions.

On the first and following places in these direct products of representations the irreducible representations contained in $[\Pi]_6$, $D_{3/2} \otimes D_{3/2}^*$, $D_{1/2} \otimes D_{1/2}^*$, and $[\varXi]_{18}$ are indicated, respectively. Bethe designations are used for the irreducible representations. The coordinates of symmetry entering the Hamiltonian (4.8), Q_2, Q_3, Q_{4x}, Q_{4y}, Q_{4z}, Q_{5x}, Q_{5y}, and Q_{5z}, correspond to the symmetry coordinates Q_3, $-Q_2$, Q_{21}, Q_{19}, Q_{18}, Q_6, Q_5, and Q_4 in the designations of van Vleck [3].

Operator $\mathcal{H}_-$ has the form

$$
\begin{aligned}
\mathcal{H}_- = &\, b_{1,\beta_1} E\big\{\big[3\big(I_{1x}^2 - I_{4x}^2\big) - 2I(I+1)\big]Q_{4x}' + \Delta_1'\big\} \\
&+ b_{2,\beta_2} E\big\{\big[\big(I_{1y}^2 - I_{4y}^2\big) - \big(I_{1z}^2 - I_{4z}^2\big)\big]Q_{5x}' + \Delta_2'\big\} \\
&+ b_{3,\beta_3}\mathbf{SQ}_4'\big(I_x^{14} + \Delta_3'\big) + b_{4,\beta_3}\big[I_x^{14}\big(3S_x Q_{4x}' - \mathbf{SQ}_4'\big) + \Delta_4'\big] \\
&+ b_{5,\beta_4}\big[I_{z36}^{y25} S_x Q_{5x}' + \Delta_5'\big] \\
&+ b_{6,\beta_3}\big[\big(I_{\overline{y}14}^{x25} S_y + I_{z14}^{x36} S_z\big)Q_{4x}' + \Delta_6'\big] \\
&+ b_{7,\beta_4}\big[\big(I_{\overline{y}14}^{x25} S_y + I_{\overline{x}36}^{z14} S_z\big)Q_{5x}' + \Delta_7'\big] \\
&+ b_{8,\beta_3}\big[\big(I_{y14}^{x25} S_y + I_{z14}^{x36} S_z\big)Q_{4x}' + \Delta_8'\big] \\
&+ b_{9,\beta_4}\big[\big(I_{y14}^{x25} S_y - I_{z14}^{x36} S_z\big)Q_{5x}' + \Delta_9'\big] \\
&+ b_{10,\beta_5} E\big(P_{y36}^{z25} Q_{5x}' + \Delta_{10}'\big) \\
&+ b_{11,\beta_6} E\big(P_{\overline{y}36}^{z25} Q_{5x}' + \Delta_{11}'\big), \tag{4.10}
\end{aligned}
$$

where $\beta_1, \beta_2, \ldots, \beta_6$ designate the products of the representations

$$
\begin{aligned}
&\Gamma_4^- \otimes \Gamma_3^+ \otimes \Gamma_1^+ \otimes \Gamma_4^-, &&\Gamma_4^- \otimes \Gamma_3^+ \otimes \Gamma_1^+ \otimes \Gamma_5^-, \\
&\Gamma_4^- \otimes \Gamma_4^+ \otimes \Gamma_4^+ \otimes \Gamma_4^-, &&\Gamma_4^- \otimes \Gamma_4^+ \otimes \Gamma_4^+ \otimes \Gamma_5^-, \\
&\Gamma_4^- \otimes \Gamma_5^+ \otimes \Gamma_1^+ \otimes \Gamma_4^-, &&\Gamma_4^- \otimes \Gamma_5^+ \otimes \Gamma_1^+ \otimes \Gamma_5^-.
\end{aligned}
$$

On the first and following places in these direct products the irreducible representations contained in $[\Pi]_6$, $D_{3/2} \otimes D_{3/2}^*$, $D_{1/2} \otimes D_{1/2}^*$, and $[\varXi]_{18}$ are indicated, respectively. The expression for Δ_i' is obtained in the same way as for Δ_i. Symmetry coordinates $Q_{4\xi}'$ and $Q_{5\xi}'$ ($\xi = x, y, z$) are transformed on the representations $^1\Gamma_4^-$ and Γ_5^-. In (4.10) it is necessary to introduce six supplementary terms (with the corresponding coupling constants b_{12,β_1}, b_{13,β_3}, b_{14,β_3}, b_{15,β_3}, b_{16,β_3}, and b_{17,β_3}), due to the representation $^2\Gamma_4^-$: $\{Q_{4x}'', Q_{4y}'', Q_{4z}''\}$.

If instead of an electron localized on the defect there is an impurity atom, then an additional $^3\Gamma_4^-$ representation is added to the representation $\varXi_{18}$. In this case (4.10) will have additional terms due to this representation.

Later, on all the constants of spin–phonon interaction from (4.8) and (4.10) we shall drop the indices showing the direct product of representation that determine the given term of the spin Hamiltonian. After calculating the matrix elements of the operator (4.7) using oscillator and spin wave functions, we then obtain the following expression for the coefficient of hypersound absorption at transitions between the states $|M_S M_{I_K}\rangle$ and $|M_S M'_{I_K}\rangle$ with spin reorientation of the nuclei in the equivalent nodes of the kth type:

$$\Gamma^{(k)}_{M_{I_k} M'_{I_k}} = \mathscr{P}_k C_k \left| \frac{\omega R}{v_s} F_{k+} + F_{k-} \right|^2, \qquad (4.11)$$

where

$$\mathscr{P}_k = \frac{2\pi n_0}{v_s k_0 T} g_k(\omega).$$

Here C_k is the spin multiplier, $g_k(\omega)$ is the function of the shape of the acoustical ENDOR line, n_0 is the number of orientationally equivalent magnetic nuclei in the unit of volume, ω and v_s are the frequency and speed of the hypersound, R is the radius of the first coordination sphere, and F_{k+} and F_{k-} are functions containing linear constants of spin–phonon coupling. The form of the functions F_{k+} and F_{k-} depends on the orientation of the magnetic field $\mathbf{H}$ relative to the crystallographic axes, the direction of propagation of the phonons, and their polarization. Thus, F_{k+} contains only constants of spin–phonon coupling that enter into $\mathscr{H}_+$, and F_{k-} only constants of coupling from $\mathscr{H}_-$.

Let us consider spin–phonon transitions with the selection rules $\Delta M_S = 0$, $\Delta M_{I_K} = \pm 1$. In this case,

$$4C_k = \begin{cases} (I + M_{I_k} + 1)(I - M_{I_k}), & \Delta M_{I_k} = +1, \\ (I - M_{I_k} + 1)(I + M_{I_k}), & \Delta M_{I_k} = -1. \end{cases}$$

Marking the multiplier at $\mathscr{P}_k$ from (4.11) through $\mathscr{F}_k$, we obtain the following expressions in some particular cases. In these expressions when we encounter the $\pm$ sign, the upper $(+)$ sign corresponds to the value of C_k at $\Delta M_{I_k} = +1$, and the lower $(-)$ sign to the value of C_k at $\Delta M_{I_k} = -1$.

A. Longitudinally polarized phonons ($\mathbf{e}_{\varkappa_l} \parallel \varkappa_l$)

(1) $\mathbf{H} \parallel \mathbf{n}_l \parallel [001]$, $\mathbf{n}_l$ is the unit vector indicating the direction of propagation of the phonons; $\varkappa_l$ and $\mathbf{e}_{\varkappa_l}$ are the phonon wave vector and the unit vector of polarization of longitudinal phonons.

$$\mathscr{F}_k = \left[M_s (b'_3 - b'_4) + \left(M_{I_k} \pm \frac{1}{2} \right) b'_{10} \right]^2, \quad k = 1, 2, 4, 5,$$

$$\mathscr{F}_3 = \mathscr{F}_6 = 0.$$

Here

$$b'_3 = b_3 + \frac{2}{\sqrt{3}} b_{13}, \qquad b'_4 = b_4 + \frac{2}{\sqrt{3}} b_{14}, \qquad b'_{10} = b_{10} + \frac{2}{\sqrt{3}} b_{17}.$$

(2) **H** $\parallel$ [001], $\mathbf{n}_l \perp$ [001]

$$\mathcal{F}_k = \left[M_s \left(b_6' + b_8' \right) + \left(M_{I_k} \pm \frac{1}{2} \right) b_{10}' \right]^2, \quad k = 3, 6,$$

$$\mathcal{F}_1 = \mathcal{F}_2 = \mathcal{F}_4 = \mathcal{F}_5 = 0,$$

where

$$b_6' = b_6 + \frac{2}{\sqrt{3}} b_{15}, \qquad b_8' = b_8 + \frac{2}{\sqrt{3}} b_{16}.$$

(3) **H** $\perp$ [001], $\mathbf{n}_l \parallel$ [001]

$$\mathcal{F}_k = (\varkappa_l R A_k \sin 2\varphi)^2 + B_k^2 \cos^2 \varphi, \quad k = 1, 4,$$

$$\mathcal{F}_p = (\varkappa_l R A_p \sin 2\varphi)^2 + B_p^2 \sin^2 \varphi, \quad p = 2, 5,$$

$$\mathcal{F}_3 = \mathcal{F}_6 = 0.$$

Here

$$A_k = \frac{1}{2} \left[M_s (\sqrt{3} a_8 - \sqrt{6} a_9 - \sqrt{2} a_{10}) + (2 M_{I_k} \pm 1) \right.$$

$$\left. \times \left(\frac{1}{\sqrt{3}} a_6 - \sqrt{6} a_7 \right) \right], \qquad A_p = -A_k \quad (k \to p),$$

$$B_1 = M_s \left(b_6' - b_8' \right) + \left(M_{I_2} \pm \frac{1}{2} \right) b_{10}', \qquad B_2 = B_1 (1 \to 2),$$

$$B_3 = B_1 (1 \to 3), \qquad B_4 = -B_1 (1 \to 4),$$

$$B_5 = -B_2 (2 \to 5), \qquad B_6 = -B_3 (3 \to 6).$$

(φ is the angle between the direction of the magnetic field and the crystallographic axis [100].)

(4) **H** $\perp$ [001], $\mathbf{n}_l \perp$ [001]

$$\mathcal{F}_k = (\varkappa_l R)^2 (A_{1k} \cos 2\phi_l \sin 2\varphi - A_{2k} \sin 2\phi_l \cos 2\varphi)^2$$

$$+ B_k^2 \cos^2 (\phi_l - \varphi), \quad k = 3, 6,$$

where

$$A_{1k} = \frac{1}{2} \sqrt{\frac{3}{2}} \left[2 M_s \left(a_2 - a_9 - \frac{1}{\sqrt{3}} a_{10} \right) + (2 M_{I_k} \pm 1)(2 a_3 - a_7) \right],$$

$$A_{2k} = \frac{1}{\sqrt{2}} \left[M_s (a_4 + a_{12} - a_{11}) + \left(M_{I_k} \pm \frac{1}{2} \right)(a_5 + a_{13}) \right],$$

$$B_3 = M_s \left(b_3' - b_4' \right) + \left(M_{I_3} \pm \frac{1}{2} \right) b_{10}', \qquad B_6 = -B_3 (3 \to 6),$$

φ_l is the angle between $\mathbf{n}_l$ and [100].

The expressions $\mathcal{F}_k$ for the rest of the nuclei are bulky in this particular case and are not given here.

B. Transverse-polarized phonons ($e_{\varkappa_t} \perp \varkappa_t$)

    (5)   **H** $\parallel [001]$, $\mathbf{n}_t \perp [001]$

$$\mathcal{F}_k = (\varkappa_t RA_{1k} \sin \phi_t)^2 + (\varkappa_t RA_{2k} \cos \phi_t + B_k)^2,$$
$$\mathcal{F}_p = (\varkappa_t RA_{2p} \cos \phi_t)^2 + (\varkappa_t RA_{1p} \sin \phi_t + B_p)^2,$$
$$\mathcal{F}_q = (\varkappa_t RA_q)^2, \quad k = 1, 4; \ p = 2, 5; \ q = 3, 6.$$

(4.12)

In expressions (4.12) the following designations are used:

$$A_{1k} = \frac{1}{\sqrt{2}}\left[M_s(a_4 - a_{11} + a_{12} + a_{14} - a_{16} + a_{17}) \right.$$
$$\left. + (2M_{I_k} \pm 1)\left(\frac{1}{2}a_5 + a_{13}\right) \right],$$

$$A_{2k} = \frac{1}{\sqrt{2}}\left[M_s(a_4 + 2a_{11} + a_{14} + 2a_{16}) \right.$$
$$\left. + \left(M_{I_k} \pm \frac{1}{2}\right)(a_5 - a_{13} - a_{15}) \right],$$

$$A_q = \frac{1}{\sqrt{2}}\left[M_s(a_4 - 2a_{11} - a_{12} + a_{14} - a_{16} - a_{17}) \right.$$
$$\left. + \left(M_{I_q} \pm \frac{1}{2}\right)(a_5 - a_{13} + a_{15}) \right],$$

$$A_{1p} = A_{2k}(k \to p), \qquad A_{2p} = A_{1k}(k \to p),$$

$$B_1 = M_s(b_3' - b_4') + \left(M_{I_1} \pm \frac{1}{2}\right)b_{10}', \qquad B_2 = B_1(1 \to 2),$$

$$B_3 = B_1(1 \to 3), \qquad B_4 = -B_1(1 \to 4),$$
$$B_5 = -B_2(2 \to 5), \qquad B_6 = -B_3(3 \to 6).$$

(4.13)

    (6)   **H** $\perp [001]$, $\mathbf{n}_t \perp [001]$

$$\mathcal{F}_k = \left[\varkappa_t R(A_{1k} \sin \phi_t \sin \varphi + A_{2k} \cos \phi_t \cos \varphi) + B_k \cos \varphi \right]^2,$$
$$\mathcal{F}_p = \left[\varkappa_t R(A_{1p} \sin \phi_t \sin \varphi + A_{2p} \cos \phi_t \cos \varphi) + B_p \sin \varphi \right]^2,$$
$$\mathcal{F}_q = \left[\varkappa_t RA_q \cos(\phi_t - \varphi) \right]^2, \quad k = 1, 4; \ p = 2, 5; \ q = 3, 6,$$

where

$$A_{1k} = \frac{1}{\sqrt{2}}\bigg[M_s(a_4 - a_{11} + a_{12} - a_{14} + a_{16} - a_{17})$$

$$+ \bigg(M_{I_k} \pm \frac{1}{2} \bigg)(a_5 - a_{13} - a_{15}) \bigg],$$

$$A_{2k} = \frac{1}{\sqrt{2}}\bigg[M_s(a_4 - a_{11} - a_{12} - a_{14} + a_{16} + a_{17})$$

$$+ \bigg(M_{I_k} \pm \frac{1}{2} \bigg)(a_5 - a_{13} - a_{15}) \bigg],$$

$$A_q = \frac{1}{\sqrt{2}}\bigg[M_s(a_4 + 2a_{11} - a_{14} - 2a_{16})$$

$$+ \bigg(M_{I_q} \pm \frac{1}{2} \bigg)(a_5 - a_{13} + a_{15}) \bigg],$$

$$A_{1p} = A_{2k}(k \to q), \qquad A_{2p} = A_{1k}(k \to p), \quad k = 1, 4; \; p = 2, 5; \; q = 3, 6.$$

B_k and B_p are the same as in the case of $A3$; Φ_t is the angle between $\mathbf{n}_t$ and [100].

The numeric estimations given in [59] show that we can experimentally observe acoustical ENDOR on the ^{39}K nuclei in the crystal KCl (with the selection rules $\Delta M_S = 0$, $\Delta M_{I_k} = \pm 1$) even at room temperature. A comparatively low acoustic power ($\sim$2 W/cm^2) is needed for the excitation of the nuclear spin transitions. Besides, from operators $\mathcal{H}_+$ and $\mathcal{H}_-$, in the acoustical ENDOR spectrum there are possible transitions with the selection rules $\Delta M_S = 0$, $\Delta M_{I_k} = \pm 2$ due to quadratic terms on the nuclear spin operators in the Hamiltonian of spin–phonon interaction.

Acoustical ENDOR, which was forecast theoretically in [59] for the first time, was observed experimentally by Golenischev-Kutuzov, Kopvillem, and Shamukov [404]. In this experiment simultaneous excitation of nuclear and electron spins was performed, respectively, by acoustic and electromagnetic vibrations in the single crystal Al$_2$O$_3$, containing 0.05 at % of Cr$_3^+$ ions. The electron spin system Cr^{3+} was used to detect the NMR of ^{27}Al nuclei. A change in the intensity of the EPR signal while presenting the acoustic field on frequencies corresponding to the transitions between the nuclear spin levels with the selection rules $\Delta M_I = \pm 1$ and $\Delta M_I = \pm 2$ was discovered. The relative change in intensity of the EPR signal in the centers of the lines of double resonance was $\sim$(10–15) % at relative deformation $\sim$3 · 10^{-7}. The shape of the lines of such resonances is of Gauss type with width $\sim$2 · 10^4 Hz [38, 404].

The availability of excitons in the crystal leads to a shift of nuclear spin levels (exciton Knight shift) and to a change in the nuclear spin–lattice relaxation rate caused by contact hyperfine interaction of the nuclei with the charge carriers forming excitons. The exchange interaction of these carriers with paramagnetic centers causes shifts in the electron spin levels and a change in the electron spin–lattice relaxation rate. These effects will be considered in Chaps. 5 and 6. Here we note only that, by these effects, the excitons influence NMR, EPR, ENDOR, acoustical EN-DOR, and all other resonances in which the spin states of the nuclei and localized electrons participate.

4.2 One-Phonon Spin–Lattice Relaxation and Acoustically Nonequivalent Nuclei

The coefficient of absorption of hypersound $\Gamma^{(k)}_{M_{I_k} M'_{I_k}}$ during acoustical ENDOR, as seen from formula (4.11), along with the usual acoustic resonance term $\sim v_s^{-3}$ contains the items $\sim v_s^{-2}$ and $\sim v_s^{-1}$, the presence of which is due to the odd representations during the consideration of symmetry properties of the spin–phonon system. At some orientations of the magnetic field and the fascicle of polarized phonons relative to the crystallographic axes, we can separate the contribution in the expression for $\Gamma^{(k)}_{M_{I_k} M'_{I_k}}$ from the symmetric coordinates Q_j, constructed from displacements of the nuclei of the nearest environment of the local center, transformed by the even and odd representations. Thus, in cases $A1$ and $A2$ the absorption coefficient of hypersound is due exceptionally to the vibrations that are described by the symmetry coordinates being transformed on the basis of odd representations. The absorption of hypersound in cases $B5$ and $B6$ for the nuclei situated on the axis [001] is affected only by displacements of the nuclei from which we can construct symmetric coordinates which are transformed on even representations. This allows us to use the acoustical ENDOR spectra to determine the character of vibrational displacements of the nuclei surrounding the local center.

We shall estimate the relative contribution to $\Gamma^{(k)}_{M_{I_k} M'_{I_k}}$ of vibrations of the nuclei described by symmetry coordinates that are transformed on the even and odd representations. For $A3$ and assuming $\phi = \pi/4$, we obtain that both types of symmetry coordinates give a similar contribution to the magnitude of the coefficient of hypersound absorption if the following correlation is applied:

$$B_k = \sqrt{2} \frac{\omega R}{v_s} A_k. \tag{4.14}$$

For F centers in the crystal KCl, substituting $R = 3.14$ Å, $v_s = 2.4 \cdot 10^5$ cm/s, and $\omega/2\pi = 10.8$ MHz, we obtain $B_k = 1.3 \cdot 10^{-5} A_k$. It follows from here that in formula (4.7) it is necessary to consider the operator $\mathcal{H}_-$, even if the constants of the spin–phonon coupling from (4.10) are some orders less than the constants of the spin–phonon interaction from (4.8). The role of the terms from (4.7) due to the odd representations grows, as is seen from (4.14), with decreasing distance between hyperfine levels.

Using the symmetry coordinates transformed on the odd representations in the spin–phonon Hamiltonian allows us to also predict the new behavior of the one-phonon spin–lattice relaxation time depending on the magnetic field and hypersound speed. We consider the relaxational transitions using only longitudinally polarized phonons. The probability of the transition conditioning the cross relaxation time $\tau_x(\Delta M_S = -1, \Delta M_{I_k} = +1)$ at $\hbar\omega < kT$ is determined by the expression

$$W^{(k)}_{1/2,M_{I_k};-1/2,M_{I_k}+1} = \frac{k_0 T}{4\pi^2\hbar^2 v_1^3 d}(I + M_{I_k} + 1)(I - M_{I_k})$$

$$\times \left[(\alpha\varepsilon_k)^2\Lambda_{k+} + \Lambda_{k-}\right],$$

where

$$\varepsilon_k = \begin{cases} g\beta H \pm a \pm b\gamma_k, & M_{I_k} = \frac{1}{2} \text{ (upper sign)}, M_{I_k} = -\frac{3}{2} \text{ (lower sign)}, \\ g\beta H, & M_{I_k} = -\frac{1}{2}, \end{cases}$$

$$\alpha = \frac{R}{\hbar v_1}, \qquad \gamma_1 = 3\sin^2\theta\cos^2\varphi - 1, \qquad \gamma_2 = 3\sin^2\theta\sin^2\varphi - 1,$$

$$\gamma_3 = 3\cos^2\theta - 1, \qquad \gamma_1 = \gamma_4, \qquad \gamma_2 = \gamma_5, \qquad \gamma_3 = \gamma_6,$$

$$\Lambda_{k+} = f_{k0}^2 + \frac{1}{5}\left\{ f_{k1}^2 + f_{k6}^2 + f_{k7}^2 + \frac{2}{3}\sum_{j=1}^{5}|f_{kj}|^2 \right.$$

$$+ 5f_{k0}(f_{k1} + f_{k6} + f_{k7}) + f_{k1}(f_{k6} + f_{k7})$$

$$\left. + 2f_{k2}(f_{k7} - f_{k6}) + f_{k6}f_{k7} \right\}, \qquad \Lambda_{k-} = \frac{1}{3}\sum_{j=8}^{10}|f_{kj}|^2.$$

Here a and b are the constants of the contact and dipole–dipole hyperfine interaction, v_1 is the speed of the propagation of the longitudinal hypersonic waves, d is the density of the crystal, and θ and φ are the polar and azimuth angles determining the direction of the vector $\mathbf{H}$. The functions f_{kj} are given in Appendix E. The index k marks the type of orientationally equivalent nuclei for which the probability of the transition between the spin levels is considered. Thus, independent of the type of relaxational transition, $W^{(1)} = W^{(4)}$, $W^{(2)} = W^{(5)}$, and $W^{(3)} = W^{(6)}$ due to the fact that the nuclei turning into one another in the process of inversion are equivalent in the magnetic field.

On approaching the high spin temperature, because the nuclear Zeeman and quadrupole energy is much less compared with the electron Zeeman energy and energies of the contact and magnetic dipole–dipole interactions, we obtain the following formula for the cross-relaxation rate [389]:

$$\frac{1}{T_x} = 2\left(\frac{1}{T_x^{(1)}} + \frac{1}{T_x^{(2)}} + \frac{1}{T_x^{(3)}}\right),$$

where

$$\frac{1}{T_x^{(k)}} = \frac{k_0 T}{2\pi^2 \hbar^2 v_l^3 d} \{\alpha^2 H^2 [5(g\beta H)^2 + 18(a + b\gamma_k)^2]\Lambda_{k+}$$

$$+ [5H^2 + 3(H_a + H_b\gamma_k)^2]\Lambda_{k-}\}\left[2H^2 + 15\left(\frac{1}{2}H_a^2 + H_b^2\right)\right]^{-1}. \quad (4.15)$$

Here $H_a = a(g\beta)^{-1}$, $H_b = b(g\beta)^{-1}$. With $a, b \ll g\beta H$ the dependence $T_x^{(k)}$ versus H is determined only by the items from (4.15) containing Λ_{k+}:

$$\frac{1}{T_x^{(k)}} = \frac{5k_0 T}{2\pi^2 \hbar^2 v_l^3 d} [(\alpha g\beta H)^2 \Lambda_{k+} + \Lambda_{k-}].$$

Thus, for the spin–phonon interaction operator, accounting for the symmetry coordinates being transformed on the odd representations leads to additional items in the expression for the one-phonon spin–lattice relaxation rate. These additional items do not change the temperature behavior of the spin–lattice relaxation time. As seen from formula (4.15), they cause a new dependence of the relaxation time on the hypersound speed ($T_x^{(k)} \sim v_l^3$).

With $a, b \ll g\beta H$ the spin–lattice relaxation described by the additional items does not depend on the magnetic field. From

$$\frac{\Lambda_{k-}}{\Lambda_{k+}} = (\alpha g\beta H)^2 \quad (4.16)$$

the additional terms give the same contribution in the expression for the spin–lattice relaxation time as the items that depend squarely on the magnetic field. The numeric estimation in formula (4.16) for the crystal KCl for $g\beta H = 6.17 \cdot 10^{-17}$ erg gives $\Lambda_{k-} = 1.5 \cdot 10^{-5}\Lambda_{k+}$. Therefore, the items that depend and do not depend on the magnetic field give comparable contributions to the expression for one-phonon spin–lattice relaxation time, even if the constants of the spin–phonon coupling determining Λ_{k-} are two orders less than the spin–phonon coupling constants determining Λ_{k+}. As in the case of hypersound absorption with acoustical ENDOR, the contribution in the expression for the spin–lattice relaxation time conditioned by odd representations increases with decreasing distance between the energy levels of the spin system.

In the absence of perturbation transformed similarly to the polar vector, the nuclei that transform into one another under the inversion operation are physically equivalent. For example, under conditions of ENDOR, these nuclei are equivalent at arbitrary orientations of the external magnetic field relative to the crystallographic axes [170, 381], because **H**, along with **S** and **I**, is a pseudovector.

Since the symmetry coordinates entering in $\mathcal{H}_-$ from (4.10) are transformed similarly to the polar vector, we expect that for acoustical ENDOR the equivalence of nuclei k and k' being transformed into one another during inversion may be violated. This "acoustical" nonequivalence of the nuclei connected by inversion actually takes place if we use the polarized phonons to excite the transitions between hyperfine levels. This effect is due to the odd vibrations; it disappears if we set the corresponding constants of the spin–phonon interaction from $\mathcal{H}_-$ equal to zero.

The acoustic nonequivalence of the nuclei is explained by the fact that $B_{k'} = -B_k$, with $M_{I_{k'}} = M_{I_k}$, and the expressions of type (4.12) contain the linear contribution from functions $B_{k'}$ and B_k. For example, in the case $B5$ with $M_{I_{k'}} = M_{I_k}$ and $M_{I_{p'}} = M_{I_p}$, we have

$$\Delta\mathcal{F}\left(k, k'\right) = \mathcal{F}_k - \mathcal{F}_{k'} = 4\varkappa_t R A_{2k} B_k \cos\Phi_t, \tag{4.17a}$$

$$\Delta\mathcal{F}\left(p, p'\right) = \mathcal{F}_p - \mathcal{F}_{p'} = 4\varkappa_t R A_{1p} B_p \sin\Phi_t. \tag{4.17b}$$

The expressions for A_{2k}, B_k, A_{1p}, and B_p are given in (4.13). Formula (4.17a) describes the effect of acoustic nonequivalence of nuclei 1 and 4 situated on the axis [100], and (4.17b) the effect of acoustic nonequivalence of nuclei 2 and 5 situated on the axis [010]. The acoustic nonequivalence of nuclei 3 and 6 situated on the axis [001] does not react during the conditions $\mathbf{H} \parallel [001]$ and $\mathbf{n}_t \perp [001]$ corresponding to the case $B5$.

Analogical results are obtained for the case $B6$:

$$\Delta\mathcal{F}\left(k, k'\right) = 4\varkappa_t R B_k(A_{1k} \sin\Phi_t \sin\varphi + A_{2k} \cos\Phi_t \sin\varphi) \cos\varphi, \tag{4.18a}$$

$$\Delta\mathcal{F}\left(p, p'\right) = 4\varkappa_t R B_p(A_{1p} \sin\Phi_t \sin\varphi + A_{2p} \cos\Phi_t \cos\varphi) \sin\varphi, \tag{4.18b}$$

where formula (4.18a) refers to the acoustic nonequivalence of nuclei 1 and 4, and (4.18b) to the acoustic nonequivalence of nuclei 2 and 5.

The difference in magnitude of the hypersound absorption coefficients for nuclei k and k' that have passed into one another in the process of inversion, as seen from (4.17a)–(4.18b), depends on the orientation of the magnetic field and the fascicle of polarized phonons relative to the crystallographic axes. Using this it may be possible to determine the contribution to the hypersound absorption coefficient from each such type of nuclei (k and k') in particular. However, because the static spin Hamiltonian, between the states of which the spin–phonon transitions are considered, has a similar form for the nuclei connected by inversion, the absorption of hypersound in both cases takes place at the same frequency, and the acoustic nonequivalence of the nuclei under these conditions exists but is not displayed. Note that during the examination of spin–lattice relaxation accounting for odd vibrations, the acoustic nonequivalence of nuclei k and k' connected by inversion is also not displayed but for another reason—it is due to the averaging of the probability of the transition on the directions of phonon propagation and polarization.

To detect the acoustic nonequivalence of nuclei k and k', their absorption of hypersound must take place at different frequencies. In particular, this happens if we consider the phonon shifts of the hyperfine levels in the case when these shifts for nuclei k and k' are not similar [385]. Another way of detecting the acoustic nonequivalence of nuclei connected by inversion consists in the electric field's influence on the nuclear spins.

The static spin Hamiltonian in the presence of a permanent electric field $\mathcal{E}$ with precision up to square terms on the nuclear spin projection operators may be presented in the form

$$\mathcal{H} = \mathcal{H}_0 + \mathcal{H}_\mathcal{E} + \mathcal{H}_{\mathcal{E}H},$$

where $\mathcal{H}_0$ is the spin Hamiltonian in the absence of the electric field.

The operator $\mathcal{H}_\mathcal{E}$ contains terms linear to the electric field and square-law on the nuclear spin operators:

$$
\begin{aligned}
\mathcal{H}_\mathcal{E} &= \alpha_0\big[(E_1 - E_4)\mathcal{E}_x + (E_2 - E_5)\mathcal{E}_y + (E_3 - E_6)\mathcal{E}_z\big] \\
&+ \alpha_1\big[\mathcal{E}_x\big(I_{1x}^2 - I_{4x}^2\big) + \mathcal{E}_y\big(I_{2y}^2 - I_{5y}^2\big) + \mathcal{E}_z\big(I_{3z}^2 - I_{6z}^2\big)\big] \\
&+ \alpha_2\big[\big(\{I_{1x}I_{1y}\} - \{I_{4x}I_{4y}\}\big)\mathcal{E}_y + \big(\{I_{1z}I_{1x}\} - \{I_{4z}I_{4x}\}\big)\mathcal{E}_z \\
&+ \big(\{I_{2y}I_{2z}\} - \{I_{5y}I_{5z}\}\big)\mathcal{E}_z + \big(\{I_{2x}I_{2y}\} - \{I_{5x}I_{5y}\}\big)\mathcal{E}_x \\
&+ \big(\{I_{3z}I_{3x}\} - \{I_{6z}I_{6x}\}\big)\mathcal{E}_x + \big(\{I_{3y}I_{3z}\} - \{I_{6y}I_{6z}\}\big)\mathcal{E}_y\big].
\end{aligned} \tag{4.19}
$$

Here α_1 and α_2 are the constants of the spin Hamiltonian, E_k is the unit 4×4 matrix, and $\{I_{k\xi}I_{k\eta}\} = I_{k\xi}I_{k\eta} + I_{k\eta}I_{k\xi}$.

The operator $\mathcal{H}_{\mathcal{E}H}$ includes the terms linear to the nuclear spin, electric, and magnetic fields:

$$
\begin{aligned}
\mathcal{H}_{\mathcal{E}H} &= \beta_1\big\{\big[(I_{1y} - I_{4y})\mathcal{E}_y + (I_{1z} - I_{4z})\mathcal{E}_z\big]H_x \\
&- \big[(I_{1y} - I_{4y})H_y + (I_{1z} - I_{4z})H_z\big]\mathcal{E}_x \\
&+ \big[(I_{2x} - I_{5x})\mathcal{E}_x + (I_{2z} - I_{5z})\mathcal{E}_z\big]H_y \\
&- \big[(I_{2x} - I_{5x})H_x + (I_{2z} - I_{5z})H_z\big]\mathcal{E}_y \\
&+ \big[(I_{3x} - I_{6x})\mathcal{E}_x + (I_{3y} - I_{6y})\mathcal{E}_y\big]H_z \\
&- \big[(I_{3x} - I_{6x})H_x + (I_{3y} - I_{6y})H_y\big]\mathcal{E}_z\big\},
\end{aligned} \tag{4.20}
$$

where β_1 is the constant of simultaneous interaction of the nuclear spin with the electric and magnetic fields.

Using operators $\mathcal{H}_\mathcal{E}$ and $\mathcal{H}_{\mathcal{E}H}$ as determined by formulas (4.19) and (4.20), we find shifts of the acoustical ENDOR frequencies for the nuclei k and k' that pass into one another at space inversion, in the case $\mathbf{H} \parallel [001]$:

$$\Delta\omega_{1/2\to3/2}(1,4) = -\Delta\omega_{-3/2\to-1/2}(1,4) = -\frac{2}{\hbar}(\alpha_1 + \beta_1 H_z)\mathcal{E}_x,$$

$$\Delta\omega_{1/2\to3/2}(2,5) = -\Delta\omega_{-3/2\to-1/2}(2,5) = -\frac{2}{\hbar}(\alpha_1 + \beta_1 H_z)\mathcal{E}_y,$$

$$\Delta\omega_{1/2\to3/2}(3,6) = -\Delta\omega_{-3/2\to-1/2}(3,6) = \frac{4}{\hbar}\alpha_1\mathcal{E}_z,$$

where

$$\Delta\omega_{M_I \to M_I'}(k, k') \equiv \omega^{(k)}_{M_I \to M_I'} - \omega^{(k')}_{M_I \to M_I'}.$$

In contrast to the case of ENDOR, the acoustical ENDOR lines split by the electric field have a different intensity due to the acoustic nonequivalence of the nuclei connected by inversion. The acoustic nonequivalence of the nuclei exists independently from the external electric field. The presence of the latter evidentiates only this nonequivalence and also leads to additional nonequivalence of nuclei k and k', influenced by the electric field on the probability of transitions.

In [393] on the basis of microscopic theory of the spin–phonon interaction and taking into account odd vibrations, it is shown that the bonding of the electron spin with displacements of the surrounding nuclei that are transformed under the Γ_4^- representation is possible only in the adiabatic approximation. This is due to the polarization of the electron shell of the ions encircling the vacant node on which the electron is localized. In the harmonic approximation, when the electron wave function does not depend on the displacements of the nuclei from the equilibrium positions, bonding of the electron spin with odd vibrations of the nuclei appears to be impossible. In this case the Hamiltonian of the dynamic hyperfine interaction contains products of similar projections of the operator of the nuclear spin and the displacements of the same nucleus. This eliminates the possibility, in the Hamiltonian, of forming combinations from the projections of the displacements of the nuclei being transformed according to odd representations.

4.3 Localized Biexcitons in the Crystal GaP:N

In Sects. 4.1 and 4.2 double resonances in the paramagnetic center with $S \geq 1/2$ were studied. We now consider the case when two electrons and two holes are localized in one center, but the center of capture of the carriers is isoelectronic. In this case it is necessary to consider the short-range interaction of the carriers with the neutral center and the Coulomb interaction between them. Due to the specificity of the binding of the charge carriers on the isoelectronic traps, the four-particle system of charges probably does not fall under the classification of multiparticle exciton–impurity complexes [273, 274] (see Chap. 1), but is rather a localized exciton molecule (localized biexciton). The first information on the observation of localized biexcitons came from Faulkner, Merz, and Dean [271], who found recombinational radiation from an exciton molecule, formed from two excitons, that was captured by the neutral isoelectron acceptor center of nitrogen in the crystal GaP.

The authors of [271, 272] showed that, during high levels of optic excitation and low temperatures in the short-wave part of the crystal GaP:N spectrum, new lines A^* and B^* appear which are attributed to the recombinational radiation from a localized biexciton captured by the isoelectron impurity of nitrogen. The observed structure of the lines A^* and B^* is interpreted by assuming that, during the formation of the captured biexciton, the spins of electrons are compensated and only the states of the bonded biexciton with total angular momentum $J_t = 0$ and 2 are realized. Thus, the level of the bonded biexciton with $J_t = 2$ is split by the crystal field into the levels Γ_3 and Γ_5, which causes the structure of lines A^* and B^*. Zeeman splitting of lines A^* and B^* in a wide interval of the change of the magnetic field (up to 50 kOe) is also interpreted on this basis.

We shall consider a cubic crystal that has a simple band of conduction of s type and a threefold degenerate valence band formed by p electrons. If the spin–orbital interaction in the valence band is sufficiently strong, then the presence of the lower valence subband, chipped off by spin–orbital interaction, is not "seen" in the optical transitions at low temperatures. In this case the four-particle wave function describing the interaction of two localized excitons must be constructed on the basis of the hole states from the upper valence subband and the electron states from the conduction band. In semiconductors of type GaP, the upper valence subband of which is degenerated fourfold, there exist 64 states of two interacting excitons captured on the isoelectronic center. They perform the basis of the reducible representation $D_{1/2} \otimes D_{1/2}^* \otimes D_{3/2} \otimes D_{3/2}^*$ and can be easily built by the methods of group theory. The four-particle wave functions related to localized biexciton have the form [170]

$$\Phi_0^0(00) = \frac{1}{2\sqrt{2}}[(+-)-(-+)]((\overset{+-}{3}\overset{}{3})-(\overset{-+}{1}\overset{}{1})-(\overset{+-}{1}\overset{}{1})-(\overset{-+}{3}\overset{}{3})),$$

$$\Phi_2^2(02) = \frac{1}{2}[(+-)-(-+)]((\overset{++}{3}\overset{}{1})-(\overset{++}{1}\overset{}{3})),$$

$$\Phi_1^2(02) = \frac{1}{2}[(+-)-(-+)]((\overset{+-}{3}\overset{}{1})-(\overset{-+}{1}\overset{}{3})),$$

$$\Phi_0^2(02) = \frac{1}{2\sqrt{2}}[(+-)-(-+)]((\overset{+-}{3}\overset{}{3})-(\overset{+-}{1}\overset{}{1})-(\overset{-+}{1}\overset{}{1})-(\overset{-+}{3}\overset{}{3})),$$

(4.21)

where $(+-)$, $(-+)$ and $\overset{++}{mm'}, \overset{+-}{mm'}, \overset{-+}{mm'}$ are used to designate the spin functions of the electrons from the conduction band and the eigenstates of the operators of the angular momenta of valence electrons.

$$(+-) = \left| S_1, \; M_{S_1} = +\frac{1}{2} \right\rangle \left| S_2, \; M_{S_2} = -\frac{1}{2} \right\rangle,$$

$$(\overset{+-}{mm'}) = \left| J_1, \; M_{J_1} = +\frac{m}{2} \right\rangle \left| J_2, \; M_{J_2} = -\frac{m'}{2} \right\rangle.$$

For each of the states $\Phi_M^J(ll')$ the numbers l and l' indicate the irreducible representations of the rotation groups $D_l \in D_{1/2} \otimes D_{1/2}$ and $D_{l'} \in D_{3/2} \otimes D_{3/2}$, on which the wave functions of the s and p electrons that are interacting between themselves and bonded on the isoelectronic center are transformed. The wave functions $\Phi_M^J(ll')$ with non-negative values M which are not indicated in (4.21) are given in Appendix F. The states to which the negative values M belong are

$$\Phi_{-M}^J(ll') = (-1)^M \left(\Phi_M^J(ll') \right)^*.$$

The functions $\overset{+}{m}, \overset{+}{m'}, \ldots$ characterize the states of the valence p electrons, but not of the holes from the valence subband. Therefore, $\Phi_M^J(ll')$ still does not describe the interaction of the excitons. The wave functions of the interacting excitons $\Psi_M^J(ll')$ can be obtained from formula (4.21) and Appendix F, if acting by the time-reversal operator on the four-particle wave functions $\Phi_M^J(ll')$:

$$\Psi_M^J(ll') = U_{J_1} U_{J_2} \mathcal{K}_0 \Phi_M^J(ll'), \tag{4.22}$$

where $\mathcal{K}_0$ is the complex conjugate operator, and U_{J_i} is the unitary operator, acting in the space of the angular momentum $J_i = 3/2$ of the valence electron from the upper valence subband, which is represented by a skew-symmetric matrix with alternating matrix elements $+1$ and -1 on the oblique diagonal and the rest of the elements equal to zero [170, 391, 398]. With the help of Wigner coefficients for point groups of symmetry [397, 405] the linear combinations were found for the wave functions $\Psi_M^J(l, l')$ that are transformed according to the irreducible representations of the symmetry group T_d of the isoelectronic center.

The experimental results on the new recombinational radiation from localized biexcitons in the crystal GaP:N [271, 272] can be easily explained by starting from the states of the interacting excitons. The interpretation of Zeeman splitting of the lines A^* and B^* on the basis of only the hole Hamiltonian of the localized biexciton, confirmed by experimental data, actually indicates the antiparallel orientation of the electron spins in the biexciton that is captured on the isoelectron acceptor. Among the 16 wave functions of interacting bond excitons that are symmetric relative to the permutation of the electrons (the spins of the electrons are antiparallel), only six of them, $\Psi_0^0(00)$ and $\Psi_M^2(02)$, are antisymmetric simultaneously relative to the permutation of holes. These states satisfy the Pauli principle and therefore are functions of the bond biexciton. This limited number of states of the bond biexciton is caused by the lack of orbital degeneration of the conduction band.

The binding energy of a separate electron to the isoelectron acceptor of nitrogen in gallium phosphide is 8 meV [266]. The binding energy of a single exciton to the isoelectronic center is ~ 11 meV, and the difference between the binding energy of the first and second excitons of the localized biexciton with an isoelectronic center N in GaP is equal to 1.88 meV [271]. Hence, we could expect the manifestation of electron–hole interaction in the bond biexciton. However, from the symmetry properties of the wave functions (4.22), this interaction, like the electron spin–spin interaction, does not perturb the biexciton levels. Therefore, the only internal interaction (besides the interaction with the crystal field) that still can perturb the biexciton states is the interaction between the holes.

4.4 Exchange and Magnetic Dipole–Dipole Interaction Between Holes in the Localized Biexciton

The Hamiltonian interaction between the holes of the localized biexciton is presented in the form

$$\mathcal{H} = \mathcal{H}_{exch} + \mathcal{H}_{dd} + \mathcal{H}', \tag{4.23}$$

where $\mathcal{H}_{\text{exch}}$ and $\mathcal{H}_{\text{dd}}$ are operators of the exchange and magnetic dipole–dipole interaction of the holes, and $\mathcal{H}'$ contains terms of the third and fourth degree on the projection operators of the holes' angular momenta. These operators are determined by the expressions:

$$\mathcal{H}_{\text{exch}} = -\mathcal{J}\mathbf{J}_1\mathbf{J}_2, \tag{4.24}$$

$$\mathcal{H}_{\text{dd}} = \frac{\gamma_h^2 \hbar^2}{r_{12}^3}\{\mathbf{J}_1\mathbf{J}_2 - 3[J_{1z}\cos\theta + \sin\theta(J_{1x}\cos\varphi + J_{1y}\sin\varphi)]$$
$$\times [J_{2z}\cos\theta + \sin\theta(J_{2x}\cos\varphi + J_{2y}\sin\varphi)]\}, \tag{4.25}$$

$$\mathcal{H}'_1 = Q_1 \sum_p J_{1p}J_{2p}(J_{1p}^2 + J_{2p}^2) + Q_2\big(\xi_{\Gamma_{3,1}}^{(1)}\xi_{\Gamma_{3,1}}^{(2)} + \eta_{\Gamma_{3,2}}^{(1)}\xi_{\Gamma_{3,2}}^{(2)}\big)$$
$$+ Q_3\xi_{\Gamma_4}^{(1)}\xi_{\Gamma_4}^{(2)}. \tag{4.26}$$

Here

$$\xi_{\Gamma_{3,1}}^{(k)} = \frac{1}{2\sqrt{3}}[3J_{kz}^2 - J(J+1)], \qquad \eta_{\Gamma_{3,2}}^{(k)} = \frac{1}{2}\big(J_{kx}^2 - J_{ky}^2\big),$$

$$\xi_{\Gamma_{4,1}}^{(k)} = \frac{1}{2}(J_{ky}J_{kz} + J_{kz}J_{ky}), \qquad \xi_{\Gamma_{4,2}}^{(k)} = \frac{1}{2}(J_{kz}J_{kx} + J_{kx}J_{kz}), \tag{4.27}$$

$$\xi_{\Gamma_{4,3}}^{(k)} = \frac{1}{2}(J_{kx}J_{ky} + J_{ky}J_{kx}); \quad p = x, y, z; \; k = 1, 2.$$

In formulas (4.23)–(4.27) $\mathcal{J}$, Q_1, Q_2, and Q_3 are the constants of the spin Hamiltonian, γ_h is the gyromagnetic ratio of the hole; r_{12}, θ, φ are the spherical coordinates of the vector $\mathbf{r}_{12} = \mathbf{r}_1 - \mathbf{r}_2$ ($\mathbf{r}_1$ and $\mathbf{r}_2$ are radius vectors of the holes of the localized biexciton); Γ_3 and Γ_4 are the irreducible representations of the symmetry point group T_d in Heine's designations [127].

The wave functions of the localized biexciton that are being transformed on the irreducible representations Γ_1, Γ_3, and Γ_5 of the symmetry group T_d have the forms

$$\Psi_{\Gamma_1} = \frac{1}{2\sqrt{2}}[(+-) - (-+)]((\overset{+-}{33}) - (\overset{+-}{11}) + (\overset{-+}{11}) - (\overset{-+}{33})),$$

$$\Psi_{\Gamma_{3,1}} = \frac{1}{2\sqrt{2}}[(+-) - (-+)]((\overset{+-}{33}) + (\overset{+-}{11}) - (\overset{-+}{11}) - (\overset{-+}{33})),$$

$$\Psi_{\Gamma_{3,2}} = \frac{1}{2\sqrt{2}}[(+-) - (-+)]((\overset{++}{31}) - (\overset{++}{13}) + (\overset{--}{13}) - (\overset{--}{31})),$$

$$\Psi_{\Gamma_{5,1}} = \frac{1}{2}[(+-) - (-+)]((\overset{+-}{13}) - (\overset{-+}{31})), \tag{4.28}$$

$$\Psi_{\Gamma_{5,2}} = \frac{i}{2\sqrt{2}}[(+-) - (-+)]((\overset{++}{31}) - (\overset{++}{13}) - (\overset{--}{13}) + (\overset{--}{31})),$$

$$\Psi_{\Gamma_{5,3}} = \frac{1}{2}[(+-) - (-+)]((\overset{+-}{31}) - (\overset{-+}{13})).$$

The operator of the exchange interaction between holes $\mathcal{H}_{\mathrm{ex}}$, determined by formula (4.24), is diagonal in the basis of the wave functions of the biexciton $\Psi_{\Gamma_\alpha,\mu_\alpha}$ ($\alpha = 1, 3, 5$; μ_α numbers the rows of the representation Γ_α):

$$\langle \Psi_{\Gamma_\alpha,\mu_\alpha} | \mathbf{J}_1 \mathbf{J}_2 | \Psi_{\Gamma_\beta,\mu_\beta} \rangle = -\frac{3}{4} \delta_{\alpha\beta} \begin{cases} 5, & \alpha = 1, \\ \delta_{\mu_\alpha \mu_\beta}, & \alpha = 3, 5. \end{cases} \tag{4.29}$$

From formula (4.29) it follows that the exchange splitting of the biexciton levels (the distance between level Γ_1 and the center of gravity of the levels Γ_3 and Γ_5) is equal to

$$\left| \Delta E_{\Gamma_1 \Gamma_{3(5)}}^{\mathrm{exch}} \right| = 3\mathcal{J}. \tag{4.30}$$

For the state Γ_1 the value of the total angular momentum of the localized biexciton is equal to zero, but for the levels Γ_3 and Γ_5 the total angular momentum is equal to 2.

We notice that in the localized biexciton the average value of the operators of electron–electron and electron–hole exchange interactions, calculated by means of the wave functions $\Psi_M^J(ll')$ or $\Psi_{\Gamma_\alpha,\mu_\alpha}$ from formulas (4.22) and (4.28), are equal to zero. Therefore, these interactions do not contribute to the splitting and shifts of biexciton levels, which is due to the properties of symmetry of the wave functions of the biexciton.

Another equivalent way of examining the exchange interaction between the holes (as for any other interactions between angular momenta) is based on applying the representation of coupled bosons for the spin operators. The two-Bose operator representation of the angular momentum was initially introduced by Schwinger [406].

In Appendix H it is shown that it is possible to come to Schwinger's two-Bose operator representation for angular momentum J, when determining some two-boson spinor operators of rank $2J$ and using their unitarity properties [170]. For a localized biexciton these two-boson spinor operators have the form

$$U_1^{\mathrm{e}} = \begin{pmatrix} a_1 \\ a_2 \end{pmatrix}, \qquad U_2^{\mathrm{e}} = \begin{pmatrix} b_1 \\ b_2 \end{pmatrix},$$

$$U_1^{\mathrm{h}} = \frac{1}{\sqrt{2}} \begin{pmatrix} \frac{1}{\sqrt{3}} A_1^3 \\ A_1^2 A_2 \\ A_1 A_2^2 \\ \frac{1}{\sqrt{3}} A_2^3 \end{pmatrix}, \qquad U_2^{\mathrm{h}} = \frac{1}{\sqrt{2}} \begin{pmatrix} \frac{1}{\sqrt{3}} B_1^3 \\ B_1^2 B_2 \\ B_1 B_2^2 \\ \frac{1}{\sqrt{3}} B_2^3 \end{pmatrix}, \tag{4.31}$$

where the upper indices e and h refer to electrons and holes; a, b, A, and B are the destruction operators of bosons of the corresponding types commuting between themselves.

After the transition to the representation of the coupled bosons, taking into account the unitarity of spinor operators (4.31), we obtain the following expression for the Hamiltonian of the exchange interaction between holes:

$$\mathcal{H}_{\text{exch}} = -\frac{\mathcal{J}}{2}\left[\frac{1}{2}\left(A_1^+ A_1 - A_2^+ A_2\right)\left(B_1^+ B_1 - B_2^+ B_2\right)\right.$$

$$\left. + A_1^+ B_2^+ A_2 B_1 + A_2^+ B_1^+ A_1 B_2\right].$$

The states of the localized biexcitons in the representation of the coupled bosons are described by the following wave functions (the symmetry group T_d):

$$\Psi_{\Gamma_1} = \frac{1}{4\sqrt{2}}\left(a_1^+ b_2^+ - a_2^+ b_1^+\right)\left\{\frac{1}{3}\left[\left(A_1^+\right)^3\left(B_2^+\right)^3 - \left(A_2^+\right)^3\left(B_1^+\right)^3\right]\right.$$

$$\left. + A_1^+ A_2^+ B_1^+ B_2^+\left(A_2^+ B_1^+ - A_1^+ B_2^+\right)\right\}|0\rangle,$$

$$\Psi_{\Gamma_3,1} = \frac{1}{4\sqrt{2}}\left(a_1^+ b_2^+ - a_2^+ b_1^+\right)\left\{\frac{1}{3}\left[\left(A_1^+\right)^3\left(B_2^+\right)^3 - \left(A_2^+\right)^3\left(B_1^+\right)^3\right]\right.$$

$$\left. + A_1^+ A_2^+ B_1^+ B_2^+\left(A_1^+ B_2^+ - A_2^+ B_1^+\right)\right\}|0\rangle,$$

$$\Psi_{\Gamma_3,2} = \frac{1}{4\sqrt{6}}\left(a_1^+ b_2^+ - a_2^+ b_1^+\right)\left[\left(A_1^+\right)^2\left(B_1^+\right)^2 + \left(A_2^+\right)^2\left(B_2^+\right)^2\right]$$

$$\times \left(A_1^+ B_2^+ - A_2^+ B_1^+\right)|0\rangle,$$

$$\Psi_{\Gamma_5,1} = \frac{1}{4\sqrt{3}}\left(a_1^+ b_2^+ - a_2^+ b_1^+\right)A_2^+ B_2^+\left[\left(A_2^+\right)^2\left(B_1^+\right)^2\right.$$

$$\left. - \left(A_1^+\right)^2\left(B_2^+\right)^2\right]|0\rangle,$$

$$\Psi_{\Gamma_5,2} = \frac{1}{4\sqrt{6}}\left(a_1^+ b_2^+ - a_2^+ b_1^+\right)\left[\left(A_2^+\right)^2\left(B_2^+\right)^2 - \left(A_1^+\right)^2\left(B_1^+\right)^2\right]$$

$$\times \left(A_1^+ B_2^+ - A_2^+ B_1^+\right)|0\rangle,$$

$$\Psi_{\Gamma_5,3} = \frac{1}{4\sqrt{3}}\left(a_1^+ b_2^+ - a_2^+ b_1^+\right)A_1^+ B_1^+\left[\left(A_2^+\right)^2\left(B_1^+\right)^2\right.$$

$$\left. - \left(A_1^+\right)^2\left(B_2^+\right)^2\right]|0\rangle.$$

Here we used the possibility of constructing the wave function of the hole, transformed on the representation Γ^*, resulting from the wave function of the electron corresponding to the representation Γ by means of the time-reversal operator.

The exchange shifts of the biexciton levels Γ_3, Γ_5, and Γ_1 are equal to

$$\delta E_{\Gamma_3(\Gamma_5)}^{\text{exch}} = -3/4\mathcal{J}, \qquad \delta E_{\Gamma_1}^{\text{exch}} = -15/4\mathcal{J},$$

from which it is seen that the value of the exchange splitting of the biexciton is determined by formula (4.30) found earlier in the spin formalism.

We account for biexciton splitting in the representation of coupled bosons because it demonstrates the fact that the simplicity of the types of wave functions and spin operators in the two-Bose operator representation makes this formalism as convenient to apply as the spin operator techniques.

From the analysis of the positions of the localized biexciton levels Γ_5, Γ_3, and Γ_1, found experimentally in [271, 272] ($E_{\Gamma_5} - E_{\Gamma_3} = 0.16$ meV; the center of gravity of the levels Γ_5 and Γ_3 is situated 0.17 meV lower than the level Γ_1), we find the value and sign of the exchange integral for the holes in the biexciton $\mathcal{J} = 0.057$ meV. Thus, the usage of only one kind of symmetry property in this analysis allows us to evidentiate the contribution of the exchange interaction between the holes, which has a ferromagnetic character, in the general splitting of the biexciton due to other effects.

For averaging of dipole–dipole interaction between holes, using dipole–dipole interaction operator $\mathcal{H}_{dd}$ from (4.25) with wave functions of the biexciton Ψ_M^2 ($M = 2, \ldots, -2$) determined by (4.22), we will complete the functions by coordinate parts and in their angular parts divide the orbital and spin variables

$$\Psi_M^2(02|1,2) = \frac{1}{\sqrt{2}}\big[(+-)-(-+)\big]\chi_M^2(2|1,2),$$

where

$$\chi_2^2(2|1,2) = \frac{1}{\sqrt{2}}\left\{\frac{1}{\sqrt{3}}Y_{-1}^1(1)Y_{-1}^1(2)(\alpha_1\beta_2-\beta_1\alpha_2)\right.$$
$$\left.+\sqrt{\frac{2}{3}}\big[Y_0^1(1)Y_{-1}^1(2)-Y_{-1}^1(1)Y_0^1(2)\big]\beta_1\beta_2\right\},$$

$$\chi_1^2(2|1,2) = -\frac{1}{\sqrt{2}}\left\{\frac{1}{\sqrt{3}}\big[Y_1^1(1)Y_{-1}^1(2)-Y_{-1}^1(1)Y_1^1(2)\big]\beta_1\beta_2\right.$$
$$\left.+\sqrt{\frac{2}{3}}\big[Y_0^1(1)Y_{-1}^1(2)\alpha_1\beta_2-Y_{-1}^1(1)Y_0^1(2)\beta_1\alpha_2\big]\right\},$$

$$\chi_0^2(2|1,2) = \frac{1}{\sqrt{2}}\left\{Y_1^1(1)Y_{-1}^1(2)\alpha_1\beta_2-Y_{-1}^1(1)Y_1^1(2)\beta_1\alpha_2\right.$$
$$+\frac{\sqrt{2}}{3}\big[Y_1^1(1)Y_0^1(2)-Y_0^1(1)Y_1^1(2)\big]\beta_1\beta_2$$
$$+\frac{\sqrt{2}}{3}\big[Y_0^1(1)Y_{-1}^1(2)-Y_{-1}^1(1)Y_0^1(2)\big]\alpha_1\alpha_2$$
$$+\frac{2}{3}Y_0^1(1)Y_0^1(2)(\alpha_1\beta_2-\beta_1\alpha_2)$$
$$\left.+\frac{1}{3}\big[Y_1^1(1)Y_{-1}^1(2)\beta_1\alpha_2-Y_{-1}^1Y_1^1(2)\alpha_1\beta_2\big]\right\}.$$

The functions $\chi^2_{-1}(2|1,2)$ and $\chi^2_{-2}(2|1,2)$ are not given here but may be found using expressions for $\chi^2_1(2|1,2)$ and $\chi^2_2(2|1,2)$; α and β are the spin functions of the holes.

After the averaging on the spin states, the matrix of the operator of dipole–dipole interaction between holes of the localized biexciton in the basis of wave functions $\{\Psi^2_2(02|1,2), \ldots, \Psi^2_{-2}(02|1,2)\}$ has the form

$$\mathcal{H}_{dd} = \gamma^2_h \hbar^2 r^{-3}_{12}$$

$$\times \begin{pmatrix} \frac{3}{2}a(\theta) & 3c^*(\theta,\varphi) & 0 & 0 & 0 \\ 3c(\theta,\varphi) & -\frac{3}{4}a(\theta) & \sqrt{\frac{3}{2}}c^*(\theta,\varphi) & 3b^*(\theta,\varphi) & 0 \\ 0 & \sqrt{\frac{3}{2}}c(\theta,\varphi) & -\frac{9}{4}a(\theta) & -\sqrt{\frac{3}{2}}c^*(\theta,\varphi) & \sqrt{6}b^*(\theta,\varphi) \\ 0 & 3b(\theta,\varphi) & -\sqrt{\frac{3}{2}}c(\theta,\varphi) & -\frac{3}{4}a(\theta) & -3c^*(\theta,\varphi) \\ 0 & 0 & \sqrt{6}b(\theta,\varphi) & -3c(\theta,\varphi) & \frac{3}{2}a(\theta) \end{pmatrix},$$

$$(4.32)$$

where

$$a(\theta) = 1 - 3\cos^2\theta, \qquad b(\theta,\varphi) = -\frac{3}{4}\sin^2\theta e^{-2i\varphi},$$

$$c(\theta,\varphi) = -\frac{3}{2}\sin^2\theta\cos^2\varphi e^{-i\varphi}.$$

Further calculation of the contribution of dipole–dipole interaction between holes in the fine structure of the energy levels of the localized biexciton is connected with the determination of the matrix elements of the operator $\mathcal{H}_{dd}$ on the orbital and radial wave functions of the biexciton.

4.5 Double Hole–Nuclear Resonance on Localized Biexcitons in the Crystal GaP:N

The peculiarities of the energy band structure of the crystal GaP, as was explained above, lead to the absence of the exchange interaction of electrons between themselves and between electrons and holes in the biexciton captured by the isoelectronic center. This is due to the absence of orbital degeneration of the conduction band, as a result of which there are a total of four different pure spin states for two interacting electrons, from which only a state with a total spin of electrons equal to zero is allowed according to Pauli's principle. This situation is characteristic for all other crystals with orbitally nondegenerate conduction bands.

Electron–electron exchange interaction can be displayed in the bond biexciton only in the presence of orbital degeneration of the conduction band. In this case it is possible to form antisymmetric combinations from electron wave functions (entering the full wave function of the biexciton) with nonzero summary angular momentum of two electrons contained in the biexciton. In connection with this, experimental works on detecting and investigating biexcitons localized on isoelectronic centers in crystals with orbitally degenerate conductivity bands are of interest. If in the point $\mathbf{K} = 0$ both the conduction and the valence bands are degenerate (for example, Γ_8^- and Γ_8^+ in the crystal Cu_2O), then the spectrum of the recombinational radiation from the localized biexciton on the isoelectronic center in the crystal will possess a richer fine structure in comparison with the structure that was found by the authors of the works [271, 272] in the crystal GaP:N.

In the optical spectra of the recombinational radiation of the localized biexciton in the crystal GaP:N, only the effects of the crystal field and exchange interaction of the holes [170, 386] were displayed; the experimental data on other types of interactions between the holes are currently absent. That is why it is advisable to estimate the values of the shifts of the energy levels of the localized biexciton affected by the magnetic dipole–dipole interaction between holes, and to discuss the possibility of their experimental detection.

We proceed from expression (4.32) for the matrix of the operator of the dipole–dipole interaction between holes $\mathcal{H}_{dd}$. For simplicity, the numeric estimations of the order of the values of dipole–dipole corrections to the energy levels of the biexciton will be performed in the case of a strong magnetic field, the direction of which is taken as the axis of quantization ($\mathbf{H} \parallel [001]$; the symmetry point group T_d is reduced up to S_4). We shall limit ourselves by examining only the secular part $\mathcal{H}_{dd}^{sec}$ of the operator $\mathcal{H}_{dd}$. On the basis of wave functions transformed according to the irreducible representations Γ'_{t1}, $^1\Gamma'_{t2}$, $^2\Gamma'_{t2}$, Γ'_{t3}, and Γ'_{t4} of the symmetry groups S_4 (the designations accepted in [397] are used), after averaging of the operator $\mathcal{H}_{dd}^{sec}$ according to the angular part of the wave function of the localized biexciton, we get

$$\langle \Gamma'_{t1} | \mathcal{H}_{dd}^{sec} | \Gamma'_{t1} \rangle = \frac{9}{4} \frac{\gamma_h \hbar^2}{r_{12}^3} (\eta - 1),$$

$$\langle {}^1\Gamma'_{t2} | \mathcal{H}_{dd}^{sec} | {}^1\Gamma'_{t2} \rangle = \langle {}^2\Gamma'_{t2} | \mathcal{H}_{dd}^{sec} | {}^2\Gamma'_{t2} \rangle = \frac{3}{2} \frac{\gamma_h \hbar^2}{r_{12}^3} (1 - 2\eta),$$

$$\langle \Gamma'_{t3} | \mathcal{H}_{dd}^{sec} | \Gamma'_{t3} \rangle = \langle \Gamma'_{t4} | \mathcal{H}_{dd}^{sec} | \Gamma'_{t4} \rangle = \frac{1}{3} \langle \Gamma'_{t1} | \mathcal{H}_{dd}^{sec} | \Gamma'_{t1} \rangle,$$

where

$$\eta = \frac{1}{5} \frac{r_1^2 + r_2^2}{r_{12}^2}.$$

The other matrix elements of the operator $\mathcal{H}_{dd}^{sec}$ in the basis of the states $\Gamma'_{t1}, \dots, \Gamma'_{t4}$ are equal to zero. Further averaging of the operator $\mathcal{H}_{dd}^{sec}$ on the radial functions of the bond biexciton $\mathcal{R}_{31}(1, 2) = \mathcal{R}_{31}(1)\mathcal{R}_{31}(2)$ leads to the following expressions for the dipole–dipole shifts of Zeeman levels of the biexciton:

$$\delta E\left(\Gamma'_{t1}\right) = 1.3 \cdot 10^2 \pi^2 \gamma_h^2 \hbar^2 a_B^{-3},$$

$$\delta E\left(^1\Gamma'_{t2}\right) = \delta E\left(^2\Gamma'_{t2}\right) = 1.88 \cdot 10\pi^2 \gamma_h^2 \hbar^2 a_B^{-3}, \qquad (4.33)$$

$$\delta E\left(\Gamma'_{t3}\right) = \delta E\left(\Gamma'_{t4}\right) = \frac{1}{3}\delta E\left(\Gamma'_{t1}\right).$$

Here a_B is the effective Bohr radius of the localized biexciton.

When deducing the formula (4.33) we limited ourselves only to the linear items in the wave functions $\mathcal{R}_{31}(r_1)$ and $\mathcal{R}_{31}(r_2)$ on r_k which give the biggest contribution to the radial integrals and, correspondingly, to the values of the shifts of the biexciton levels. Assuming in (4.33) $a_B = 45$ Å, we obtain $\delta E(\Gamma'_{t1}) = 0.53 \cdot 10^{-6}$ eV, $\delta E(\Gamma'_{t2}) = -0.77 \cdot 10^{-6}$ eV [170]. The value of the gyromagnetic ratio of the hole γ_h necessary for performing this numeric estimation was found from the experimental data [271, 272] on the Zeeman splitting of the lines of the recombinational radiation of the bond biexciton in GaP:N ($\gamma_h \simeq 8 \cdot 10^6$ s$^{-1} \cdot$ Oe^{-1}).

Such small shifts of the energy levels cannot be seen in optic spectroscopy, although they probably can be determined by the beat resonance method. In the work [407], in order to carry out the nuclear resonance beats under conditions of optical orientation by the transversal fascicle of light, a sufficiently small value of the constant magnetic field intensity (0.07 Oe) was used so that the magnetic splitting of the excited state did not exceed the natural width of the level, with a value of $\sim 0.28 \cdot 10^{-8}$ eV.

Magnetic dipole–dipole shifts of the localized biexciton levels result in the form of weak anisotropic effects on the spectrum of the hole paramagnetic resonance (HOPR) of the localized biexciton. The weaker interactions between the holes (in contrast to the exchange and dipole–dipole interactions) lead to still smaller shifts of the biexciton energy levels. The experimental manifestation of these shifts depends on the value of the spin Hamiltonian $\mathcal{H}'$ constants, determined by formula (4.26). The operator $\mathcal{H}'$ causes, in particular, the following shift of the biexciton level Γ_1:

$$\delta E(\Gamma_1) = \frac{123}{8}Q_1 - \frac{1}{192}Q_2\left[32J^2(J+1)^2 - 120J(J+1) - 417\right] + \frac{51}{16}Q_3.$$

Since the nuclei spins of the nitrogen atom on which the biexciton is localized differ from zero ($I = 1$), there exists a hyperfine interaction of this spin with the total hole angular momentum $J = 2$ ($D_2 \in D_{3/2} \otimes D_{3/2}$) that causes the magnetic momentum of the biexciton. The energy level with $J = 2$ is the lowest level of the biexciton in the absence of a crystal field. This level is split by a crystal field of tetrahedral symmetry into doublet Γ_3 and triplet Γ_5, the doublet Γ_3 being the ground state of the localized biexciton. The hole Zeeman energy is higher in comparison with the energy of the hole–nuclear hyperfine interaction; therefore, neglecting the crystal field splitting the biexciton level with $J = 2$, we can represent the wave function of the biexciton–nuclear system in the form

$$|J, M_J; I, M_I\rangle = |J, M_J\rangle |I, M_I\rangle,$$

where $|J, M_J\rangle$ is the eigenfunction of the J_z operator ($J = 2$) and $|I, M_I\rangle$ is the eigenfunction of the I_z operator ($I = 1$). Taking into account that for the biexciton captured by the nitrogen trap in GaP crystal $\gamma_h > 0$ [181, 271] (γ_h is the hole gyromagnetic ratio), carrying out only the Fermi hole–nuclear hyperfine interaction and neglecting the nuclear Zeeman interaction, which is assumed to be much smaller than hole Zeeman splitting, we obtain:

$$E_{2,\pm1} = E_0 - 2\delta \pm 2a_J, \qquad E_{\pm2,0} = E_0 \mp 2\delta,$$

$$E_{1,\pm1} = E_0 - \delta \pm a_J, \qquad E_{\pm1,0} = E_0 \mp \delta,$$

$$E_{-1,\pm1} = E_0 + \delta \mp a_J, \qquad E_{0,\pm1} = E_{0,0} = E_0,$$

$$E_{-2,\pm1} = E_0 + 2\delta \mp 2a_J,$$

where E_0 is the energy of the biexciton level of D_2 symmetry unperturbed by the crystal and magnetic fields, $\delta = g_h \beta H$ is the hole Zeeman splitting, and a_J is the constant of Fermi hole–nuclear hyperfine interaction. There are transitions between hyperfine levels with frequency $\omega_1 = a_J/\hbar$ ($\omega_{1,-1\leftrightarrow0} = \omega_{1,1\leftrightarrow0} = \omega_{-1,1\leftrightarrow0} = \omega_{-1,0\leftrightarrow-1} = a_J/\hbar$) and with frequency $\omega_2 = 2\omega_1$ ($\omega_{2,-1\leftrightarrow0} = \omega_{2,0\leftrightarrow1} = \omega_{-2,1\leftrightarrow0} = \omega_{-2,0\leftrightarrow-1} = 2a_J/\hbar$). For all these quantum transitions the selection rules $\Delta M_J = 0$, $\Delta M_I = \pm1$ apply.

As a result of strong delocalization of the holes ($a_B \simeq 45$ Å) we should expect a small value of the constant of hyperfine interaction a_J of the hole in the biexciton with the nucleus of the impurity nitrogen atom. The constant a_J can be measured while observing the transitions between the hole–nuclear hyperfine levels under conditions of saturation of the magnetic dipole transitions between the biexciton states with the change of the projection of the hole angular moment (the hole–nuclear double resonance, HONDOR). The HONDOR in the localized biexciton takes place on the frequencies $\omega = a_J/\hbar$ and 2ω with the selection rules $\Delta M_J = 0$, $\Delta M_I = \pm1$.

Thus, the small concentration of nitrogen impurity atoms in GaP crystal does not allow one to observe the influence of the localized biexcitons on the ^{14}N NMR spectrum because of the low sensitivity of the NMR method. The interaction of the holes of the localized biexciton with the nitrogen nuclei can be examined only by two methods: the hole paramagnetic resonance (HOPR) method or the more precise HONDOR one.

4.6 Double Radio–Optical Exciton Resonance

The optical pumping of an atomic gas in the presence of resonance radio-frequency field leads to the broadening and shifts of Zeeman levels of energy as well as to the partial transfer of the radio-frequency coherency in the pumping cycle [408–413]. Similar effects are also characteristic for an exciton gas on account of its specificity [388].

Let us consider the radio-frequency coherency transfer in the system of excitons and photons under the following assumptions:

1. The ground state of the relative motion of the electron–hole in the exciton has a purely spin or spin–orbital degeneracy.
2. The values of Zeeman's splittings of the ground and excited states of the relative motion of the exciton are less than those of the corresponding binding energy in these states. The gyromagnetic ratios of the exciton in the ground and excited states are different.
3. The optical band of the excitation represents by itself the resonance curve with maximum on the frequency of the optical transition $\omega_{\Gamma_\mu} \to \omega_{\Gamma_m}$ and halfwidth Δ (Γ_μ and Γ_m are the irreducible representations of the symmetry point group on which the wave functions of the ground and excited states of the exciton are transformed). The optical excitations are performed by N photons with wave vectors $\mathbf{K}_1, \mathbf{K}_2, \ldots, \mathbf{K}_N$, that are parallel between themselves. The polarization of the photons is determined by the unit vector $\mathbf{e}_{\lambda_0}$.
4. $\Delta \gg \tau^{-1}$ (τ is the radiational lifetime of the excited state of the relative movement of the exciton), $\Delta \gg \omega_{\Gamma_\mu \Gamma'_\mu}$, $\Delta \gg \omega_{\Gamma_m, \Gamma'_m}$.

In the absence of exciton–photon interaction, the wave function of the system in the time momentum $t = t_0$ is represented by the state vector

$$|\Psi(t_0)\rangle = \sum_\mu a_{\Gamma_\mu}(t_0)|\Gamma_\mu\rangle. \tag{4.34}$$

Coherence between the Zeeman components of the ground state of the exciton exists if the nondiagonal elements of the density matrix are different from zero:

$$\langle a_{\Gamma_{\mu'}}(t_0)a^*_{\Gamma_\mu}(t_0)\rangle \neq 0.$$

Taking into account the exciton–photon interaction, the state (4.34) in the time momentum t has the form

$$|\Psi(t)\rangle = \sum_\mu a_{\Gamma_\mu}(t)|\Gamma_\mu; \mathbf{K}_1, \mathbf{K}_2, \ldots, \mathbf{K}_N, \mathbf{e}_{\lambda_0}\rangle$$

$$+ \sum_{mi} a_{\Gamma_m, -\mathbf{K}_i}(t)|\Gamma_m; \mathbf{K}_1, \mathbf{K}_2, \ldots, \mathbf{K}_{i-1}, \mathbf{K}_{i+1}, \ldots, \mathbf{K}_N, \mathbf{e}_{\lambda_0}\rangle$$

$$+ \sum_{\mu,i,\mathbf{K},\lambda} a_{\Gamma_\mu, -\mathbf{K}_i, \mathbf{K}, \lambda}(t)|\Gamma_\mu; \mathbf{K}_1, \mathbf{K}_2, \ldots, \mathbf{K}_{i-1}, \mathbf{K}_{i+1}, \ldots, \mathbf{K}_N, \mathbf{e}_{\lambda_0}; \mathbf{K}, \mathbf{e}_\lambda\rangle,$$

$$\tag{4.35}$$

where $\mathbf{K}$ and $\mathbf{e}_\lambda$ are the wave vector and the unit vector of polarization of the reradiated photon. The first sum in (4.35) corresponds to the exciton in the ground state in the presence of N photons, and the second sum corresponds to the exciton in the excited state of electron–hole relative motion in the presence of $N - 1$ of the excited photons (the photon with wave vector $\mathbf{K}_i$ and polarization $\mathbf{e}_{\lambda_0}$ was absorbed). The third sum in (4.35) corresponds to the exciton in the ground state in the presence of $N - 1$ excited photons and a reradiated one. The state (4.35) describes three phases of the optical pumping of excitons assuming that the time necessary for the realization of the complete cycle is less than the lifetime of the exciton relative to the process of electron–hole annihilation.

The concrete calculation will be performed further for the case of excitons in cubic crystals of the type CuCl with symmetry point group T_d. In the crystal CuCl in a low homogeneous magnetic field ($H \lesssim 10$ kOe) the Zeeman components of the level $1S$ (Γ_5) of the orthoexciton are practically equidistant [388]; therefore, the interaction of the rotating magnetic field $\mathbf{H}_1(\mathbf{H}_1 \perp \mathbf{H})$ with the excitons may be represented by the operators of the effective spins $J = 1$:

$$\mathcal{H}_1 = \frac{\omega_1}{2}\left(J_+ e^{-i\omega t} + J_- e^{i\omega t}\right),$$

where ω is the frequency of the alternative field, $\omega_1 = \gamma H_1$, and γ is the gyromagnetic ratio for the exciton in the $1S(\Gamma_5)$ state.

In the presence of the magnetic field $\mathbf{H} \parallel [001]$ the reduction of the symmetry group $T_d \Rightarrow S_4$ takes place, while $\Gamma_5 = \Gamma_2 + \Gamma_3 + \Gamma_4$ (for the irreducible representations Heine's designations are used [127]).

The external alternative magnetic field leads to the appearance of the radio-frequency coherence in the system of Zeeman sublevels $1S(\Gamma_2, \Gamma_3, \Gamma_4)$, and for the time momentum t_0 the wave function of the system is described by the expression

$$|\Psi(t_0)\rangle = \sum_{\mu=2}^{4} a_{\Gamma_\mu}(t_0)|\Gamma_\mu\rangle,$$

and for the polarization Γ_3 of the magnetic component of the radio-frequency field $a_{\Gamma_4}(t_0) = 0$.

By means of the unitary operator

$$U(\omega, t) = E + iJ_z \sin \omega t + J_z^2(\cos \omega t - 1),$$

where E is the unit 3×3 matrix, we shall pass to the rotating system of coordinates for the excitons in the state $1S(\Gamma_2, \Gamma_3, \Gamma_4)$ of relative motion, but the excitons in the state $2P(\Gamma_1, \Gamma_3, \Gamma_4)$ of the relative motion and the radiation field will be considered in the representation of the interaction. Then in the case of polarization Γ_3 of the light that excites the transitions $1S(\Gamma_2, \Gamma_3, \Gamma_4) \rightarrow 2P(\Gamma_1, \Gamma_3, \Gamma_4)$ we obtain the following equations of motion for the components of the density matrix of excitons from the band $1S(\Gamma_2, \Gamma_3, \Gamma_4)$:

$$\frac{d\rho_{\Gamma_2\Gamma_2}}{dt} = -\frac{1}{T_p}|D_{23}^0|^2(1 - |D_{23}|^2)\rho_{\Gamma_2\Gamma_2} + \frac{1}{T_p}|D_{41}^0|^2|D_{21}|^2\rho_{\Gamma_4\Gamma_4},$$

$$\frac{d\rho_{\Gamma_4\Gamma_4}}{dt} = -\frac{1}{T_p}|D_{41}^0|^2(1 - |D_{41}|^2)\rho_{\Gamma_4\Gamma_4}, \qquad (4.36)$$

$$\frac{d\rho_{\Gamma_3\Gamma_2}}{dt} = \left[-\frac{1}{2T_p} + i|D_{23}^0|^2(\delta - \varepsilon)\right]\rho_{\Gamma_3\Gamma_2},$$

where

$$D_{\mu m} = \langle \Gamma_\mu | \mathbf{e}_\lambda \mathbf{D} | \Gamma_m \rangle, \qquad D^0_{\mu m} = \langle \Gamma_\mu | \mathbf{e}_{\lambda_0} \mathbf{D} | \Gamma_m \rangle,$$

$\mathbf{D}$ is the operator of the electric dipole moment of the exciton in dimensionless units, T_p is the average interval of time separating the acts of the consequent absorption of two photons by the same exciton, $\hbar\delta$ is the shift of exciton level $1S(\Gamma_5)$ due to the exciton–photon interaction, and $\varepsilon = \omega - \omega_{\Gamma_3\Gamma_2}$ is the deviation from the magnetic resonance frequency.

If the radio-frequency coherence was introduced to the time momentum $t = t_0$ in the system of Zeeman sublevels $1S(\Gamma_2, \Gamma_3)$, then the solutions of (4.36) have the form

$$\rho_{\Gamma_3\Gamma_2} = \frac{\omega_1}{\alpha^2}\left(\varepsilon \sin\frac{1}{2}\alpha t_0 - \frac{i\alpha}{2}\sin\alpha t_0\right)\exp\left\{i\left(\tilde{\Omega} + \frac{i}{2T_p}\right)t\right\},$$

$$\rho_{\Gamma_2\Gamma_2} = \frac{1}{|D^0_{41}|^2(1 - |D_{41}|^2) + |D^0_{23}|^2(|D_{23}|^2 - 1)}$$

$$\times \left\{\left(\frac{\varepsilon^2}{\alpha^2}\sin^2\frac{1}{2}\alpha t_0 + \cos^2\frac{1}{2}\alpha t_0\right)\left[|D^0_{23}|^2(|D_{23}|^2 - 1)\right.\right.$$

$$\left.+ |D^0_{41}|^2(1 - |D_{41}|^2)\right] + |D_{21}|^2|D^0_{41}|^2\right\}$$

$$\times \exp\left[-\frac{1}{T_p}|D^0_{41}|^2(1 - |D_{41}|^2)t\right],$$

$$\rho_{\Gamma_4\Gamma_4} = 0,$$

where

$$\alpha = \sqrt{\varepsilon^2 + \omega_1^2}, \qquad \tilde{\Omega} = |D^0_{23}|^2(\delta - \varepsilon).$$

The introduction of the radio-frequency coherence in the system of Zeeman sublevels $1S(\Gamma_2, \Gamma_3)$ leads to amplitude modulation of the luminescence $2P(\Gamma_1, \Gamma_3, \Gamma_4) \to 1S(\Gamma_2, \Gamma_3, \Gamma_4)$. The depth of the modulation also depends, in particular, on the frequency of the external alternative magnetic field and on the distance between the Zeeman components Γ_3 and Γ_4 of the exciton in $1S$ states:

$$\mathcal{I}_{2P\to 1S} \sim \frac{1}{\tau T_p} \cdot \frac{1}{\tau^{-2} + \omega^2} \cdot |D^0_{23}|^2|D_{23}|^2$$

$$\times \exp\left\{-\frac{t}{2T_p}\right\}\left\{\frac{\omega_1}{\alpha}\left(\frac{1}{2\tau}\sin\alpha t_0 - \frac{\varepsilon\omega}{\alpha}\sin\frac{1}{2}\alpha t_0\right)\right.$$

$$\left.\times \sin\Omega t + \frac{\omega_1}{\alpha}\left(\frac{\varepsilon}{\alpha\tau}\sin\frac{1}{2}\alpha t_0 + \frac{\omega}{2}\sin\alpha t_0\right)\cos\Omega t\right\},$$

$$\Omega = \tilde{\Omega} - \omega. \tag{4.37}$$

Expression (4.37) is true for the intensity of the exciton luminescence in the infrared region of the spectrum in the presence of the alternative magnetic field of the resonance frequency, if along with the specific assumptions at the beginning of this section, the inequalities $\tau \ll T_p$, $\omega_{1S(\Gamma_5)2P(\Gamma_4)} \gg \tau^{-1}$, $\omega_{\Gamma_2\Gamma_3} > \gamma H_1$, and $T_p < \tau_s$ (τ_s is the lifetime of the exciton in the ground $1S(\Gamma_5)$ state) are applied.

As seen from formula (4.37), the circularly polarized luminescence relative to the magnetic field being modulated on the intensity is damped exponentially with time. Since Zeeman exciton subbands are broadened due to exciton–phonon interaction, and direct detection of the magnetic resonance is difficult, an optical method of detecting such a resonance, based on the effect of the modulation of the circularly polarized exciton luminescence with double radio–optical resonance, is of interest. However, the short lifetime of the exciton in the ground and excited states means that the optical pumping cycle discussed above will probably present experimental difficulties.

For the first time the magnetic resonance of excitons in semiconductors was optically detected by Morigaki, Dawson, and Cavenett [414], who observed the circular dichroism of the optical radiation in the region of the magnetic resonance of triplet excitons bound on ionized donors in GaSe crystal. They determined the resonance by using the orientation of the magnetic field parallel to the c axis of the crystal. The spectrum of the magnetic resonance of the excitons in this case is described by the spin Hamiltonian of axial symmetry with the parameters $g_{\text{ex}\|} = 1.85 \pm 0.03$ and $D = 0.11 \pm 0.04$ cm^{-1}. Radio–optical double resonance (RODOR) on the excitons was also observed in the semiconductor LiYF$_4$ [415]. Thus, with the appearance of the first experimental works in the field of RODOR on excitons in semiconductors, this method has become practically important for the investigation of exciton spin structure [416].

Chapter 5
Investigation of Excitons by NMR Spectroscopy Methods

The excitons in crystals may influence the nuclear magnetic resonance (NMR) in all those cases when hyperfine interaction of the charge carriers forming the exciton with nuclear spins is essential. The modulation by the vibrations of the lattice of this interaction causes an influence of the excitons on the nuclear spin relaxation rate, but local magnetic fields that are created by excitons on the nuclei lead to energy shifts of the nuclear spin levels and, thus, to shifts of the NMR frequencies (the exciton Knight shift). These shifts essentially depend on the type of statistics which the exciton gas is subject to, a fact that may be used practically for diagnostic purposes of the exciton gas. The influence of excitons on NMR in semiconductors is considered in Sects. 5.1–5.3.

In Sect. 5.4 it is shown that the nuclear spin relaxation rate is also influenced by orthobiexcitons. The biexciton contribution to the nuclear spin–lattice relaxation rate may be separated from the exciton one, and the influence of the excitons on nuclear spin relaxation, in turn, may be separated from the influence of the free charge carriers. Thus, the states of the free carriers, excitons, and biexcitons may influence the relaxation of the nuclear spins differently. This fact may be used to investigate these states by NMR methods.

Along with investigating the influence of the magnetic fields created by the excitons and biexcitons on the positions of the NMR lines and the nuclear spin relaxation rates, in Sect. 5.5 a method of narrowing the exciton light absorption lines broadened due to exciton–exciton interaction at high levels of optical excitation of crystals is discussed. This method is similar to the method of narrowing NMR lines under the interaction of multipulse sequences known in high-resolution NMR spectroscopy [68, 69].

The results of this chapter are published in [170, 378, 417–420].

5.1 Relaxation of Nuclear Spin via Triplet Excitons

In semiconductors nuclear spin relaxation via free charge carriers is due to Fermi hyperfine interaction of s electrons in the conduction band with the spins of the nu-

I. Geru, D. Suter, *Resonance Effects of Excitons and Electrons*,
Lecture Notes in Physics 869, DOI 10.1007/978-3-642-35807-4_5,
© Springer-Verlag Berlin Heidelberg 2013

clei [29]. When the free semiconductor carriers move in crossed electric and magnetic fields, the contact hyperfine interaction also influences the nuclear spin–lattice relaxation rate [421]. In semiconductors and dielectrics, if a sufficient number of electron–hole pairs is coupled in the excitons and the populated bands have orbital or spin degeneration, then at high density of the exciton gas we should expect an effective relaxation of the nuclear spin via excitons.

The nondiagonal part of the operator of the hyperfine interaction of electrons and holes in the second quantization representation using the model for a crystal with simple energy bands has the form

$$\mathcal{H}'_{IS} = \frac{4\pi}{3}\gamma_n\hbar^2 \sum_{\mathbf{k},\mathbf{k}'}\left[\gamma_e\mathcal{U}^*_{\mathbf{k}',e}(0)\mathcal{U}_{\mathbf{k},e}(0)\left(I_-a^+_{\mathbf{k}',\uparrow}a_{\mathbf{k},\downarrow} + I_+a^+_{\mathbf{k}',\downarrow}a_{\mathbf{k},\uparrow}\right)\right.$$

$$\left. + \gamma_h\mathcal{U}^*_{\mathbf{k}',h}(0)\mathcal{U}_{\mathbf{k},h}(0)\left(I_-b^+_{\mathbf{k}',\uparrow}b_{\mathbf{k},\downarrow} + I_+b^+_{\mathbf{k}',\downarrow}b_{\mathbf{k},\uparrow}\right)\right], \qquad (5.1)$$

where $I_\pm = I_x \pm iI_y$ (I_x and I_y are the nuclear spin projection operators), $\mathcal{U}_{\mathbf{k},e}(0)$ and $\mathcal{U}_{\mathbf{k},h}(0)$ are the periodic parts of the Bloch electron functions in the conduction band and the hole in the valence band at the location of the nucleus, $a^+_{\mathbf{k},s}$ and $a_{\mathbf{k},s}$ are the creation and destruction operators of the electron in the conduction band with wave vector $\mathbf{k}$ and spin projection s, $b^+_{\mathbf{k}',s'}$ and $b_{\mathbf{k}',s'}$ are the creation and destruction operators of the hole in the valence band with wave vector $\mathbf{k}'$ and spin projection s', and γ_e, γ_h, and γ_n are the gyromagnetic ratios for the electron, hole, and nucleus, respectively. Here and throughout the chapter, the volume of the system is assumed to be $\mathcal{V} = 1$.

The states of the triplet exciton are determined by the expressions

$$|n\mathbf{K}, 1\rangle = \sum_{\mathbf{p}}\varphi_n(\mathbf{p} - \beta\mathbf{K})a^+_{\mathbf{K}-\mathbf{p},\uparrow}b^+_{\mathbf{p},\uparrow}|0\rangle,$$

$$|n\mathbf{K}, 0\rangle = \frac{1}{\sqrt{2}}\sum_{\mathbf{p}}\varphi_n(\mathbf{p} - \beta\mathbf{K})\left(a^+_{\mathbf{K}-\mathbf{p},\uparrow}b^+_{\mathbf{p},\downarrow} + a^+_{\mathbf{K}-\mathbf{p},\downarrow}b^+_{\mathbf{p},\uparrow}\right)|0\rangle, \qquad (5.2)$$

$$|n\mathbf{K}, -1\rangle = \sum_{\mathbf{p}}\varphi_n(\mathbf{p} - \beta\mathbf{K})a^+_{\mathbf{K}-\mathbf{p},\downarrow}b^+_{\mathbf{p},\downarrow}|0\rangle.$$

Here $\varphi_n(\mathbf{p} - \beta\mathbf{K})$ is the Fourier transform of the function of relative motion of the electron and hole in the exciton, $\mathbf{K}$ is the exciton wave vector, $\beta = m_h/m_{ex}$, $m_{ex} = m_e + m_h$, and m_e and m_h are the effective masses of the electron and hole.

We shall consider the nonelastic scattering of Wannier–Mott triplet excitons with spin reorientation in the external magnetic field due to the condition that the energy of the nuclear spin–spin interaction is much less than the Zeeman energy of the nucleus. In the weak-coupling approximation of excitons with the nucleus, for the full probability of the transition in the unit of time between the spin states $|IM_I\rangle$ and $|IM'_I\rangle$ of the nucleus in a Maxwellian distribution of the excitons on quasi-momenta, we obtain

$$
\begin{aligned}
W_{M_I M_I'} &= \frac{8}{9\pi^2} \Big\{ \big[I(I+1) - M_I'(M_I'+1) \big] \delta_{M_I, M_I'+1} \\
&\quad + \big[I(I+1) - M_I'(M_I'-1) \big] \delta_{M_I, M_I'-1} \Big\} \\
&\quad \times \gamma_n^2 \left(\frac{m_{ex}}{\hbar} \right)^3 \Delta k_0 T F(\sigma, T) \exp\left\{ \frac{\mu - E_g + G}{k_0 T} \right\} ch \frac{\Delta}{k_0 T},
\end{aligned}
$$

where

$$
\begin{aligned}
F(\sigma, T) &= \frac{\gamma_e^2 \eta_e^2}{1 + \frac{4}{\sigma} \frac{k_0 T}{G}} \mathcal{K}_1\left[\frac{\Delta}{2k_0 T} \left(1 + \frac{4}{\sigma} \frac{k_0 T}{G} \right) \right] \\
&\quad + \frac{\gamma_h^2 \eta_h^2}{1 + 4\sigma \frac{k_0 T}{G}} \mathcal{K}_1\left[\frac{\Delta}{2k_0 T} \left(1 + 4\sigma \frac{k_0 T}{G} \right) \right] \\
&\quad - \frac{\gamma_e \gamma_h \eta_e \eta_n}{1 + 2\frac{1+\sigma^2}{\sigma} \frac{k_0 T}{G}} \mathcal{K}_1\left[\frac{\Delta}{2k_0 T} \left(1 + 2\frac{1+\sigma^2}{\sigma} \frac{k_0 T}{G} \right) \right].
\end{aligned}
\tag{5.3}
$$

Here $\sigma = m_e/m_h$, $\Delta = \gamma_{ex} \hbar H$ is Zeeman splitting of the triplet exciton, E_g is the semiconductor energy gap, G is the potential of the exciton ionization, μ is the chemical potential of the excitons, and $\mathcal{K}_1(z)$ is a modified Bessel function of the third kind (McDonald's function). We assume that $|\mathcal{U}_{\varkappa, j}(0)|^2$ has the value η_j, not depending on the energy of the electron (hole) (see [29]). We also consider the excitons to be in a state of relative motion of $1s$ type, and the Fourier transform of this state is approximated by the exponential function [422]

$$
\varphi(p) \approx 8\sqrt{\pi} a_{ex}^{3/2} \exp\{-2a_{ex}^2 p^2\},
\tag{5.4}
$$

where a_{ex} is the exciton Bohr radius. The approximation (5.4) is true due to the condition $a_{ex}^2 p^2 \ll 1$, which is well satisfied for thermal excitons in semiconductors. For large values of p the function (5.4) tends to zero more rapidly than the exact expression for the Fourier transform of the function of relative motion of the electron and hole in the exciton. The use of approximation (5.4) finally leads to an insignificant increase in the time relaxation value.

In the approximation of high spin and lattice temperatures, taking into account that

$$
\exp\left\{ \frac{\mu - E_g + G}{k_0 T} \right\} = \left(\frac{2\pi}{m_{ex} k_0 T} \right)^{3/2} \frac{\hbar^3}{1 + 2ch\frac{\Delta}{k_0 T}} n_{ex},
$$

where n_{ex} is the exciton concentration, we obtain the expression for the longitudinal relaxation time of the nuclear spin via the orthoexcitons [378]:

$$
\frac{1}{T_{1,ex}} = \frac{32}{9} \sqrt{\frac{2}{\pi k_0 T}} \gamma_n^2 m_{ex}^{3/2} \Delta \frac{ch\frac{\Delta}{2k_0 T}}{1 + 2ch\frac{\Delta}{k_0 T}} n_{ex} F(\sigma, T).
\tag{5.5}
$$

If the inequalities

$$1 + 4\xi \frac{k_0 T}{G} \ll \frac{2k_0 T}{\Delta}; \quad \xi = \sigma, \frac{1}{\sigma}, \frac{1+\sigma^2}{2\sigma} \tag{5.6}$$

are satisfied, then for the function $F(\sigma, T)$ from (5.5) we have

$$F(\sigma, T) = \frac{2k_0 T}{\Delta} \left[\frac{\gamma_e^2 \eta_e^2}{(1 + \frac{4}{\sigma} \frac{k_0 T}{G})^2} + \frac{\gamma_h^2 \eta_h^2}{(1 + 4\sigma \frac{k_0 T}{G})^2} - \frac{\gamma_e \gamma_h \eta_e \eta_h}{(1 + 2\frac{1+\sigma^2}{\sigma} \frac{k_0 T}{G})^2} \right]. \tag{5.7}$$

The range of temperature variation in (5.5)–(5.7) is limited from above by the condition $k_0 T < G$, which eliminates the thermal decay processes of excitons. In another extreme case (the inverted inequalities of (5.6)) the function $F(\sigma, T)$ contains an exponentially small multiplier $\sim \exp[-\Delta/2k_0 T]$, which leads to a sharp reduction in the relaxation rate. Formula (5.5) is not applicable when $\Delta > G$, because in this case a significant change of the exciton wave functions in the magnetic field takes place, and in calculating the transition probabilities it is not possible to proceed from the free exciton states [86, 378].

We note that the wave functions (5.2) describe the triplet states of "direct" excitons; therefore, formula (5.5) determines the relaxation rate of the nuclear spin via the direct triplet excitons.

Let us evaluate the nuclear spin relaxation rate via excitons in the CuCl crystal [170, 378] in which the extremes of the lowest conduction band (Γ_1) and the upper valence band (Γ_{15}) correspond to the point $\mathbf{k} = 0$. The band Γ_{15} is split into Γ_7 and Γ_8 bands due to the spin–orbital interactions. During the binding of the electron from the band Γ_1 (Γ_6, taking the spin into account) with the hole from the subband Γ_7, a $1s$-type exciton ($\Gamma_6 \otimes \Gamma_7 = \Gamma_2 \oplus \Gamma_{15}$) is formed. The lowest level is level Γ_2 of the paraexciton. The distance between level Γ_2 and level Γ_{15} of the orthoexciton is equal to 50 cm^{-1} at $T = 4.2$ K [170, 378]; therefore, we can ignore state Γ_2 during the consideration of Zeeman splitting of level Γ_{15}. The results of the optical spectra of reflection and Röentgen spectra of photoemission show that, in the formation of the hole band Γ_7, the electrons of s type do not participate [378]. Therefore, in expression (5.5) only the first term from (5.3) contributes. Figure 5.1 shows the temperature dependence of the inverse spin–lattice relaxation time of the ^{63}Cu nuclei in CuCl at $n_{ex} = 10^{19}$ cm^{-3},[1] $m_e = 0.4 m_0$ (m_0 is the mass of the free electron), $m_h = 20 m_0$, $\gamma_e = 1.76 \cdot 10^7$ Oe^{-1} s^{-1}, $\gamma_n = 7.3 \cdot 10^3$ Oe^{-1} s^{-1}, and $\eta_e \sim 10^2$ [170, 378].

From the energy conservation law, transitions with the selection rules $\Delta M_s = +1$ ($M_s = -1, 0$) and $\Delta M_s = -1$ ($M_s = +1, 0$) are forbidden for excitons with wave vectors that are enclosed in a sphere of radius $k_c = \hbar^{-1}(2m\Delta)^{1/2}$, and the spin projections are equal to M_s and $M_s - 1$, respectively. Therefore, at $T \to 0$ the relaxation rate is $T_{1,ex}^{-1} \to 0$. The shape of the curve $T_{1,ex}^{-1}(T)$ is determined by the

---

[1] According to [423], the maximal concentration of excitons in the CuCl crystal at laser excitation may be $\sim 4 \cdot 10^{20}$ cm^{-3}.

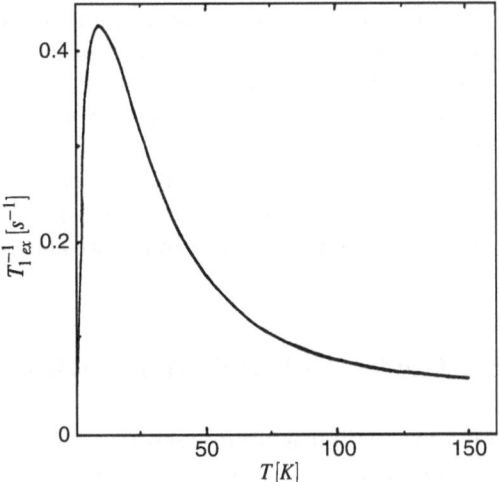

Fig. 5.1 Temperature dependence of the nuclear spin relaxation rate of ^{63}Cu nuclei via excitons in the CuCl crystal

function of the relative motion of electrons and holes in the exciton and the function of the distribution of the excitons in the band. The position of the maximum and the shape of the curve $T_{1,ex}^{-1}(T)$ is not changed practically with a change of the magnetic field H intensity in the interval $5 \cdot 10^2$–$5 \cdot 10^4$ Oe.

In the work [378] it was shown that in the CuCl crystal at $n_{ex} \sim 10^{19}$ cm^{-3} the concentration of the free charge carriers, affected by nonelastic exciton–exciton collisions, is $n_e \lesssim 10^{16}$ cm^{-3}. In the range of temperatures corresponding to the maximum of the curve $T_{1,ex}^{-1}(T)$ in Fig. 5.1, the electron gas with the concentration $n_e \sim 10^{16}$ cm^{-3} is not degenerated, and the relaxation rate of the nuclear spin conditioned by free electrons is determined by the formula

$$\frac{1}{T_{1,e}} = \frac{16}{9}\sqrt{\frac{2\pi}{k_0 T}}\gamma_e^2 \gamma_n^2 \eta_e^2 m_e^{3/2} \Delta_e n_e \frac{ch\frac{\Delta_e}{2k_0 T}}{ch\frac{\Delta_e}{k_0 T}} K_1\left(\frac{\Delta_e}{2k_0 T}\right), \qquad (5.8)$$

where Δ_e is electron Zeeman splitting.

At $\Delta_e \ll k_0 T$ formula (5.8) is transformed into the usual expression for $T_{1,ex}^{-1}$ corresponding to nuclear spin relaxation via free electrons [29, 378]. The estimation according to formula (5.8) at $\Delta_e = 0.272$ cm^{-1}, $n_e = 10^{16}$ cm^{-3}, and $T = 7$ K gives $T_{1,e} \sim 3 \cdot 10^5$ s. A comparison of this value for $T_{1,e}$ with the magnitude found in Fig. 5.1 shows that the influence of free carriers compared with that of excitons on the processes of nuclear spin–lattice relaxation can be ignored. At high exciton concentrations the spin–lattice relaxation of the nuclei via excitons becomes effective enough.

The large spin relaxation rate of the nucleus via excitons is connected with the short-range nature of the potential of the contact hyperfine interaction and is conditioned by the high values of the Bloch function modules of the free carriers at the location of the nucleus. When we neglect the "splashes" of wave functions of the free carriers in the nodes of the crystal lattice, the rate of nuclear spin relaxation via excitons is reduced by four orders of magnitude [420].

The temperature dependence with a sharp maximum for the rate of nuclear spin relaxation via excitons, as obtained in the work [378], was not previously known for other mechanisms of nuclear spin relaxation in solids. A similar anomaly in the temperature dependence is also characteristic for the spin relaxation rate of paramagnetic centers via excitons for short-range exchange scattering of the excitons on paramagnetic impurities with spin reorientation (see Chap. 6).

5.2 Exciton Knight Shift of NMR Lines

It is known [29] that NMR is sensitive to the presence of paramagnetic centers and free charge carriers creating an additional local field in the location of the nucleus. In semiconductors or dielectrics, if a sufficient number of electron–hole pairs are coupled into excitons and the lowest exciton band has an orbital or spin degeneracy, then the effect of the exciton Knight shift of the NMR lines can possibly occur, the magnitude of which is proportional to the concentration of excitons. In the model for a crystal with simple energy bands, the diagonal part of the operator of the hyperfine interaction of the electrons and holes with the nucleus in the second quantization representation has the form

$$\mathcal{H}^0_{IS} = \frac{4\pi}{3}\gamma_n\hbar^2 I_z \sum_{\mathbf{k},\mathbf{k}'}\left[\gamma_e U^*_{\mathbf{k}',e}(0)U_{\mathbf{k},e}(0)\left(a^+_{\mathbf{k}',\uparrow}a_{\mathbf{k},\uparrow} - a^+_{\mathbf{k}',\downarrow}a_{\mathbf{k},\downarrow}\right)\right.$$
$$\left. + \gamma_h U^*_{\mathbf{k}',h}(0)U_{\mathbf{k},h}(0)\left(b^+_{\mathbf{k}',\uparrow}b_{\mathbf{k},\uparrow} - b^+_{\mathbf{k}',\downarrow}b_{\mathbf{k},\downarrow}\right)\right],$$

where I_z is the operator of the zth projection of the nuclear spin, and the other designations coincide with those of formula (5.1).

The correction in the spin energy of the nucleus due to hyperfine interaction of excitons with the nucleus in first order perturbation theory (exciton–nuclear interaction is assumed to be weak) is described by the expression

$$\overline{\mathcal{H}^0}_{IS} = -\frac{8\pi}{3}\gamma_n\hbar^2 I_z(\gamma_e\eta_e + \gamma_h\eta_h)\frac{sh(\Delta/k_0 T)}{1 + 2ch(\Delta/k_0 T)}n_{ex} \tag{5.9}$$

Here the averaging is performed according to the probability of finding the orthoexcitons at the given temperature in states with different wave vectors and projections of the spin. The wave functions (5.2) and Maxwellian distribution of excitons on quasi-momenta were used when averaging. It was use also the usual assumption that $|U_{\mathbf{k},e(h)}(0)|^2$ is equal to the value $\eta_{e(h)}$, which does not depend on the electron (hole) energy [29, 417]. In this approximation the Knight shift caused by the excitons does not depend on the type of wave function of the relative motion of the electron and hole in the exciton.

The formula (5.9) can be presented as

$$\overline{\mathcal{H}_{IS}^0} = -\gamma_n \hbar I_z \Delta H_{ex},$$

where the additional magnetic field ΔH_{ex} created by the triplet excitons on the nucleus coincides with the external magnetic field H in direction and is determined by the expression

$$\Delta H_{ex} = \frac{8\pi}{3}\hbar(\gamma_e \eta_e + \gamma_h \eta_h)\frac{sh(\Delta_0/k_0 T)}{1 + 2ch(\Delta_0/k_0 T)} n_{ex}, \qquad (5.10)$$

where $\Delta_0 = \gamma_{ex}\hbar H$ is Zeeman splitting of the triplet exciton.

To estimate the exciton influence on the position of the NMR line according to formula (5.10), the ^{63}Cu NMR frequency shift in the CuCl crystal affected by the orthoexcitons from band Γ_{15} was found. Since the conduction band Γ_6 in CuCl is formed from the states of s electrons, then $\eta_e \sim 10^2$ [170, 417], and, as in the formation of the hole subband Γ_7, the valence electrons of s type do not participate (see Sect. 5.1), thus $\eta_h = 0$. The temperature dependence of the exciton Knight shift of the NMR line for the ^{63}Cu nuclei in the CuCl crystal at $\Delta_0 = 0.136$ cm^{-1} and the other values of the parameters given in Sect. 5.1 is presented in Fig. 5.2, where the exciton gas phase corresponds to $T > 3$ K. At an exciton concentration of $n_{ex} = 10^{19}$ cm^{-3}, the average distance between excitons has a magnitude of $\sim$50 Å, which is much more than the radius of a $1s$-type exciton in the CuCl crystal ($a_{ex} = 7$ Å); therefore, the excitons still do not lose their individuality.

Because of the large magnitude of the band gap of the CuCl crystal, the influence of free charge carriers on the NMR frequency shift may be ignored. The possible contribution from the free electrons created in the process of laser pumping may be eliminated by applying the resonance method for creating excitons. We can also ignore the contribution to the Knight shift of the NMR line from free electron–hole pairs generated by exciton–exciton collisions. We determine the concentration of such n_{e-h} pairs from the system of kinetic equations

$$\frac{dn_{ex}}{dt} = G - \frac{n_{ex}}{\tau} - \alpha n_{ex}^2,$$

$$\frac{dn_{e-h}}{dt} = \alpha n_{ex}^2 - \frac{n_{e-h}}{\tau_1},$$

where G is the concentration of excitons generated per unit of time during pumping, τ is the lifetime of the exciton, τ_1 is the lifetime of the electron–hole pair with respect to the binding in the exciton, and α is the constant of the process of exciton nonelastic pair collisions. Under stationary conditions, for n_{e-h} we have

$$n_{e-h} = \frac{1}{2}\frac{\tau_1}{\alpha\tau^2}\left(1 + G\alpha\tau^2 - \sqrt{1 + 4G\alpha\tau^2}\right). \qquad (5.11)$$

For a pumping concentration of $G = 2 \cdot 10^{28}$ cm^{-3} s^{-1} corresponding to an exciton concentration of $n_{ex} = 10^{19}$ cm^{-3} in CuCl, $\alpha = 10^{-10}$ cm^3 s^{-1}, $\tau \lesssim 10^{-9}$ s, and $\tau_1 = 10^{-12}$ s [97, 171, 417], according to (5.11) we find $n_{e-h} \sim 10^{16}$ cm^{-3}. In this case the estimation of the contribution of the free charge carriers to the Knight shift of the NMR line is $\Delta H_e \sim 10^{-2}$ Oe at $T \sim 4$ K, which is much less than the exciton Knight shift.

5.3 NMR Evidence of Bose–Einstein Condensation of Excitons

In Sect. 1.4 it was shown that the optical indication of the exciton (biexciton) Bose–Einstein condensation phenomenon at high levels of the excitation of the crystals in practice may be insufficient to recognize this phenomenon. The reason is that there are different causes which may lead to a narrow line of radiation in a short-wavelength part of the exciton luminescence spectrum at a frequency near the transition frequency of the exciton state.

One of the causes hindering the recognition of the phase transition of the exciton gas to the state of Bose–Einstein condensation by means of optical methods is the narrowing of exciton radiation lines at laser generation. Another cause is the sufficiently strong narrowing of the radiation line during the decay of a biexciton in which one exciton annihilates radiatively and the other remains in the polariton state. In the latter case in the emission spectrum of the exciton molecule state the polariton doublet $L-T$ [255] appears, and one of the components of the doublet is narrowed in comparison with the other. The spectral positions of these components correspond to a short-wave narrow peak with a wider long-wave wing.

All of this hinders the experimental investigation of the phenomenon of Bose–Einstein condensation of excitons and biexcitons on the background of other phenomena, influenced by the interaction of the excitons between themselves and with the field of radiation. Therefore, along with an optical indication of the phenomenon of Bose–Einstein exciton condensation, we want to find another feature of this phenomenon, on the basis of which we can independently determine the presence (or absence) of the transition of excitons to the Bose–Einstein condensed phase in the quasi-momenta space. Such a new indication of the existence of exciton Bose–Einstein condensation, as will be shown later, is the sharp temperature dependence of the exciton Knight shift of the NMR line near the critical point of phase transition. This occurs because the magnitude and temperature dependence of the exciton Knight shift of the NMR line essentially depends on the type of distribution of the excitons on quasi-momenta.

We shall consider the general case when some of the excitons on a crystal are in the gas phase, and the others are in the Bose condensed state. Let us assume that Bose–Einstein condensation takes place in the state with $\mathbf{K} = 0$ for the excitons from the lowest subband of the Zeeman triplet. Then the exciton distribution function has the form

$$f(\varepsilon_{\mathbf{K}}, M_s) = n_{\mathbf{K}}^{ex} \delta_{\mathbf{K},0} \delta_{M_s,-1} + \left[\xi^{-1} \exp\left(\frac{\varepsilon_{\mathbf{K}} + \Delta_0 M_s}{k_0 T}\right) - 1\right]^{-1}. \qquad (5.12)$$

Here $\xi = \exp(\frac{\mu}{k_0 T})$, μ is the chemical potential, and $n_{\mathrm{K}}^{\mathrm{ex}}$ is the concentration of excitons in the condensate.

Taking into account the distribution function (5.12), we find that the additional magnetic field on the nucleus ΔH_{tot}, due to both the exciton Bose condensate and the overcondensed excitons, is determined by the formula

$$\Delta H_{\mathrm{tot}} = \frac{4\pi}{3} \hbar (\gamma_e \eta_e + \gamma_h \eta_h) \left\{ n_{\mathrm{ex}} \left[1 - \left(\frac{T}{T_c} \right)^{3/2} \right] \right.$$
$$\left. - \frac{1}{\hbar^3} \left(\frac{m k_0 T}{2\pi} \right)^{3/2} \left[g_{3/2} \left(\xi e^{-\frac{\Delta_0}{k_0 T}} \right) - g_{3/2} \left(\xi e^{\frac{\Delta_0}{k_0 T}} \right) \right] \right\}, \qquad (5.13)$$

where $g_p(x)$ is the Riemann zeta function [424], n_{ex} is the total exciton concentration (the number of condensed and overcondensed excitations on the unit of volume), T_c is the critical temperature, which satisfies the equation

$$T_c = \frac{2\pi \hbar^2}{k_0 m} n_{\mathrm{ex}}^{2/3} \left[g_{3/2} \left(e^{-2\frac{\Delta_0}{k_0 T_c}} \right) + g_{3/2} \left(e^{-\frac{\Delta_0}{k_0 T_c}} \right) + g_{3/2}(1) \right]^{-2/3}, \qquad (5.14)$$

and

$$T_c(\Delta_0 \to \infty) / T_c(\Delta_0 \to 0) = 3^{2/3},$$

which indicates the growth of the critical temperature with increasing external magnetic field H.

In the range of values of the argument $x \gg 1$ for the function $g_{3/2}(x)$, the following approximate expression is true:

$$g_{3/2}(e^{-x}) \approx e^{-x}, \quad e^{-x} \ll 1. \qquad (5.15)$$

Therefore, in the case when all the excitons are in the gaseous phase at temperatures far from the critical temperature $(T \gg T_c)$, expression (5.13) for the exciton Knight shift ΔH_{tot}, taking into account approximation (5.15), is transformed into (5.10).

In the condensed phase $(T \lesssim T_c)$ in the case of weak magnetic fields $(\Delta_0/k_0 T \ll 1)$, using the approximate expansion of the zeta function [171]

$$g_{3/2}(e^{-x}) \approx g_{3/2}(1) - 2\sqrt{\pi x},$$

for the Knight shift conditioned by the condensed and overcondensed excitons we obtain

$$\Delta H_{\mathrm{tot}} = \frac{4\pi}{3} \hbar (\gamma_e \eta_e + \gamma_h \eta_h) \left\{ n_{\mathrm{ex}} \left[1 - \left(\frac{T}{T_c} \right)^{3/2} \right] + \frac{m^{3/2}}{\pi \hbar^3} k_0 T \Delta_0^{1/2} \right\}. \qquad (5.16)$$

As we see from (5.10), (5.13), and (5.16), the passage through the critical point is characterized by a qualitative change of the dependence of the exciton Knight shift on the magnetic field and temperature. The numerical estimations show that

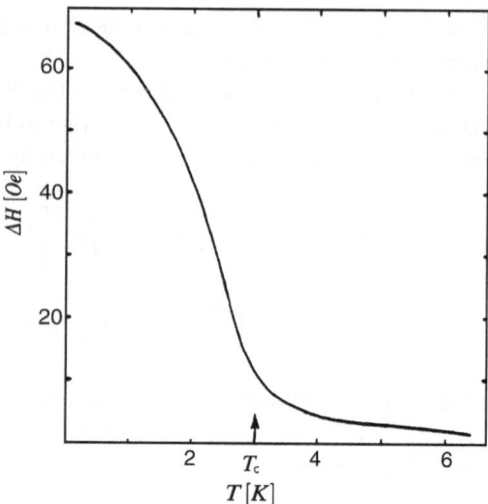

Fig. 5.2 Temperature dependence of the exciton Knight shift of ^{63}CuCl crystal near the critical temperature of phase transition to the state of exciton Bose–Einstein condensation

there is also a sharp increase in the Knight shift below the critical temperature T_c. In Fig. 5.2 the Bose-condensed exciton phase corresponds to the temperature range $T \leq 3$ K (calculated by the formula (5.15) for $n_{ex} = 10^{19}$ cm^{-3} in CuCl crystal at $\Delta_0 = 0.136$ cm^{-1}, the critical temperature is $T_c \approx 3$ K). As can be seen from Fig. 5.2, the critical temperature T_c which satisfies (5.14) corresponds to the inflection point in the temperature dependence of the exciton Knight shift. Below the critical temperature, the Knight shift is sharply increasing, which is a new feature of manifestation of the phase transition in a system of excitons (a radio-frequency feature of the phenomenon of Bose–Einstein condensation of excitons).

The sharp increase in the exciton Knight shift of the NMR line near T_c specifies the phase transition of excitons to the Bose-condensed state under the condition that one can ignore the contribution to the Knight shift due to free electron–hole pairs created in exciton–exciton collisions. In Sect. 5.2 the concentration of these pairs was estimated, and it was shown that the contribution of free carriers to the Knight shift of the NMR line compared with that of overcondensed excitons may be ignored. The free carrier contribution may be neglected even more in the case of the Knight shift of the NMR line affected by the exciton Bose-condensate.

Thus, it is of interest to perform experiments on evidentiating the exciton Bose-condensate by investigating the effects of the influence of excitons on NMR along with the exciton luminescence during Bose–Einstein condensation [171, 260]. These effects can be displayed at high levels of optical excitation of crystals, not only in the case of pure triplet spin exciton states, but also in those cases when the exciton bands are orbitally degenerate and the state of s electrons contributes to the wave functions of free charge carriers ($\eta_{e,h} \neq 0$).

The high value of the exciton Knight shift of the NMR line (as well as the large value of the rate of nuclear spin relaxation via excitons) is due to the short-range character of the potential of contact hyperfine interaction and is caused by large values of the Bloch function modules of free carriers in the location of the nucleus. When the splashes of the wave functions of the free charge carriers in the nodes of a crystal lattice are ignored, the exciton Knight shift is reduced by two orders of magnitude [420].

Note that a shift similar to the Knight shift is also characteristic for electron paramagnetic resonance (EPR) lines in the case of short-range exchange interaction of charge carriers coupled to excitons with paramagnetic centers. A sharp increase in the EPR line shift influenced by the Bose–Einstein condensation of the excitons may also be used to detect this phenomenon.

5.4   Relaxation of Nuclear Spin via Orthobiexcitons

The effects of excitons on the nuclear spin relaxation rate are displayed, as was shown in Sect. 5.2, for high exciton concentrations. However, in this case, the interaction between excitons is essential and, in particular, it can lead to biexciton formation. We will consider further the influence of biexcitons on nuclear spin relaxation.

The wave function of the ground state of a biexciton is antisymmetric relative to the permutation of the electrons (holes) comprising the excitons [166, 168]. Since the spins of electrons and holes in the biexciton are compensated, the biexcitons in the ground state do not influence the nuclear spin relaxation rate. However, relaxation of the nuclear spin via biexcitons is possible if the latter are in an excited state.

Let us consider the nonelastic scattering of the biexcitons on nuclei with nuclear spin reorientation in an external magnetic field. We shall neglect the processes of the transformation of orthobiexcitons into parabiexcitons during the scattering on the nuclei, since the value of exchange splitting of the biexciton is much higher than its Zeeman energy in the excited state (for example, in the CuCl crystal the distance between the ortho- and parabiexciton states is ~ 6.5 meV [203, 419], while $\varepsilon_1^{(0,1)} - \varepsilon_0^{(0,1)} = \varepsilon_0^{(0,1)} - \varepsilon_{-1}^{(0,1)} \simeq 0.035$ meV for the magnetic field $H \sim 3 \cdot 10^3$ Oe. Here the upper indices of ε correspond to the vibrational and rotational quantum numbers, and the lower ones correspond to the projections of the orthobiexciton spin). Then the reorientation of the nuclear spin due to the interaction with the biexcitons is affected by the processes of the interband scattering of orthobiexcitons on the nucleus with the selection rules $\Delta M_S = \pm 1$ (M_S is the projection of the orthobiexciton spin).

Let us suppose that the gyromagnetic ratios for the electrons (holes) forming the biexciton are similar. Then in the model for a crystal with simple energy bands, the nondiagonal part of the Hamiltonian of the contact hyperfine interaction of the nucleus with the electrons and holes composing the biexciton has the form (5.1). The wave functions of the orthobiexciton are determined by the formulas

$|1, 1\rangle$

$$= \frac{1}{2\sqrt{2}} \sum_{\mathbf{p},\mathbf{k}_n,\mathbf{k}_v} g(\mathbf{p})\varphi_1\left[\mathbf{k}_n + \beta\left(\mathbf{p} - \frac{1}{2}\mathbf{K}\right)\right]\varphi_2\left[\mathbf{k}_v - \beta\left(\mathbf{p} + \frac{1}{2}\mathbf{K}\right)\right]$$

$$\times \left(a^+_{\frac{\mathbf{K}}{2}-\mathbf{p}-\mathbf{k}_n,\uparrow} b^+_{\mathbf{k}_n,\downarrow} a^+_{\frac{\mathbf{K}}{2}+\mathbf{p}-\mathbf{k}_v,\uparrow} b^+_{\mathbf{k}_v,\uparrow} - a^+_{\frac{\mathbf{K}}{2}-\mathbf{p}-\mathbf{k}_n,\uparrow} b^+_{\mathbf{k}_n,\uparrow} a^+_{\frac{\mathbf{K}}{2}+\mathbf{p}-\mathbf{k}_v,\uparrow} b^+_{\mathbf{k}_v,\downarrow}\right)|0\rangle,$$

$|1, 0\rangle$

$$= \frac{1}{4} \sum_{\mathbf{p},\mathbf{k}_n,\mathbf{k}_v} g(\mathbf{p})\varphi_1\left[\mathbf{k}_n + \beta\left(\mathbf{p} - \frac{1}{2}\mathbf{K}\right)\right]\varphi_2\left[\mathbf{k}_v - \beta\left(\mathbf{p} + \frac{1}{2}\mathbf{K}\right)\right]$$

$$\times \left(a^+_{\frac{\mathbf{K}}{2}-\mathbf{p}-\mathbf{k}_n,\uparrow} b^+_{\mathbf{k}_n,\downarrow} a^+_{\frac{\mathbf{K}}{2}+\mathbf{p}-\mathbf{k}_v,\downarrow} b^+_{\mathbf{k}_v,\uparrow} - a^+_{\frac{\mathbf{K}}{2}-\mathbf{p}-\mathbf{K}_n,\uparrow} b^+_{\mathbf{k}_n,\uparrow} a^+_{\frac{\mathbf{K}}{2}+\mathbf{p}-\mathbf{k}_v,\downarrow} b^+_{\mathbf{k}_v,\downarrow}\right.$$

$$\left. - a^+_{\frac{\mathbf{K}}{2}-\mathbf{p}-\mathbf{k}_n,\downarrow} b^+_{\mathbf{k}_n,\uparrow} a^+_{\frac{\mathbf{K}}{2}+\mathbf{p}-\mathbf{k}_v,\uparrow} b^+_{\mathbf{k}_v,\downarrow} + a^+_{\frac{\mathbf{K}}{2}-\mathbf{p}-\mathbf{k}_n,\downarrow} b^+_{\mathbf{k}_n,\downarrow} a^+_{\frac{\mathbf{K}}{2}+\mathbf{p}-\mathbf{k}_v,\uparrow} b^+_{\mathbf{k}_v,\uparrow}\right)|0\rangle,$$

$|1, -1\rangle$

$$= \frac{1}{2\sqrt{2}} \sum_{\mathbf{p},\mathbf{k}_n,\mathbf{k}_v} g(\mathbf{p})\varphi_1\left[\mathbf{k}_n + \beta\left(\mathbf{p} - \frac{1}{2}\mathbf{K}\right)\right]\varphi_2\left[\mathbf{k}_v - \beta\left(\mathbf{p} + \frac{1}{2}\mathbf{K}\right)\right]$$

$$\times \left(a^+_{\frac{\mathbf{K}}{2}-\mathbf{p}-\mathbf{k}_n,\downarrow} b^+_{\mathbf{k}_n,\uparrow} a^+_{\frac{\mathbf{K}}{2}+\mathbf{p}-\mathbf{k}_v,\downarrow} b^+_{\mathbf{k}_v,\downarrow} - a^+_{\frac{\mathbf{K}}{2}-\mathbf{p}-\mathbf{k}_n,\downarrow} b^+_{\mathbf{k}_n,\downarrow} a^+_{\frac{\mathbf{K}}{2}+\mathbf{p}-\mathbf{k}_v,\downarrow} b^+_{\mathbf{k}_v,\uparrow}\right)|0\rangle.$$

$$(5.17)$$

Here $g(\mathbf{p})$ is the Fourier transform of the function of relative motion of excitons in the biexciton, $\varphi_1[\mathbf{k}_n + \beta(\mathbf{p} - \frac{1}{2}\mathbf{K})]$ and $\varphi_2[\mathbf{k}_v - \beta(\mathbf{p} + \frac{1}{2}\mathbf{K})]$ are Fourier transforms of the functions of relative motion of electrons and holes in excitons 1 and 2, forming the biexciton, $\mathbf{k}_n$ and $\mathbf{k}_v$ are the wave vectors of electrons and holes in the biexciton, and $\mathbf{K}$ is the wave vector of the biexciton.

The representation of the wave function of the orthobiexciton in the form (5.17) containing the function of the relative motion of two excitons $g(\mathbf{p})$ is possible in the adiabatic approximation when the motion of one exciton with respect to the other in the biexciton is much slower in relation to the motion of the electron and hole in the exciton. The spin structure of the individual excitons is not preserved and, as we see from (5.17), the levels of the orthobiexciton are characterized by the projections of its total spin. For certainty, we consider the case when the total spin of the ortho-biexciton is affected by the parallel orientation of the electron spins at antiparallel orientation of the hole spins. Further calculations apply when the biexciton, before the collision with the nucleus, in an excited state with $n = 0$, $l = 1$, $m_l = 0$ (n, l, and m_l are the vibrational, rotational, and magnetic quantum numbers), is transferred to the state with $n = 0$, $l = 1$ without changing the projection of the orbital angular momentum. Fourier transforms of the functions of the relative motion of excitons in the biexcitons $g(\mathbf{p})$ and electrons and holes in the exciton $\varphi_1(\mathbf{q})$ and $\varphi_2(\mathbf{q})$ for the biexcitons with $n = 0$ and $l = 1$ have the form

$$g(\mathbf{p}) = \frac{i2^7(2\pi a_{\mathrm{B}}^5)^{1/2}}{(1 + 4a_{\mathrm{B}}^2\mathbf{p}^2)^3}p_z, \qquad \varphi_{1,2}(\mathbf{q}) = \varphi_{1S}(\mathbf{q}) = \frac{8\sqrt{\pi}a^{3/2}}{(1 + a^2\mathbf{q}^2)^2}, \qquad (5.18)$$

where $a_{\mathrm{B}} = 2a_{\mathrm{ex}}$ ($a = a_{\mathrm{ex}}$ is the exciton Bohr radius).

In the approximation of weak bonding of orthobiexcitons with the nucleus, for the total probability of the transition per unit time between the nuclear spin states $|I, m\rangle$ and $|I, n\rangle$ we obtain

$$\begin{aligned}
W_{mn} = \frac{2\pi}{\hbar}\sum_{\mathbf{K}} &\left|F(\mathbf{K}, \mathbf{K}')\right|^2\left[I(I+1) - n(n+1)\right] \\
&\times\left[f_{\mathrm{B}}(\varepsilon_{\mathbf{K}}, 0) + f_{\mathrm{B}}(\varepsilon_{\mathbf{K}}, -1)\right]\delta_{m,n+1}\delta(\varepsilon_{\mathbf{K}} - \varepsilon_{\mathbf{K}'} - \Delta) \\
&+ \left[I(I+1) - n(n-1)\right]\left[f_{\mathrm{B}}(\varepsilon_{\mathbf{K}}, 0) + f_{\mathrm{B}}(\varepsilon_{\mathbf{K}}, 1)\right] \\
&\times\delta_{m,n-1}\delta(\varepsilon_{\mathbf{K}} - \varepsilon_{\mathbf{K}'} + \Delta),
\end{aligned} \qquad (5.19)$$

where

$$\begin{aligned}
F(\mathbf{K}, \mathbf{K}') = \frac{\sqrt{2\pi}}{3}&\gamma_e\gamma_n\hbar^2 n_e \sum_{\mathbf{p},\mathbf{k}_n,\mathbf{k}_v} g^*(\mathbf{p}) \\
&\times \varphi_{1S}^*\left[\mathbf{k}_n + \beta\left(\mathbf{p} - \frac{1}{2}\mathbf{K}\right)\right]\varphi_{1S}^*\left[\mathbf{k}_v - \beta\left(\mathbf{p} + \frac{1}{2}\mathbf{K}\right)\right] \\
&\times\left\{g\left[\frac{1}{2}(\mathbf{K} - \mathbf{K}') - \mathbf{p}\right]\varphi_{1S}\left[\mathbf{k}_v + \beta\left(\frac{1}{2}\mathbf{K} - \mathbf{K}' - \mathbf{p}\right)\right]\right. \\
&\times\varphi_{1S}\left[\mathbf{k}_n - \beta\left(\frac{1}{2}\mathbf{K} - \mathbf{p}\right)\right] - g\left[\frac{1}{2}(\mathbf{K} - \mathbf{K}') + \mathbf{p} + \mathbf{k}_n - \mathbf{k}_v\right] \\
&\times\varphi_{1S}\left[\alpha\mathbf{k}_v + \beta(\mathbf{p} + \mathbf{k}_n) + \beta\left(\frac{1}{2}\mathbf{K} - \mathbf{K}'\right)\right]\varphi_{1S}\left[\alpha\mathbf{k}_n\right. \\
&\left.+ \beta\left(\mathbf{k}_v - \mathbf{p} - \frac{1}{2}\mathbf{K}\right)\right] + g\left[\frac{1}{2}(\mathbf{K} - \mathbf{K}') - \mathbf{p} - \mathbf{k}_n + \mathbf{k}_v\right] \\
&\times\varphi_{1S}\left[\alpha\mathbf{k}_n + \beta\left(\frac{1}{2}\mathbf{K} - \mathbf{K}' + \mathbf{k}_v - \mathbf{p}\right)\right]\varphi_{1S}\left[\alpha\mathbf{k}_v + \beta\left(\mathbf{p} + \mathbf{k}_n - \frac{1}{2}\mathbf{K}\right)\right] \\
&- g\left[\frac{1}{2}(\mathbf{K} - \mathbf{K}') + \mathbf{p}\right]\varphi_{1S}\left[\mathbf{k}_v - \beta\left(\frac{1}{2}\mathbf{K} + \mathbf{p}\right)\right]\varphi_{1S}\left[\mathbf{k}_n\right. \\
&\left.+ \beta\left(\frac{1}{2}\mathbf{K} - \mathbf{K}' + \mathbf{p}\right)\right] - g\left[\frac{1}{2}(\mathbf{K} - \mathbf{K}') + \mathbf{p}\right]\varphi_{1S}\left[\mathbf{k}_n + \beta\left(\mathbf{p} - \frac{1}{2}\mathbf{K}\right)\right] \\
&\times\varphi_{1S}\left[\mathbf{k}_v - \beta\left(\mathbf{K}' - \frac{1}{2}\mathbf{K} + \mathbf{p}\right)\right] + g\left[\frac{1}{2}(\mathbf{K}' - \mathbf{K}) - \mathbf{p} + \mathbf{k}_v - \mathbf{k}_n\right]
\end{aligned}$$

$$
\times \varphi_{1S}\left[\alpha \mathbf{k}_n + \beta\left(\mathbf{k}_v - \mathbf{p} - \frac{1}{2}\mathbf{K}\right)\right] \times \varphi_{1S}\left[\alpha \mathbf{k}_v - \beta\left(\mathbf{K}' - \frac{1}{2}\mathbf{K} - \mathbf{k}_n\right)\right]
$$

$$
+ g\left[\frac{1}{2}(\mathbf{K} - \mathbf{K}') - \mathbf{p}\right]
$$

$$
\times \varphi_{1S}\left[\mathbf{k}_v - \beta\left(\frac{1}{2}\mathbf{K} + \mathbf{p}\right)\right]\varphi_{1S}\left[\mathbf{k}_n - \beta\left(\mathbf{K}' - \frac{1}{2}\mathbf{K} - \mathbf{p}\right)\right]
$$

$$
- g\left[\frac{1}{2}(\mathbf{K}' - \mathbf{K}) + \mathbf{p} - \mathbf{k}_n - \mathbf{k}_v\right]\varphi_{1S}\left[\mathbf{k}_v + \beta\left(\mathbf{p} - \frac{1}{2}\mathbf{K} + \mathbf{k}_n - \mathbf{k}_v\right)\right]
$$

$$
\times \varphi_{1S}\left[\mathbf{k}_n - \beta\left(\mathbf{K}' - \frac{1}{2}\mathbf{K} + \mathbf{p} + \mathbf{k}_n - \mathbf{k}_v\right)\right]\Bigg\}. \tag{5.20}
$$

Here $f_B(\varepsilon_{\mathbf{K}}, M_s)$ is the function of the distribution of biexcitons with kinetic energy $\varepsilon_{\mathbf{K}}$ and spin projection M_s, Δ is the Zeeman splitting of the orthobiexciton with $n = 0$, $l = 1$, and total spin $S = 1$ ($S_{e,\text{tot}} = 1$, $S_{h,\text{tot}} = 0$), $\alpha = 1 - \beta$, and η_e is determined in Sect. 5.1.

The calculation of the total probability of the transition W_{mn} is cumbersome; therefore, we shall calculate W_{mn} approximately, assuming that $\alpha = 0$ and $\beta = 1$. This corresponds to the case when the mass of the electron is much smaller than the mass of the hole. Using the approximation (5.4) and taking into account (5.18), we obtain the following expression for $F(\mathbf{K}, \mathbf{K}')$ from (5.20):

$$
F(\mathbf{K}, \mathbf{K}') = -\frac{2^{10}}{3\sqrt{2\pi}}\gamma_e\gamma_n\eta_e\hbar^2\left\{\frac{1}{6^{5/2}}\left[12a^2(K_z - K_z')^2 - 1\right]\right.
$$

$$
\left. \times \exp\left[-7a^2(\mathbf{K} - \mathbf{K}')^2\right] + \frac{1}{(78)^{5/2}}\exp\left[-a^2(\mathbf{K} - \mathbf{K}')^2\right]\right\}. \tag{5.21}
$$

Substituting (5.21) into (5.19) and after the corresponding integration on quasi-momenta, we obtain the total probability of the transition W_{mn}, based on which we can calculate the nuclear spin–lattice relaxation rate. In the approximation of high spin and lattice temperatures, the rate of nuclear spin relaxation via biexcitons is

$$
\frac{1}{T_{1B}} = \frac{2^{24}}{9(2\pi^5)^{1/2}}\gamma_e^2\gamma_n^2\eta_e^2 m_B^{3/2}(k_0 T)^{3/2}\frac{ch\frac{\Delta}{2k_0 T}}{1 + 2ch\frac{\Delta}{k_0 T}}n_B
$$

$$
\times \left[\left(a_1 x^{-1} + a_2 x^{-2}\right)\mathcal{K}_0(\xi x) + \left(a_3 x^{-1} + a_4 x^{-2} + a_5 x^{-3}\right)\mathcal{K}_1(\xi x)\right.
$$

$$
\left. + a_6 y^{-1}\mathcal{K}_0(\xi y) + \left(a_7 y^{-1} + a_8 y^{-2}\right)\mathcal{K}_1(\xi y) + a_9\mathcal{K}_1(\xi z)\right], \tag{5.22}
$$

where $m_B = 2m_{ex}$ is the biexciton mass. Here $\mathcal{K}_p(\xi Q)$ is McDonald's function $(p = 0, 1; Q = x, y, z)$. The other designations are as follows:

$$x = 1 + 56\frac{m_{ex}^2}{m_e m_h}\frac{k_0 T}{G}, \qquad y = 1 + 32\frac{m_{ex}^2}{m_e m_h}\frac{k_0 T}{G},$$

$$z = 1 + 56\frac{m_{ex}^2}{m_e m_h}\frac{k_0 T}{G}, \qquad a_1 = -\frac{1}{3^5 2^3}c\xi,$$

$$a_2 = \frac{2}{15}c^2, \qquad a_3 = \frac{2}{3^3}\xi\left(\frac{c^2}{5} + \frac{1}{2^8 3^2}\right), \qquad a_4 = -\frac{1}{2 \cdot 3^5}c, \qquad (5.23)$$

$$a_5 = \frac{4c^2}{15\xi}, \qquad a_6 = \frac{1}{2^2 \cdot 117^{5/2}}c\xi, \qquad a_7 = -\frac{1}{2^6 \cdot 117^{5/2}}\xi,$$

$$a_8 = \frac{1}{2 \cdot 117^{5/2}}c, \qquad a_9 = \frac{1}{2 \cdot 39^5}\xi; \qquad \xi = \frac{\Delta}{2k_0 T}, \qquad c = \frac{m_{ex}a_{ex}^2\Delta}{\hbar^2}.$$

In formulas (5.23) G is the potential of ionization of the exciton.

The range of the temperature changes in the processes of nuclear spin–lattice relaxation via biexcitons is limited from above by the condition $k_0 T < G_B$ (G_B is the potential of the biexciton dissociation), which eliminates the processes of thermal disintegration of biexcitons. Similarly to the case of nuclear spin relaxation via excitons, formula (5.22) is not applicable in strong magnetic fields ($\Delta > G_B$).

Let us consider nuclear spin relaxation due to the interaction with orthobiexcitons in the CuCl crystal. For this crystal the approximation noted above is well satisfied (according to [423] for CuCl $\alpha = 0.02$, $\beta = 0.98$). In Fig. 5.3 the temperature dependence of the nuclear spin relaxation rate via orthobiexcitons (curve 1) for ^{63}Cu nuclei in CuCl at $n_B = 10^{18}$ cm^{-3} is presented. The shape of the curve $T_{1B}^{-1} = f(T)$ is determined by the functions of relative motion for the excitons in the biexciton, by the function of relative motion of electrons and holes in the excitons (from which the biexciton is formed), and by the function of the distribution of biexcitons on quasi-momenta in the biexciton band. For comparison, Fig. 5.3 also shows the temperature dependence of the nuclear spin relaxation rate via excitons (curve 2) calculated at the same values of the parameters.

An analysis of curves 1 and 2 in Fig. 5.3 leads to the following conclusions:

- The rate of nuclear spin relaxation via biexciton excited states is two orders of magnitude lower than the rate of nuclear spin relaxation via excitons under similar conditions.
- The maximum of the temperature dependence of the nuclear spin relaxation rate via biexcitons is shifted to lower temperatures relative to the analogous maximum of nuclear spin relaxation rate via excitons.
- The temperature gradient of the nuclear spin relaxation rate via biexcitons is greater than that via excitons, and the graph of the function $T_1^{-1} = f(T)$ is much narrower for nuclear spin relaxation via biexcitons than via excitons.

These particularities of the temperature behavior of the nuclear spin relaxation can be used for evidentiating contributions from excitons and biexcitons among the

Fig. 5.3 Comparison of the
temperature dependencies of
nuclear spin relaxation rate
via biexcitons (*1*) and
excitons (*2*) for ^{63}Cu nuclei
in the CuCl crystal

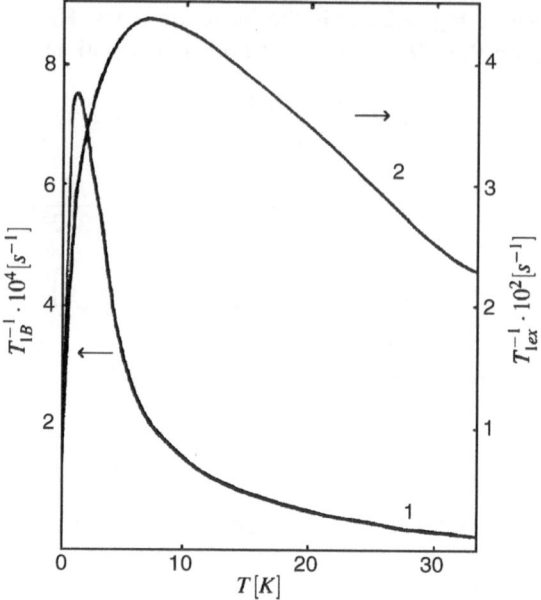

other mechanisms of nuclear spin relaxation in solids. We may also conclude that even in a high-density exciton gas the effects of the influence of exciton–exciton interaction of the attractive type on the nuclear spin relaxation are not determining.

5.5 Partial Averaging of the Exciton–Exciton Interaction Under Influence of Terasound

This section studies the possibility of applying multipulse methods of high-resolution NMR spectroscopy [68, 69] to the optical spectroscopy of excitons. Note in particular that to realize this possibility it is not quite necessary to consider only the triplet (magnetic) excitons. We consider the case in which the exciton band is orbitally nondegenerate, the spin structure of the excitons is not taken into consideration, there is no external constant magnetic (or other) field, and the unique actual processes are intraband exciton–exciton scattering and interaction of excitons with terasound or with hypersonic phonons. This situation is realized at low temperatures and high levels of optical excitation of the crystals, when the frequency of exciton–exciton collisions is higher than the frequency of collisions of excitons with thermal phonons, and the main contribution to the width of the exciton absorption or exciton photoluminescence line is caused by exciton–exciton interaction. For a high exciton concentration, as was noted in Sect. 1.4, a shift of the exciton level due to exciton–exciton interaction takes place.

In the GaSe crystal this shift takes place in the long-wavelength part of the spectrum [425–427]. In [427] it was experimentally proved that the red shift of the exciton absorption line was primarily conditioned by exciton–exciton collisions and is not connected with the heating of the sample by laser irradiation, which, as was shown in [428], also leads to a shift of the exciton level on the long-wavelength side. In the ZnTe crystal, with increased optical pumping, in certain limits, the exciton level is also shifted to the long-wavelength part of the spectrum, which is confirmed by the broadening and long-wavelength shift of the maximum of the $P_{n=2}$ band caused by the inelastic scattering of excitons. In this case the radiation line of electron–hole plasma is observed from the long-wavelength side line of radiation of the free excitons. This line is shifted to shorter wavelengths at 1.5 meV during the change of the excitation energy E_{exc} from the values $E_{exc} > \mu$ (μ is the chemical potential corresponding to $R_{exc} = 15$ MW/cm^2) to $E_{exc} < \mu$ [429]).

In this section we discuss the narrowing of the exciton lines of absorption whose widths are due to intraband exciton–exciton collisions, under the influence of a sequence of four terasound pulses, similar to the WAugh, HUber, and HAeberlen (WAHUHA) sequence of four electromagnetic pulses used in multipulse solid state NMR spectroscopy [68, 69]. Let us consider a crystal containing high-density excitons, taking into account only one (the lowest) exciton band. We consider only those exciton concentrations for which the exciton–exciton interaction is manifested, but does not yet manifest collective effects in the exciton system. As noted above, it is assumed that the exciton ground state is nondegenerate in the orbital and spin quantum numbers. The interaction between excitons is caused by elastic exciton–exciton scattering within a nondegenerate exciton band. In this case, the width of the optical absorption line related to the quantum transitions from the ground state of the crystal in a nondegenerate exciton band will be determined by the processes of intraband exciton–exciton collisions. The Hamiltonian of the exciton system in the occupation number representation is given by [147]

$$\mathcal{H} = \sum_{\mathbf{k}} [E_{ex}(\mathbf{k}) - \mu] A_{\mathbf{k}}^{+} A_{\mathbf{k}} + \frac{1}{2\mathcal{V}} \sum_{\mathbf{k}_1, \mathbf{k}_2, \mathbf{q}} V_{\mathbf{q}} A_{\mathbf{k}_1}^{+} A_{\mathbf{k}_2}^{+} A_{\mathbf{k}_2 - \mathbf{q}} A_{\mathbf{k}_1 + \mathbf{q}}, \qquad (5.24)$$

where $A_{\mathbf{k}}^{+}$ and $A_{\mathbf{k}}$ are the exciton operators of creation and destruction, $E_{ex}(\mathbf{k})$ is the energy of exciton formation with wave vector $\mathbf{k}$ and quadratic dispersion law, $V_{\mathbf{q}}$ is the Fourier transform of the energy of interaction between two excitons, μ is the chemical potential of the system, and $\mathcal{V}$ is the volume of the crystal.

The form of Hamiltonian (5.24) differs from that of the Hamiltonian which describes the Zeeman interaction of nuclear spins with a constant magnetic field, and the secular part of the operator of magnetic dipole–dipole interactions between nuclear spins [68]

$$\mathcal{H}_{spin} = \mathcal{H}_Z + \mathcal{H}_d^0, \qquad (5.25)$$

$$\mathcal{H}_Z = \gamma H_0 \sum_j I_{Zj}, \qquad (5.26)$$

$$\mathcal{H}_d^0 = \gamma^2 \hbar \sum_i \sum_{< j} r_{ij}^{-3} P_2(\cos\theta_{ij})(\mathbf{I}_i \mathbf{I}_j - 3 I_{Zi} I_{Zj}), \qquad (5.27)$$

where γ is the gyromagnetic ratio of the studied nuclei, I_{Zi} and I_{Zj} are operators of the projections of nuclear spin on the Z axis (the direction of the static magnetic field $\mathbf{H}_0$), θ_{ij} is the angle between this direction and the direction of a vector between spins i and j, and $P_2(X) = (1/2)(3X^2 - 1)$ is the Legendre polynomial.

To implement the narrowing of NMR lines in solids under the influence of a sequence of WAHUHA electromagnetic pulses, it is essential to depict the main contribution to the width of the NMR lines by the secular part $\mathcal{H}_d^0$ of the operator of magnetic dipole–dipole interactions at high magnetic fields, which commutes with the operator $\mathcal{H}_Z$:

$$\left[\mathcal{H}_Z, \mathcal{H}_d^0 \right] = 0.$$

This leads to a separation of two thermodynamic reservoirs that are not related to each other: the Zeeman reservoir, due to the interaction with a constant magnetic field H_0, and the dipole–dipole reservoir, due to the secular part of the dipole–dipole Hamiltonian $\mathcal{H}_d^0$ [68]. In this case, the energy of each system separately is an integral of motion, and one might obtain a situation when the temperatures of each reservoir will be different.

A comparison of Hamiltonians $\mathcal{H}$ and $\mathcal{H}_{\text{spin}}$, given by (5.24) and (5.25), has not shown if the exciton system can be divided into two independent thermodynamic reservoirs similar to the case of interacting nuclear spins in a strong magnetic field. However, the analogy between $\mathcal{H}$ and $\mathcal{H}_{\text{spin}}$, and, accordingly, the possibility of separation of two independent thermodynamic reservoirs in the exciton system, can be brought out, if the Schwinger representation [406] for angular momentum is used in the Hamiltonian (5.24).

We now introduce the effective spin operator $\mathbf{S}_{\mathbf{k}_1 \mathbf{k}_2}$, the projection of which in the directions X, Y, Z can be presented in the form [430]

$$S_{\mathbf{k}_1 \mathbf{k}_2}^X = \frac{1}{2} \left(A_{\mathbf{k}_1}^+ A_{\mathbf{k}_2} + A_{\mathbf{k}_1} A_{\mathbf{k}_2}^+ \right),$$

$$S_{\mathbf{k}_1 \mathbf{k}_2}^Y = \frac{1}{2i} \left(A_{\mathbf{k}_1}^+ A_{\mathbf{k}_2} - A_{\mathbf{k}_1} A_{\mathbf{k}_2}^+ \right), \tag{5.28}$$

$$S_{\mathbf{k}_1 \mathbf{k}_2}^Z = \frac{1}{2} \left(A_{\mathbf{k}_1}^+ A_{\mathbf{k}_1} - A_{\mathbf{k}_2}^+ A_{\mathbf{k}_2} \right).$$

After some transformations using (5.28), the Hamiltonian (5.24) can then be presented as follows:

$$\mathcal{H} = \mathcal{H}_0 + \mathcal{H}_{\text{ex}}^Z,$$

where

$$\mathcal{H}_0 = \sum_{\mathbf{k}_1} \left[E_{\text{ex}}(\mathbf{k}_1) - \mu \right] S_{\mathbf{k}_1, \mathbf{k}_1 + \mathbf{q}}^Z - \sum_{\mathbf{k}_2} \left[E_{\text{ex}}(\mathbf{k}_2) - \mu \right] S_{\mathbf{k}_2 - \mathbf{q}, \mathbf{k}_2}^Z$$

$$+ \frac{1}{2} \sum_{\mathbf{k}} \left[E_{\text{ex}}(\mathbf{k}_1) - \mu \right], \tag{5.29}$$

$$\mathcal{H}_{ex}^Z = \frac{1}{2\mathcal{V}} \sum_{\mathbf{k}_1 \mathbf{k}_2 \mathbf{q}} V_q \left(\mathbf{S}_{\mathbf{k}_1, \mathbf{k}_1+\mathbf{q}} \mathbf{S}_{\mathbf{k}_2-\mathbf{q}, \mathbf{k}_2} - S_{\mathbf{k}_1, \mathbf{k}_1+\mathbf{q}}^Z S_{\mathbf{k}_2-\mathbf{q}, \mathbf{k}_2}^Z \right). \tag{5.30}$$

When deriving the Hamiltonian $\mathcal{H}_0$ from (5.29), it was taken into account that the operators $A_{\mathbf{k}_1+\mathbf{q}}^+ A_{\mathbf{k}_1+\mathbf{q}}$ and $A_{\mathbf{k}_2-\mathbf{q}}^+ A_{\mathbf{k}_2-\mathbf{q}}$ which realize the two-boson representation of the z-component for the $\mathbf{S}_{\mathbf{k}_1+\mathbf{q}, \mathbf{k}_2-\mathbf{q}}$ operator, i.e.,

$$S_{\mathbf{k}_1+\mathbf{q}, \mathbf{k}_2-\mathbf{q}}^Z = \frac{1}{2} \left(A_{\mathbf{k}_1+\mathbf{q}}^+ A_{\mathbf{k}_1+\mathbf{q}} - A_{\mathbf{k}_2-\mathbf{q}}^+ A_{\mathbf{k}_2-\mathbf{q}} \right),$$

satisfy the condition

$$A_{\mathbf{k}_1+\mathbf{q}}^+ A_{\mathbf{k}_1+\mathbf{q}} + A_{\mathbf{k}_2-\mathbf{q}}^+ A_{\mathbf{k}_2-\mathbf{q}} = 1.$$

The Hamiltonian $\mathcal{H}_{ex}^Z$ from (5.30) was obtained neglecting the rare processes of pair collisions during which two excitons, having pre-collision impulses of $\mathbf{k}_1$ and $\mathbf{k}_2$ (in $\hbar$ units), have the impulses $\mathbf{k}_1' = \mathbf{k}_2$ and $\mathbf{k}_2' = \mathbf{k}_1$, respectively, after collision. Taking these processes into account, we find that the averaged energy (on the exciton occupation numbers) of interaction between excitons decreases by a value of

$$\delta E = \frac{1}{2} n_{ex} \sum_{\mathbf{q}} V_{\mathbf{q}}.$$

Comparing (5.29) and (5.30) with (5.26) and (5.27), respectively, we can see that there is a deep analogy between the problem of elastic intraband exciton–exciton collisions in semiconductors and the problem of the magnetic dipole–dipole interactions between nuclear spins in a strong magnetic field. In a system of high-density excitons, as well as in the case of interacting nuclear spins in a strong magnetic field, there are two unrelated thermodynamic reservoirs, similar to the Zeeman and dipole–dipole reservoirs in the case of nuclear spin systems. However, there is a significant difference between exciton–exciton and dipole–dipole reservoirs: the absence of the factor 3 in the operator $S_{\mathbf{k}_1, \mathbf{k}_1+\mathbf{q}}^Z S_{\mathbf{k}_2-\mathbf{q}, \mathbf{k}_2}^Z$ of (5.30) and the presence of this factor in the operator $I_{Zi} I_{Zj}$ (see (5.27)). As will be shown below, this leads to the fact that, in contrast to spin systems for which the magnetic dipole–dipole interactions are averaged to zero under the influence of a sequence of four WAHUHA electromagnetic pulses, the impact of a sequence of four WAHUHA terasound pulses leads only to a partial averaging of the exciton–exciton interactions for exciton systems.

In contrast to the system of interacting nuclear spins for which the dipole–dipole reservoir can be identified only in the presence of a strong magnetic field, the "exciton–exciton" reservoir with the corresponding Hamiltonian $\mathcal{H}_{ex}^Z$ always exists, because the energy of exciton–exciton interaction is much smaller than the eigenvalues of the operator $\mathcal{H}_0$, which is an analog of the Hamiltonian for the interactions of nuclei with a magnetic field. Due to a quasi-discrete energy spectrum of excitons and a finite reduced Brillouin zone, the projection operators of the total effective spins $S_1^Z = \sum_{\mathbf{k}_1} \sum_{\mathbf{q}} S_{\mathbf{k}_1, \mathbf{k}_1+\mathbf{q}}^Z$ and $S_2^Z = \sum_{\mathbf{k}_2} \sum_{\mathbf{q}} S_{\mathbf{k}_2-\mathbf{q}, \mathbf{k}_2}^Z$ have a large, but finite number of eigenvalues. Therefore, the formalism developed for systems with a finite number of degrees of freedom can be applied to them.

Let us consider now the impact of a sequence of four WAHUHA 90-degree tera-sound pulses on the system of effective spins periodically repeated with a period $t_c < T_2$ (T_2 is the transverse exciton relaxation time). For pure direct-band semiconductors $T_2 \sim 10^{-9}$–10^{-7} s [97]; therefore, in order to perform experiments of WAHUHA type in the case of high-density excitons in direct-band semiconductors, when the defining processes are intraband scattering of excitons on excitons, tera-sound impulses are required. For indirect excitons, for which the value of T_2 is three orders of magnitude higher, one can also use a sequence of hypersound pulses.

In accordance with the criterion of applicability of the method of coherent averaging [68, 69], we assume that the energy of the exciton–exciton interaction is a small perturbation in comparison with the energy of interaction of coherent tera-sound with excitons. The interaction Hamiltonian of terasound with excitons can be presented in the form

$$\mathcal{H}_1 = \sqrt{N_{\mathbf{q}\sigma}} \sum_{\mathbf{k}} \left[g_\sigma(\mathbf{q}) S^+_{\mathbf{k},\mathbf{k}+\mathbf{q}} \exp(-i\omega_{\mathbf{q},\sigma} t) + \text{H.c.} \right], \qquad (5.31)$$

where $S^+_{\mathbf{k},\mathbf{k}+\mathbf{q}} = S^X_{\mathbf{k},\mathbf{k}+\mathbf{q}} + i S^Y_{\mathbf{k},\mathbf{k}+\mathbf{q}}$, $g_\sigma(\mathbf{q})$ is the exciton–phonon coupling constant, and $N_{\mathbf{q}\sigma}$ is the number of terasonic phonons with frequency $\omega_{\mathbf{q}\sigma}$, wave vector $\mathbf{q}$, and polarization σ. Equation (5.31) corresponds to the interaction of excitons with a monochromatic terasonic wave of high amplitude (macrocompleted mode) when the corpuscular structure of the terasonic field is neglected ($\langle b_{\mathbf{q},\sigma} \rangle = \langle b^+_{\mathbf{q},\sigma} \rangle \approx \sqrt{N_{\mathbf{q}\sigma}}$).

The evolution operator of the system (propagator) at the time moment corresponding to the end of the cycle ($t_c = 6\tau$) is given by

$$U_{\text{int}}(t_c) = \exp\left(-\frac{i}{\hbar} \mathcal{H}^Z_{\text{ex}} \tau \right) P_X \exp\left(-\frac{i}{\hbar} \mathcal{H}^Z_{\text{ex}} \tau \right) P_{-Y} \exp\left(-2\frac{i}{\hbar} \mathcal{H}^Z_{\text{ex}} \tau \right)$$

$$\times P_Y \exp\left(-\frac{i}{\hbar} \mathcal{H}^Z_{\text{ex}} \tau \right) P_{-X} \exp\left(-\frac{i}{\hbar} \mathcal{H}^Z_{\text{ex}} \tau \right),$$

which, considering the result of the action of the operators

$$P_{\pm\xi} = \exp\left(\mp i\frac{\pi}{2} \sum_{\mathbf{k}_1 \mathbf{k}_2} S^\xi_{\mathbf{k}_1 \mathbf{k}_2} \right), \quad \xi = X, Y$$

on $\exp(-\frac{i}{\hbar} \mathcal{H}^Z_{\text{ex}} \tau)$, takes the form

$$U_{\text{int}}(t_c) = \exp\left(-\frac{i}{\hbar} \mathcal{H}^Z_{\text{ex}} \tau \right) \exp\left(-\frac{i}{\hbar} \mathcal{H}^Y_{\text{ex}} \tau \right) \exp\left(-2\frac{i}{\hbar} \mathcal{H}^X_{\text{ex}} \tau \right)$$

$$\times \exp\left(-\frac{i}{\hbar} \mathcal{H}^Y_{\text{ex}} \tau \right) \exp\left(-\frac{i}{\hbar} \mathcal{H}^Z_{\text{ex}} \tau \right), \qquad (5.32)$$

where $\mathcal{H}^X_{\text{ex}}$ and $\mathcal{H}^Y_{\text{ex}}$ are obtained from $\mathcal{H}^Z_{\text{ex}}$ by replacing Z by X and Y, respectively. Here, τ is the time during which the system develops under the influence of the effective "exciton–exciton" Hamiltonian (free development); then the first tera-sound impulse is applied to it.

After applying the Magnus expansion to the operator $U_{int}(t_c)$ in (5.32) (see, e.g., [69]), the following expression for the average Hamiltonian is derived in the zeroth approximation [430]:

$$\overline{\mathcal{H}_{ex}^{Z(0)}} = \frac{1}{3\mathcal{V}} \sum_{\mathbf{k}_1, \mathbf{k}_2, \mathbf{q}} V_{\mathbf{q}} S_{\mathbf{k}_1, \mathbf{k}_1 - \mathbf{q}} S_{\mathbf{k}_2 + \mathbf{q}, \mathbf{k}_2}. \tag{5.33}$$

Equation (5.33) shows that, in contrast to the spin systems with magnetic dipole–dipole interaction, the exciton–exciton interaction is not averaged to zero. A scalar part of the interaction diminished by one and half times remains. This is due to the different forms for the effective Hamiltonian $\mathcal{H}_{ex}^Z$ and the truncated Hamiltonian of the magnetic dipole–dipole interactions.

Thus, by introducing the projection operators of the effective spin (pseudospin), the Hamiltonian of intraband exciton–exciton scattering can be reduced to a form which is almost similar to the secular part of the operator of magnetic dipole–dipole interactions in a spin system. In this pseudospin formalism two independent thermodynamic reservoirs can be separated in the exciton system, by analogy with the Zeeman and dipole–dipole reservoirs in a spin system, with energies as integrals of motion. In contrast to the magnetic dipole–dipole reservoir, which can be separated only in the presence of a strong static magnetic field, the "exciton–exciton" reservoir exists even in the absence of interaction with any external constant fields.

For all elementary excitations in solids obeying Bose–Einstein statistics, the paired processes of elastic collisions, in which the quasi-particles remain within the same energy band, can be described by similar Hamiltonians. Our analysis for excitons can then be extended to other elementary excitations of bosonic type (biexcitons, polaritons, and magnons). In each of these cases it is possible to introduce a pseudospin and the selection of two nonconnected thermodynamic reservoirs, and, in the framework of coherent averaging techniques, one can use multipulse methods of NMR spectroscopy of solids for narrowing of the corresponding spectral lines of emission or absorption. A similar analysis can be also done in regard to the processes of inelastic scattering of quasi-particles with a turn to other branches of elementary excitation spectra.

According to the averaging Krylov–Bogolyubov–Mitropolsky method, it was shown in the article [431] that the expression for the average Hamiltonian, based on the Magnus expansion, is nonsecular in higher orders of the perturbation theory. The authors of the article [432] have developed a method of canonical transformations which allows us to study the dynamics of spin systems in a variety of multipulse experiments, and the thermodynamics effects in multipulse NMR spectroscopy of solids have been studied based on this method. These generalizations of the average Hamiltonian method should also be taken into account when the effects of multipulse sequences on excitons and other quasi-particles of bosonic type are considered. In this framework of the average Hamiltonian method, we have shown that multipulse techniques developed in solid state NMR spectroscopy can be applied, in particular, to the optical spectroscopy of excitons.

Chapter 6
Interaction of Excitons with Paramagnetic Centers

This chapter discusses the exchange interaction between the electron and hole forming a free exciton with paramagnetic centers (PCs). During this interaction the spin of the PC is reoriented, and the exciton is transferred from one state of relative motion to another with a change in both the magnitude and direction of the wave vector.

In Sects. 6.1 and 6.2 we study the cross-section of the exchange scattering of excitons on deep PCs at singlet-triplet exciton transitions and transitions between Zeeman components of the triplet exciton as well as during scattering of $\Gamma_6 \otimes \Gamma_8 \otimes \Gamma_1$ excitons on PCs in cubic semiconductors. An important particularity for practical applications of the exchange interaction of the free excitons with PCs is the possibility to reconstruct the energy spectrum of excitons, including optically forbidden inactive exciton levels, on the basis of exciton exchange cross-sections. The presence of the effective interaction between the excitons and PCs also allows us to determine, on the basis of electron paramagnetic resonance (EPR) spectra, the concentration of excitons in crystals with a very low density of the exciton gas ($n_{ex} \lesssim 10^8$ cm^{-3}), at which optical methods of exciton detection are not applied.

The Coulomb interaction between electrons and holes in excitons is displayed in the process of the exchange scattering of excitons on PCs in the form of the effect of exciton reduction of the spin relaxation time of PCs, which we discovered and have considered in Sect. 6.3. This effect consists in shortening of the PC spin relaxation time during the Coulomb binding of the electron and hole into the exciton in comparison with the relaxation conditioned by the same carriers in the free (not bound) state.

In Sect. 6.4 the indirect exchange of the PC via a field of excitons is considered, and the dependence of the isotropic effective exchange integral on the temperature and distance between paramagnetic ions is determined.

The exchange interaction of charge carriers bound into free excitons with paramagnetic ions of high concentration causes giant spin splittings of the exciton states. In Sect. 6.5 we consider the effect of giant spin splitting of the exciton band in semiconductors that are heavily doped by paramagnetic ions. We discuss how the energy spectrum of the $\Gamma_6 \otimes \Gamma_8 \otimes \Gamma_1$ excitons in cubic crystals is reconstructed during their

I. Geru, D. Suter, *Resonance Effects of Excitons and Electrons*,
Lecture Notes in Physics 869, DOI 10.1007/978-3-642-35807-4_6,
© Springer-Verlag Berlin Heidelberg 2013

exchange interaction with a system of paramagnetic ions of high concentration in a magnetic field.

Finally, Sect. 6.6 provides the experimental data on the giant spin splitting into six components of the $1S$ exciton line in the reflection spectrum and on the anomalous large magnetic circular dichroism of the reflection and giant Faraday rotation in the domain of the $1S$ exciton band in the crystal CdTe:Mn^{2+}. These experimental results were first obtained by Rabchenko et al. in 1977 and interpreted theoretically by one of us in [71–73]. The anomalous large magneto–optical effects affected by the giant splitting of the $1S$ exciton band in the exchange field of paramagnetic ions of high concentration are considered in this section. They appear to be the first examples of the experimental manifestation of the exchange interaction between charge carriers and impurity centers of high concentration in semiconductors in the exciton range of the spectrum. The large magnitudes of the indicated effects, which surpass by two–three orders of magnitude similar effects in pure semiconductors and semiconductors with low doping by paramagnetic impurities, determine the perspective of this new direction of investigation. The results of this chapter are published in [71–73, 170, 433–435].

6.1 Spin Relaxation of Deep Centers in Semiconductors via Singlet and Triplet Excitons

The study of the cross-section of the exchange scattering of free carriers on a PC by the method of spin relaxation gives important information about the carrier-impurity interaction and some parameters of the band structure of the semiconductors [436–439]. Thus, the experimental investigation by this method of neutral impurity Fe^0 in Si [437] allowed to establish the constant of the exchange interaction, to estimate the square modulus of the Bloch wave function of the bottom of the conduction band in the interstitial Fe^0 in Si [438] and explain the shortening of the spin relaxation of Fe^0 at an electron concentration of all $n_e \sim 10^8$ cm^{-3}. One expects the high efficiency of the spin relaxation of the PC to also be preserved in the process of exchange scattering of the charge carriers bound in excitons, as it is for relaxation of the nuclear spin via contact hyperfine interaction with excitons [170, 378, 418] (see Chap. 5).

Although the Hamiltonians of the exchange and contact hyperfine interactions coincide with the designated precision, the application of the results obtained for nuclear spin relaxation via excitons [170, 378, 418] to the electron spin relaxation of PCs is limited by the following circumstances. The transitions between the singlet and triplet exciton states were not taken into account in considering the spin relaxation of the nuclei via excitons, since the nuclear Zeeman energy is somewhat smaller than the magnitude of the exchange splitting of the exciton. However, for spin relaxation of PCs, the exciton transitions between the triplet and singlet states, as will be shown, can play a determining role in a number of cases. The transitions between states of relative motion of electrons and holes in exitons were also not taken into consideration. It corresponds only to the case of low temperatures.

Let us consider the spin relaxation of PCs, influenced by the exchange interaction with the singlet and triplet excitons. We shall proceed from the Γ_6-electron and Γ_7-hole states representing the $1S$ exciton in a cubic crystal split ($\Gamma_6 \otimes \Gamma_7 = \Gamma_2 \oplus \Gamma_{15}$) by the electron–hole exchange interaction $\mathcal{H}_0 = \Delta \mathbf{S} \mathbf{J}$ ($\mathbf{S}$ is the electron spin operator, $\mathbf{J}$ is the operator of the hole effective angular momentum $J = 1/2$ in the band Γ_7). Correspondingly, the exciton states $|\Gamma\gamma, \mathbf{K}\rangle$ will be characterized by the irreducible representation Γ ($\Gamma = \Gamma_2, \Gamma_{15}$), its row γ, and wave vector $\mathbf{K}$. The sum of the operators of the exchange interaction of PCs with electrons and holes obtained in [439] for the bands Γ_6 and Γ_7 is the Hamiltonian $\mathcal{H}$ describing the interaction of the exciton with the PC. The operator $\mathcal{H}$ in the $\Gamma_2 \oplus \Gamma_{15}$ representation takes the form [433]

$$\mathcal{H}^{\mathrm{ex}} = -\frac{1}{V} \sum_{\Gamma\gamma\mathbf{K}} \sum_{\Gamma'\gamma'\mathbf{K}'} \mathcal{H}^{\mathrm{ex}}_{\Gamma\gamma,\Gamma'\gamma'}(\mathbf{K} - \mathbf{K}') A^+_{\Gamma\gamma\mathbf{K}} A_{\Gamma'\gamma'\mathbf{K}'}, \tag{6.1}$$

where $A^+_{\Gamma\gamma\mathbf{K}}$ is the exciton creation operator in the state $|\Gamma\gamma, \mathbf{K}\rangle$ and V is the volume of the crystal. Approximating the Fourier transform of the function of relative motion of the electron and hole in the exciton normalized on a unit by the exponential function [422, 433], for the matrix elements in (6.1) we find

$$\mathcal{H}^{\mathrm{ex}}_{\Gamma_{15},\Gamma_{15}}(\mathbf{K} - \mathbf{K}') = b_+ \mathbf{J}_1 \mathbf{S}, \qquad \mathcal{H}^{\mathrm{ex}}_{\Gamma_2,\Gamma_{15}}(\mathbf{K} - \mathbf{K}') = b_- \left(-\frac{S_+}{\sqrt{2}}, S_z, \frac{S_-}{\sqrt{2}} \right),$$

$$\mathcal{H}^{\mathrm{ex}}_{\Gamma_2,\Gamma_2}(\mathbf{K} - \mathbf{K}') = 0, \qquad b_\pm = I_e e^{-a^2_{\mathrm{ex}}\beta^2(\mathbf{K}-\mathbf{K}')^2} \pm I_h e^{-a^2_{\mathrm{ex}}(1-\beta)^2(\mathbf{K}-\mathbf{K}')^2}, \tag{6.2}$$

where $\mathbf{J}_1$ is the angular momentum operator $J_1 = 1$, determined on the triplet state Γ_{15}; I_e and I_h are the constants of the exchange interaction of the PC with the electron and hole [439].

The probability of the relaxational transition of the PC, $M \to M'$, will be expressed through cross-sections (averaged on thermalized excitons) of the exchange scattering $\overline{\sigma}_{M,M'}$, the exciton concentration n_{ex}, and their average thermal speed:

$$W_{M,M'} = n_{\mathrm{ex}} \bar{v}_{\mathrm{ex}} \overline{\sigma}_{M,M'}; \qquad \bar{v}_{\mathrm{ex}} = \sqrt{\frac{8k_0 T}{\pi m_{\mathrm{ex}}}}. \tag{6.3}$$

In the Born approximation in the case when the Zeeman splittings of the exciton and PC are less than $k_0 T$, for $\overline{\sigma}_{M,M'}$, by means of (6.1) and (6.2), we have

$$\overline{\sigma}_{M,M'} = \frac{m^2_{\mathrm{ex}}}{\pi \hbar^4} I^2_0 \left[(S_+)^2_{M,M-1} \delta_{M',M-1} + (S_-)^2_{M,M+1} \delta_{M',M+1} \right], \tag{6.4}$$

$$I^2_0 = \left(1 + 3e^{-x} \right)^{-1} \left[F(x) I^2_-(y) + e^{-x} I^2_+(0) \right], \tag{6.5}$$

$$I^2_\pm(y) = I^2_e \lambda \left(\frac{1}{v}, y \right) + I^2_h \lambda(v, y) \pm 2 I_e I_h \lambda \left(\frac{1+v^2}{2v}, y \right), \tag{6.6}$$

$$\lambda(\xi, y) = e^{-2\xi|y|}/(1 + 8\xi k_0 T/G); \quad x = \frac{\Delta}{k_0 T}, \quad y = \frac{\Delta}{G}, \quad v = m_e/m_h. \quad (6.7)$$

Here $F(x) = \frac{1}{2}|x|e^{-x/2}\mathcal{K}_1(\frac{|x|}{2})$, where $\mathcal{K}_1$ is McDonald's function (the integral representation and the graph of F_x are given in [438]), and $G = e^2/2a_{ex}$ is the energy of ionization of the $1S$ exciton. The property of the function $F(x)$ is shown in the Appendix G.

The coefficients $\lambda(\xi, y)$ from (6.6) and (6.7) show that the relative motion of the carriers leads to a decrease in $\overline{\sigma}_{MM'}$ at the expense of two effects: the processes of inelastic scattering at transitions of the exciton between the singlet and triplet levels, and the decrease of the probability of the exchange scattering at high $\mathbf{K}$. Note that such dependencies are qualitatively observed during the electron collisions in the atoms.

The temperature dependence $\overline{\sigma}_{M,M'}$ is determined by the expressions (6.4)–(6.7) and depends on the magnitude and the sign of Δ. The case $\Delta < 0$, $|\Delta| \gg k_0 T$ corresponds to the situation considered in [378]. The given inequalities allow us to introduce the approximations

$$\lambda(\xi, y) \simeq 1, \qquad I_0^2 = \frac{1}{3}(I_e + I_h)^2.$$

Here we see that, during the relaxation via the triplet excitons, interference of the scattering amplitudes of the electron and hole takes place. Thus, if $I_e = -I_h$ (as, for example, in CdSe:Mn [440]), this channel of relaxation is switched off, but the relaxation is intensified during the transition of the exciton to the Γ_2 state with the temperature dependence $W_{M,M'} \sim \exp(-|\Delta|/k_0 T)$. If the singlet is the lowest state ($\Delta > 0$), but as earlier $k_0 T \ll \Delta$, then the processes of inelastic scattering ($\sim F(x)$ in (6.5)) may dominate if $\lambda(\xi, y)$ is not very small. In the case $\lambda \simeq 1$ we have

$$I_0^2 \simeq \frac{1}{2}\sqrt{\pi x}e^{-x}(I_e - I_h)^2,$$

so that $W_{M,M'} \sim \exp(-\Delta/k_0 T)$. For high temperatures ($x \ll 1$, $F(x) \simeq 1$) the interference part in (6.5) decreases if $1 + v^2/y < 1$ and the result does not depend on the signs of Δ and $\eta = I_e/I_h$. Assuming for simplicity $\lambda(\xi, y) = 1$, we find $I_0^2 \simeq \frac{1}{2}(I_e^2 + I_h^2)$. Thus, the graph of the temperature dependence $\overline{\sigma}_{M,M'}$ has the form of a "step" in the domain $k_0 T \sim |\Delta|$, if $\Delta < 0$. The height of the step is determined by the ratio $\frac{3}{2}(1 + \eta^2)(1 + \eta)^{-2}$. In the case $\Delta > 0$ the slope of the curve $\overline{\sigma}_{M,M'}(T)$ in the domain $k_0 T \sim \Delta$ is rapidly decreased, passing from the exponent to a magnitude not depending on the temperature. These effects may be used for the determination of the exchange constants I_e, I_h, and Δ. With the increase of temperatures when $k_0 T > G/8\xi$, the exchange cross-section decreases, but $W_{M,M'}$ is diminished as $T^{-1/2}$, passing through the maximum at T_{max} ($Gv/8 < k_0 T_{max} < G/8v$).

The quantitative estimations at $I_e = -I_h \simeq 1$ eV, $m_e = m_h = m_0$ (m_0 is the mass of the free exciton) lead to a magnitude of time relaxation of PCs via excitons of $\tau_{ex} \simeq 10^{-2}$ s for the easily experimentally achieved concentration of excitons $n_{ex} = 10^8$ cm^{-3}.

6.2 Relaxation of Paramagnetic Centers via $\Gamma_6 \otimes \Gamma_8 \otimes \Gamma_1$ Excitons in Cubic Crystals

Let us consider the influence of the isotropic and anisotropic electron–hole exchange interaction on the temperature dependence of the PC relaxation rate affected by $1S$ excitons in cubic crystals. Near the bottom of the lowest conduction band and the top of the upper valence band, the electron and hole states of these crystals are transformed correspondingly according to irreducible representations Γ_6 and Γ_8 (excitons of the $\Gamma_6 \otimes \Gamma_8 \otimes \Gamma_1$ type). We obtain the Hamiltonian of the exchange interaction of excitons with PCs using the operator of the exchange interactions of PCs with the conduction electrons in the band Γ_6,

$$\mathcal{H}^c = -\frac{I_e}{V} \sum_{k\sigma k'\sigma'} (2\mathbf{s}_c\mathbf{S})_{\sigma\sigma'} a_{\sigma k}^+ a_{\sigma' k'}, \tag{6.8}$$

and with the electrons of the valence band Γ_8 of the cubic semiconductor [441],

$$\mathcal{H}^v = -\frac{I_v}{V} \sum_{nkn'k'} \sum_{n_1 n_2} G_{n_1 n_2}^{nn'}(\mathbf{k}, \mathbf{k}')(2\mathbf{J}\mathbf{S})_{n_1 n_2} a_{nk}^+ a_{n'k'}. \tag{6.9}$$

Here I_e and I_v are the constants of the exchange interaction of PCs with the conduction and valence electrons, s_c is the spin of the conduction electron, S is the spin of paramagnetic center, and J is the total angular momentum of the valence electron ($J = 3/2$); a_{nk}^+ and a_{nk} are the creation and destruction operators of the band electrons in the state $|n, \mathbf{k}\rangle$, n is the number of the band and $\mathbf{k}$ is the wave vector. The dependence on $\mathbf{k}$ and $\mathbf{k}'$ of the matrix elements of the operator $\mathcal{H}^v$ is determined by the expression

$$G_{n_1 n_2}^{nn'}(\mathbf{k}, \mathbf{k}') = C_{nn_1}^*(\mathbf{k}) C_{n'n_2}(\mathbf{k}'),$$

where $C_{nn_1}(\mathbf{k})$ are the coefficients of decomposition of the functions $|n, \mathbf{k}\rangle$ on the basis $|n_1, 0\rangle$ ($n_1, n_2 = \pm 3/2, \pm 1/2; \sigma, \sigma' = \pm 1/2$).

The transition from the electron to the hole representation in the operator (6.9) is fulfilled by substituting a_{nk}^+ by $b_{\hat{\mathcal{K}}n, -k}$ (b is the destruction operator of the hole) due to the fact that

$$\hat{\mathcal{K}}|n_1\rangle = (-1)^{3/2 - n_1} |-n_1\rangle,$$

where $\hat{\mathcal{K}}$ is the time-reversal operator. As a result we find a hole Hamiltonian $\mathcal{H}^h$ with a form that fully coincides with (6.9), if we consider that the operators $\mathbf{J}$ and a_{nk}^+ act on the hole states and $I_h = I_v$.

Further, we shall consider only those spin relaxation processes with exciton transitions only between the components of the multiplets of the ground $1S$ state. The state of the exciton will be characterized again by the wave vector $\mathbf{K}$ and the row γ of the irreducible representation Γ of the point group of symmetry of the crystal. The wave function of the exciton is expressed via the Luttinger basis $\psi_{\sigma k_1}^e$ and $\psi_{nk_2}^h$

for the electron and hole bands Γ_6 and Γ_8:

$$\Psi_{\Gamma\gamma\mathbf{K}} = \sum_{\sigma n} \sum_{\mathbf{k}_1\mathbf{k}_2} B_{\sigma\mathbf{k}_1,n\mathbf{k}_2}^{\Gamma\gamma\mathbf{K}} \psi_{\sigma\mathbf{k}_1}^{e} \psi_{n\mathbf{k}_2}^{h},$$

$$B_{\sigma\mathbf{k}_1,n\mathbf{k}_2}^{\Gamma\gamma\mathbf{K}} = \delta_{\mathbf{k}_1,\mathbf{K}-\mathbf{k}_2} D_{\sigma n}^{\Gamma\gamma} \varphi_{1S}(\mathbf{k}_2 - \beta\mathbf{K}). \tag{6.10}$$

Here $\varphi_{1S}(\mathbf{K})$ is the Fourier transform of the function of the relative motion of the carriers for $1S$ excitons, and $D_{\sigma n}^{\Gamma\gamma}$ are the components of the vectors of the effective spin Hamiltonian [122]:

$$\mathcal{H}(\mathbf{H}) = -\Delta_1 \mathbf{J}\mathbf{S} - \Delta_2 \left(S_x J_x^3 + S_y J_y^3 + S_z J_z^3\right) + \mathcal{H}_Z(\mathbf{H}), \tag{6.11}$$

where Δ_1 and Δ_2 are the constants of the isotropic and anisotropic exchange interactions,[1] and $\mathcal{H}_Z(\mathbf{H})$ includes the operators of the Zeeman energies of the electrons and holes.

The Hamiltonian (6.11) describes the isotropic exchange splitting on the exciton states with total angular momenta $J = 1, 2$, which in the cubic field are transformed into Γ_4 and $\Gamma_3 \oplus \Gamma_5$ or into Γ_5 and $\Gamma_3 \oplus \Gamma_4$. Since both cases lead to similar results, we shall suppose that the first will be realized.

The peculiarities of the fine structure of the exciton may be displayed in the spin relaxation of the PC, if $\langle \mathcal{H}_Z \rangle \ll \Delta_{1,2}$, which corresponds to a choice of a quite weak magnetic field $\mathbf{H}$. Therefore, it is convenient to represent the interaction operators (6.8) and (6.9) in the basis $|\Gamma\gamma, \mathbf{K}\rangle$ where $\Gamma = \Gamma_4, \Gamma_5, \Gamma_3$, and γ numbers the correct wave functions of the zero approximation on $\mathcal{H}_Z$ in the limits of each multiplet. For the case $\mathbf{H} \parallel [001]$ the correct functions of representations Γ_4, Γ_5, and Γ_3 are given, for example, in [122]. In this basis the operators $\mathcal{H}^e$ and $\mathcal{H}^h$, because they have similar structures, are convenient to unite. After summation over the vectors $\mathbf{k}_1$ and $\mathbf{k}_2$ from (6.10) and using the approximation $\varphi_{1S}(\mathbf{K})$ by the exponential function [422, 434], the Hamiltonian of the exchange interaction of the excitons with PCs, $\mathcal{H}^{ex} = \mathcal{H}^e + \mathcal{H}^h$, has the form (6.1). For $\Gamma_6 \otimes \Gamma_8 \otimes \Gamma_1$ excitons (unlike the case of excitons of type $\Gamma_6 \otimes \Gamma_7 \otimes \Gamma_1$) the spin operators $\mathcal{H}^{ex}_{\Gamma\gamma,\Gamma'\gamma'}(\mathbf{K} - \mathbf{K}')$ in formula (6.1) are depicted by 8×8 matrices, which in the basis of the functions of the representations Γ_4, Γ_5, and Γ_3 have the following structures [434]:

$$\mathcal{H}^{ex}_{\Gamma_4,\Gamma_4}(\mathbf{K} - \mathbf{K}') = a\mathbf{J}_1\mathbf{S}, \qquad \mathcal{H}^{ex}_{\Gamma_5,\Gamma_5}(\mathbf{K} - \mathbf{K}') = -c\mathbf{J}_1\mathbf{S},$$

$$\mathcal{H}^{ex}_{\Gamma_3,\Gamma_3}(\mathbf{K} - \mathbf{K}') = 0,$$

$$\mathcal{H}^{ex}_{\Gamma_4,\Gamma_5}(\mathbf{K} - \mathbf{K}') = \sqrt{\frac{3}{2}} b \begin{pmatrix} 0 & S_+ & \sqrt{2}S_z \\ S_- & 0 & S_+ \\ \sqrt{2}S_z & -S_- & 0 \end{pmatrix}, \tag{6.12}$$

---

[1] Δ_1 and Δ_2 from (6.11) differ by the coefficients from Δ_1 and Δ_2 used in the work [442].

$$\mathcal{H}^{\mathrm{ex}}_{\Gamma_4, \Gamma_3}(\mathbf{K} - \mathbf{K}') = b \begin{pmatrix} -\frac{1}{\sqrt{2}} S_- & \sqrt{\frac{3}{2}} S_+ \\ 2 S_z & 0 \\ -\frac{1}{\sqrt{2}} S_+ & \sqrt{\frac{3}{2}} S_- \end{pmatrix}.$$

The matrix $\mathcal{H}^{\mathrm{ex}}_{\Gamma_5, \Gamma_3}(\mathbf{K} - \mathbf{K}')$ is obtained from $\mathcal{H}^{\mathrm{ex}}_{\Gamma_4, \Gamma_3}(\mathbf{K} - \mathbf{K}')$ if in the latter case the places of the columns are changed, the signs at S_- are changed into the opposite ones, and instead of b we write c. In (6.12) we introduce the designations

$$a = \frac{1}{2}(5\tilde{I}_{\mathrm{h}} - \tilde{I}_{\mathrm{e}}), \qquad b = \frac{1}{2}(\tilde{I}_{\mathrm{e}} - \tilde{I}_{\mathrm{h}}), \qquad c = \frac{1}{2}(\tilde{I}_{\mathrm{e}} + 3\tilde{I}_{\mathrm{h}}),$$

$$\tilde{I}_{\mathrm{e}} = I_{\mathrm{e}} \exp\{-a_{\mathrm{ex}}^2 \beta^2 (\mathbf{K} - \mathbf{K}')^2\}, \tag{6.13}$$

$$\tilde{I}_{\mathrm{h}} = I_{\mathrm{h}} \exp\{-a_{\mathrm{ex}}^2 (1 - \beta)^2 (\mathbf{K} - \mathbf{K}')^2\}.$$

Here $\mathbf{J}_1$ is the operator of the effective momentum ($J_1 = 1$) determined on triplet states Γ_4 or Γ_5.

As seen from (6.13), at $a_{\mathrm{ex}}|\mathbf{K} - \mathbf{K}'| \ll 1$ (low temperatures, elastic scattering processes) we obtain $\tilde{I}_{\mathrm{e}} = I_{\mathrm{e}}$, $\tilde{I}_{\mathrm{h}} = I_{\mathrm{h}}$; therefore, the relative motion of the carriers in the exciton do not influence the spin relaxation of the PC.

Let us remark that the matrices of the Zeeman Hamiltonians of the electron and hole in the basis Γ_3, Γ_5, Γ_4 with precision up to the dimensional factors have the form (6.12), if S_μ is substituted by H_μ, $\mu = x, y, z$. It follows that the g-factors of the triplet states Γ_4 and Γ_5 are isotropic. This conclusion comes from the sequence of the cubic symmetry of the systems with an even number of electrons [5], but it contradicts the conclusion of the work [122] on the strong anisotropy of the g-factor for Γ_5 excitons. The correct expression for the g-factor in [122] is given for the case $\mathbf{H} \parallel [100]$.

We shall consider that the excitons are found in thermal equilibrium with the lattice (the lifetime of the exciton is longer than the relaxation time). The probability of the relaxation transition of PCs with spin reorientation is connected with the cross-section of the exchange scattering $\bar{\sigma}_{M, M'}$ (see formula (6.3)), averaged on filling exciton numbers. The calculation of $\bar{\sigma}_{M, M'}$ will be performed in Born's approximation and high temperature limit, when the energy of the thermal movement $k_0 T$ is much higher than the Zeeman splitting of the PC and the exciton. In this approximation in $\bar{\sigma}_{M, M'}$ we can evidentiate the sums $\sum_{\gamma \in \Gamma} \sum_{\gamma' \in \Gamma'} |\mathcal{H}^{\mathrm{ex}}_{\Gamma\gamma, \Gamma'\gamma'}(\mathbf{K} - \mathbf{K}')|^2$ which are invariant relative to any unitary transformations of the basis functions in the limits of every multiplet Γ_3, Γ_5, and Γ_4, including the case of an arbitrary change of the direction of magnetic field $\mathbf{H}$ relative to the axis [001]. Thus, the spin relaxation of a PC is an isotropic process. Besides, as is seen from (6.12), the selection rules at the exchange scattering of excitons are similar to those of the free electrons [438] and holes [439], $M - M' = \pm 1$. For low exciton concentrations n_{ex} (when the Boltzmann statistics is used), during the calculations of $\bar{\sigma}_{M, M'}$ the integrals $I(s, t)$ appear. Using the approximation $I(s, t)$ through the modified Bessel function $\mathcal{K}_1(z)$ (see Appendix G) in the whole interval of temperatures ($k_0 T < G$) considered, we find that

$$\bar{\sigma}_{M,M\pm1} = \frac{m_{ex}^2}{\pi\hbar^4}(S_{\mp})_{M,M\pm1}^2 I^2(T),\qquad(6.14)$$

where

$$I^2(T) = \left[2 + 3\left(e^{-x_4} + e^{-x_5}\right)\right]^{-1}\left\{F(x_5)I_3^2(y_5)\right.$$

$$\left. + F(x_4)I_{-1}^2(y_4) + \frac{1}{4}\left[e^{-x_5}I_3^2(0) + e^{-x_4}I_{-5}^2(0)\right]\right.$$

$$\left. + \frac{3}{2}e^{-x_5}F(x_4 - x_5)I_{-1}^2(y_4 - y_5)\right\},\qquad(6.15)$$

$$I_p^2(y) = I_e^2\lambda\left(\frac{1}{\nu}, y\right) + p^2 I_h^2\lambda(\nu, y) + 2p I_e I_h\lambda\left(\frac{\nu + \nu^{-1}}{2}, y\right).$$

Here $x_i = E(\Gamma_i)/k_0 T$, $y_i = E(\Gamma_i)/G$, $\nu = m_e/m_h$, and $\lambda(\xi, y)$ while $\xi = \nu, \nu^{-1}, \frac{1}{2}(\nu + \nu^{-1})$ are found according to formula (6.6). The function F (continuous and smooth) is determined by the expression

$$F(x) = e^{-x\theta(x)}\int_0^\infty \sqrt{t(t + |x|)}e^{-t}\,dt = \frac{1}{2}|x|e^{-x/2}\mathcal{K}_1\left(\frac{1}{2}|x|\right),$$

where $\theta(x)$ is the Heaviside step function and $\mathcal{K}_1$ is McDonald's function.

If we select $E(\Gamma_3)$ as the beginning of the reading of energy, then

$$E(\Gamma_4) = 2\Delta_1 + 5\Delta_2, \qquad E(\Gamma_5) = \frac{3}{2}\Delta_2.$$

The coefficient $\lambda(\xi, y)$ from (6.15) describes the influence of the relative movement of the electron and hole in the exciton on the spin relaxation of the PC. The exponent in the numerator $\lambda(\xi, y)$ (formula (6.7)) determines the ratio of cross-sections of the elastic and nonelastic scattering on PCs and is transformed in the unit, if the scattering is not connected with the exchange of the kinetic energy of the exciton by spin one. The denominator $\lambda(\xi, y)$ takes into consideration the decrease of the cross-section of the exchange scattering with the increase of the exciton kinetic energy.

6.3 Shortening of Spin Relaxation Time of Paramagnetic Centers due to Interaction with Excitons

We shall discuss in detail the influence of the Coulomb binding of the carriers into excitons on the PC relaxation rate. If $\nu < 1$ then the relative motion of the carriers in the exciton will first affect the contribution of the electron part ($\sim I_e^2$) in relaxation. If thereby $I_e \gg I_h$, as is expected for the positively charged PC, then in the domain

Fig. 6.1 Temperature dependence of the exchange cross-section at scattering of excitons on PCs at $\eta = -0.75$, $|\Delta_2| = \Delta_1/20$, $\nu = 1$. 1—$\Delta_2 < 0$, 2—$\Delta_2 > 0$, 3—$G \gg 30\Delta_1$, 4—$G = 30\Delta_1$

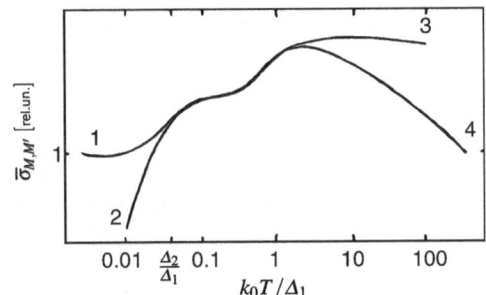

of high temperatures, $k_0T > \nu G/8$, the cross-section $\overline{\sigma}_{M,M'} \sim T^{-1}$ and the probability $W_{M,M'} \sim T^{-1/2}$. At low temperatures, $k_0T < \nu G/8$, the dependence $\overline{\sigma}_{M,M'}$ on T is determined by the ratio $E_{\Gamma\Gamma'}/k_0T$. If $|E_{\Gamma\Gamma'}|/k_0T < 1$ (small constants Δ_1 and Δ_2), then $F(E_{\Gamma\Gamma'}/k_0T) = 1$ and $\overline{\sigma}_{M,M'} = \text{const}$. In this case $W_{M,M'} \sim T^{1/2}$. Thus, while passing from low to high temperatures, the cross-section $\overline{\sigma}_{M,M'}(T)$ passes via a maximum, whose position in the case here is determined from the condition

$$k_0 T_{\max} = \frac{1}{8}\nu G.$$

We shall analyze the temperature dependencies (6.14) in the case of considerable Δ_1 and Δ_2 for $\Delta_1 > 0$ (ferromagnetic electron–hole exchange), $|\Delta_2| \ll \Delta_1$, and $\lambda(\xi, y) = 1$. In dependence on the relation between the temperature T and Δ_2, Δ_1, formula (6.14) admits the following approximations:

(A) $\Delta_2 > 0, k_0T < \Delta_2$. This case corresponds to the maximal population of the doublet Γ_3—the ground state of the exciton in the limits of which, in correspondence with (6.12), exchange scattering with spin reorientation does not take place. Therefore, the temperature dependence of $\overline{\sigma}_{M,M'}$ is determined by the relaxation transitions between the states Γ_3 and Γ_5, and $\overline{\sigma}_{M,M'}$ exponentially goes out with the decrease of temperature ($\overline{\sigma}_{M,M'} \sim e^{-E(\Gamma_5)/k_0T}$).

(B) $\Delta_2 < 0, k_0T \ll |\Delta_2|$. The triplet Γ_5 is the ground state of the exciton. Ignoring the transitions to the Γ_3 state, we find

$$I^2(T) = \frac{1}{12}(I_e + 3I_h)^2, \tag{6.16}$$

so that $\overline{\sigma}_{M,M'} = \text{const}$ and $W_{M,M'} \sim T^{1/2}$.

(C) $|\Delta_2| < k_0T < \Delta_1$. Relaxation transitions of the excitons during scattering on PCs take place as if there were no cubic anisotropy ($\Delta_2 = 0$) in the Hamiltonian (6.11). The result for $I^2(T)$ is distinguished from (6.16) by the coefficient $1/4$ instead of $1/12$. Therefore, during the transition from situation B to situation C, the "stepped" dependence $\overline{\sigma}_{M,M'}$ in the region $k_0T \sim |\Delta_2|$ is realized, but during the transition from A to B we have the rapid decrease of the slope of the temperature curve $\overline{\sigma}_{M,M'}$ (Fig. 6.1).

(D) $\Delta_1 < k_0 T < \nu G/8$. In this temperature domain we have

$$I^2(T) = \frac{1}{2}(I_e^2 + 5I_h^2).\tag{6.17}$$

The transition to this case is also accompanied by a "stepped" change of the cross-section $\overline{\sigma}_{M,M'}$. The position of the "step" on the graph of the temperature dependence $\overline{\sigma}_{M,M'}(T)$ corresponds to $k_0 T \sim \Delta_1$, and the height of the step determines the relation

$$\alpha = \frac{1}{2} \cdot \frac{(I_e + 3I_h)^2}{I_e^2 + 5I_h^2}.$$

The parameter α is connected with the ratio of electron and hole exchange integrals $\eta = I_e/I_h$ as follows:

$$\alpha = \frac{1}{2} \cdot \frac{(\eta + 3)^2}{\eta^2 + 5}.$$

The analysis shows that $\eta > 0$ while $0.9 \le \alpha \le 1.4$ and $\eta < 0$ if $0 \le \alpha < 0.5$. In the domain $0.5 < \alpha < 0.9$ the sign η is not determined uniquely, but it can be established if the order of the magnitude $|\eta|$ is known.

As is seen, in the temperature domain $k_0 T < E(\Gamma_4)$ the relaxation rate of PCs via excitons depends on the values and signs of the constants I_e and I_h. In particular, when $\eta = -3$ complete compensation of the amplitude of scattering from the electron and hole coupled into the exciton takes place. In the domain of high temperatures $k_0 T > E(\Gamma_4)$ the contribution to the spin relaxation of the PC from the electron and hole is additive and does not depend on the signs of the constant of the exchange interaction. Thus, the contribution from I_e in (6.14) for the D case with precision up to the coefficient $(m_{ex}/m_e)^2 > 1$ coincides with the cross-section of the exchange scattering of free electrons $\overline{\sigma}_{M,M'}^e$. The contribution of the holes in (6.14) for this case is connected with $\overline{\sigma}_{M,M'}^h$—the cross-section of the scattering of free holes on the PC. This contribution is not only realized through one coefficient $(m_{ex}/m_h)^2$, since for the free holes partial compensation of the amplitude of intra- and interband scattering of light and heavy holes takes place. The last effect leads to the appearance of the factor 1.5 [439] instead of 2.5 in (6.17). The binding of the carriers into excitons may intensify the relaxation of the PC (in comparison with the relaxation via the free carriers) by the factor

$$r_{\Gamma_6 \otimes \Gamma_8 \otimes \Gamma_1} = (\eta^2 + 5)\frac{1 + \nu^{3/2}}{1.5 + \nu^{3/2}\eta^2} > 1\tag{6.18}$$

at high temperatures (D case) or may lead to a slowing down of the spin relaxation at low temperatures for $\eta = -3$ (cases B and C) or for $\Delta > 0$ (case A).

A similar acceleration of the relaxation of the PC during Coulomb binding of the charge carriers into excitons takes place for the $\Gamma_6 \otimes \Gamma_7 \otimes \Gamma_1$ excitons, considered in Sect. 6.2. In this case the factor of the exciton reduction of the PC relaxation time

$r_{\Gamma_6 \otimes \Gamma_7 \otimes \Gamma_1}$ is determined by the expression

$$r_{\Gamma_6 \otimes \Gamma_7 \otimes \Gamma_1} = \left(\eta^2 + 1\right) \frac{(1+\nu)^{3/2}}{1 + \nu^{3/2}\eta^2} > 1.$$

For the crystal CdTe:Mn^{2+} ($\eta = -0.75$ [71–73], $\Delta_1 = 0.3$ meV [122], $a_{ex} = 60$ Å) the position of the "step" on the graph of the temperature dependence $\overline{\sigma}_{M,M'}(T)$ corresponds to $T \sim \Delta_1/k_0 \simeq 3.5$ K (Fig. 6.1). At the same concentrations of free electrons and holes, in the domain $T > \Delta_1/k_0$ using formula (6.18), we obtain $r_{\Gamma_6 \otimes \Gamma_8 \otimes \Gamma_1} \simeq 5$. This witnesses the efficiency of the mechanism of PC relaxation during the exchange interaction with the excitons in the CdTe crystal.

The large spin relaxation rate of PCs via excitons is affected by a strong electron–impurity and hole–impurity exchange interaction (the electrons and holes forming free excitons are involved simultaneously in Coulomb and exchange interactions). The intraexciton electron–hole exchange interaction determines the splittings and the symmetry of the ground state of the exciton and leads to the particularities in the temperature dependence of exchange scattering cross-section $\overline{\sigma}_{M,M'}$. If the ground state is the nonmagnetic doublet Γ_3, then the relaxation is realized by means of transitions into the above Γ_5 state and determines the exponential temperature dependence of $\overline{\sigma}_{M,M'}$. The "stepped" temperature dependence of $\overline{\sigma}_{M,M'}$ in the domain $k_0 T \sim \Delta_1 > 0$ witnesses the introduction of a new relaxation channel connected with Γ_4 states. We draw attention to the fact that this method of investigation with Δ_1 and Δ_2 may be more effective than that using optical methods, since the Γ_3 and Γ_5 exciton states are optically forbidden.

6.4 Indirect Interaction of the Paramagnetic Centers via Excitons

The development of the Vonsovsky c–l model [443] (c denotes the free electron and l denotes the localized spin) and the creation on its basis of the Ruderman–Kittel–Kasuya–Yosida (RKKY) theory allowed one to explain a number of regularities in the physics of electron and nuclear magnetism influenced by indirect exchange interaction between localized spins [17, 23, 303, 444–446]. The indirect exchange between the PCs in semiconductors by means of a nondegenerate electron gas (subject to the same statistics as the excitons) was considered by Karpenko and Berdyshev [447] and Kochelaev [448], who showed that this interaction decreases monotonously with increasing distance between the PCs. The shielding radius of the exchange interaction depends on the temperature and in a wide interval of the change of m_e and T exceeds the magnitude of the interatomic distance.

Among other mechanisms of the indirect exchange we note the exchange of the spins of the PCs by virtual phonons. The strict theory of this mechanism in the ignorance of delay was developed by Aminov and Kochelaev [449], but a theory accounting for the effect of delay was developed by Orbach and Tachiki [450].

Deigen, Kashirina, and Suslin [451] calculated the indirect interaction of the PC via the phonon field into ionic and atomic crystals, taking into account the permutation symmetry of the wave function, and showed that this interaction according to the order of magnitude is comparable with the Coulomb exchange. This type of indirect spin–spin exchange via the phonon field (as well as plasmons), with a nonrelativistic origin, was studied in [451–455].

Let us consider the exchange interaction of the paramagnetic ions via the triplet excitons in the model for a crystal having simple energy bands with extremes in the center of the Brillouin zone. The Hamiltonian of interaction of the electrons and holes forming the excitons with the paramagnetic impurity centers is represented in the form

$$
\mathcal{H}_{\text{int}} = -\frac{1}{2N} \sum_{\mathbf{k}_e, \mathbf{k}_e'} \sum_n \mathcal{I}_e(\mathbf{k}_e, \mathbf{k}_e') \exp[i(\mathbf{k}_e - \mathbf{k}_e')\mathbf{R}_n]
$$

$$
\times \left[S_n^+ a_{\mathbf{k}_e', \downarrow}^+ a_{\mathbf{k}_e, \uparrow} + S_n^- a_{\mathbf{k}_e', \uparrow}^+ a_{\mathbf{k}_e, \downarrow} + S_n^z \left(a_{\mathbf{k}_e', \uparrow}^+ a_{\mathbf{k}_e, \uparrow} - a_{\mathbf{k}_e', \downarrow}^+ a_{\mathbf{k}_e, \downarrow} \right) \right]
$$

$$
-\frac{1}{2N} \sum_{\mathbf{k}_h, \mathbf{k}_h'} \sum_n \mathcal{I}_h(\mathbf{k}_h, \mathbf{k}_h') \exp[i(\mathbf{k}_h - \mathbf{k}_h')\mathbf{R}_n]
$$

$$
\times \left[S_n^+ b_{\mathbf{k}_h', \downarrow}^+ b_{\mathbf{k}_h, \uparrow} + S_n^- b_{\mathbf{k}_h', \uparrow}^+ b_{\mathbf{k}_h, \downarrow} + S_n^z \left(b_{\mathbf{k}_h', \uparrow}^+ b_{\mathbf{k}_h, \uparrow} - b_{\mathbf{k}_h', \downarrow}^+ b_{\mathbf{k}_h, \downarrow} \right) \right],
$$

where S_n is the spin operator of the paramagnetic ion, $\mathcal{I}_e(\mathbf{k}_e, \mathbf{k}_e')$ and $\mathcal{I}_h(\mathbf{k}_h, \mathbf{k}_h')$ are the integrals of the exchange interaction of the PC with the electron and the hole of the free exciton, N is the number of paramagnetic ions in the crystal, and $\mathbf{R}_n$ is the radius vector of the PC.

The energy of the indirect exchange interaction of the PC via excitons will be determined by second order perturbation theory. After the corresponding summation on the exciton spin states determined by formula (5.2) and averaging on the probabilities of finding the excitons in the states with different wave vectors at given temperature, the operator of the indirect exchange interaction of the PC via excitons may be represented in the form of the Dirac exchange Hamiltonian

$$
H'' = -\sum_{mn} \mathcal{I}_{\text{eff}}(\mathbf{R}_{mn}) \mathbf{S}_m \mathbf{S}_n, \tag{6.19}
$$

where the effective exchange integral $\mathcal{I}_{\text{eff}}(\mathbf{R}_{mn})$ has the form

$$
\mathcal{I}_{\text{eff}}(\mathbf{R}_{mn}) = \frac{1}{2} \left(\frac{\mathcal{I}}{N} \right)^2 \sum_{\mathbf{K}, \mathbf{K}'} \frac{B e^{-\gamma \mathbf{K}^2}}{\varepsilon_{\mathbf{K}} - \varepsilon_{\mathbf{K}'}} e^{i(\mathbf{K} - \mathbf{K}')\mathbf{R}_{mn}}
$$

$$
\times \left\{ \left[1 + \frac{a_{\text{ex}}^2 \alpha^2}{4} (\mathbf{K} - \mathbf{K}')^2 \right]^{-4} + \left[1 + \frac{a_{\text{ex}}^2 \beta^2}{4} (\mathbf{K} - \mathbf{K}')^2 \right]^{-4} \right.
$$

$$
\left. + 2 \left[1 + \frac{a_{\text{ex}}^2 \alpha^2}{4} (\mathbf{K} - \mathbf{K}')^2 \right]^{-2} \times \left[1 + \frac{a_{\text{ex}}^2 \beta^2}{4} (\mathbf{K} - \mathbf{K}')^2 \right]^{-2} \right\}. \tag{6.20}
$$

In (6.20) $\varepsilon_{\mathbf{K}}$ is the kinetic energy of the $1S$ exciton with wave vector $\mathbf{K} = \mathbf{k}_e + \mathbf{k}_h$, $\beta = m_h/m_{ex}$, $\alpha = 1 - \beta$, where B and γ are designated as

$$B = n_{ex}(4\pi\gamma)^{3/2}, \quad \gamma = \hbar^2(2m_{ex}k_0T)^{-1}.$$

For formula (6.19) it was assumed that $\mathcal{I}_e(\mathbf{k}_e, \mathbf{k}_e')$ and $\mathcal{I}_h(\mathbf{k}_h, \mathbf{k}_h')$ do not depend on the wave vectors of the electron and hole (similarly to the situation for the contact hyperfine interaction of free carriers with nuclei [29, 378, 417]), which corresponds to the short-range exchange interaction of the electrons and holes with PCs. For simplicity, it was assumed that the electron and hole interact similarly with the PC ($\mathcal{I}_e = \mathcal{I}_h = \mathcal{I}$). However, this assumption may not be satisfied if we take into account the real structure of the bands of the crystal, as occurs, for example, for the conduction band Γ_6 and valence band Γ_8 in cubic crystals (see Sect. 6.2).

The first term in (6.20) is influenced by the contribution from the exchange interaction of the electrons with the PCs. Taking into account the correlation

$$\left[1 + \frac{a_{ex}^2\alpha^2}{4}(\mathbf{K} - \mathbf{K}')^2\right]^4 = \int d\mathbf{r}_1 \, \varphi^2(\mathbf{r}_1)e^{i\alpha\mathbf{r}_1(\mathbf{K}-\mathbf{K}')}$$

$$\times \int d\mathbf{r}_2 \, \varphi^2(\mathbf{r}_2)e^{i\alpha\mathbf{r}_2(\mathbf{K}-\mathbf{K}')},$$

where $\varphi(\mathbf{r})$ is the wave function of the relative electron–hole motion in the $1S$ exciton, this term may be represented in the form

$$\mathcal{I}_{eff}^{(e)}(\mathbf{R}_{mn}) = \frac{1}{2}\left(\frac{\mathcal{I}}{N}\right)^2 \frac{m_{ex}B}{\hbar^2(2\pi)^6} \iint d\mathbf{r}_1 \, d\mathbf{r}_2 \, \varphi^2(\mathbf{r}_1)\varphi^2(\mathbf{r}_2)$$

$$\times \int d\mathbf{K} \, e^{-\gamma K^2 + i\mathbf{K}[\mathbf{R}_{mn}+\alpha(\mathbf{r}_1+\mathbf{r}_2)]} \int d\mathbf{K}' \left(K^2 - K'^2\right)^{-1} \qquad (6.21)$$

$$\times e^{-i\mathbf{K}'[\mathbf{R}_{mn}+\alpha(\mathbf{r}_1+\mathbf{r}_2)]}.$$

The principal value of the integral on $\mathbf{K}'$ in (6.21) is equal to

$$\mathcal{P}\int \frac{d\mathbf{K}'}{K^2 - K'^2}e^{i\mathbf{K}'[\mathbf{R}_{mn}+\alpha(\mathbf{r}_1+\mathbf{r}_2)]} = 2\pi^2\frac{\cos K|\mathbf{R}_{mn} + \alpha(\mathbf{r}_1 + \mathbf{r}_2)|}{|\mathbf{R}_{mn} + \alpha(\mathbf{r}_1 + \mathbf{r}_2)|}. \qquad (6.22)$$

Further integration of the expression (6.20) is rather bulky, and the final result depends on the parameters of the system. In the particular case $\alpha = \beta$ ($m_e = m_h$) for $\mathcal{I}_{eff}(\mathbf{R}_{mn})$ we have

$$\mathcal{I}_{eff}(\mathbf{R}_{mn}) = \frac{1}{8}\left(\frac{\mathcal{I}}{n}\right)^2 \frac{m_{ex}n_{ex}}{\hbar^2\alpha\pi^{1/2}pa_{ex}}$$

$$\times \frac{g}{\alpha}\left\{\frac{5}{4}\alpha^2\left[1 - \frac{2p}{\alpha} - \frac{6g}{5\alpha^2} + \frac{8p^2}{5\alpha} - \frac{8p}{5\alpha^3}\right.\right.$$

Fig. 6.2 Dependence of
effective exchange integral
$\mathcal{I}_{\text{eff}}$ on distance between PCs
at their indirect interaction
through excitons

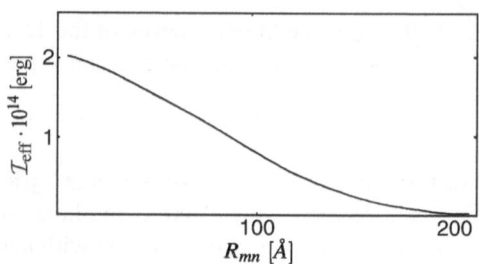

$$\times \left(\frac{p^2}{3} - \frac{3}{2}g^2 \right) - \frac{8g}{5\alpha^4} \left(p^2 - \frac{g^2}{2} \right) - \frac{8pg^4}{5\alpha^5}$$

$$- \frac{8}{15} \frac{g^6}{\alpha^6} \right] \exp\left(\frac{g^2}{\alpha^2} + \frac{2p}{\alpha} \right) \left[1 - \Phi\left(\frac{p}{g} + \frac{g}{2} \right) \right]$$

$$- \frac{5\alpha^3}{4} \left[1 + \frac{2p}{\alpha} - \frac{6g^2}{5\alpha^2} + \frac{8p^2}{5\alpha^2} + \frac{8p}{5\alpha^3} \left(\frac{p^2}{3} - \frac{3}{2}g^2 \right) \right.$$

$$- \frac{8g^2}{5\alpha^4} \left(p^2 - \frac{g^2}{2} \right) + \frac{8pg^4}{5\alpha^5} - \frac{8}{15} \frac{g^6}{\alpha^6} \right] \exp\left(\frac{g^2}{\alpha^2} - \frac{2p}{\alpha} \right)$$

$$\times \left[1 - \Phi\left(\frac{p}{g} - \frac{g}{\alpha} \right) - 2\Phi\left(\frac{g}{\alpha} \right) \right] + \exp\left(-\frac{p^2}{g^2} \right)$$

$$\times \left[1 + \frac{4}{15\alpha^2} (p^2 - 2g^2) + \frac{4g^4}{15\alpha^4} \right] - \frac{5\alpha g}{\sqrt{\pi}} \exp\left(-\frac{2p}{\alpha} \right)$$

$$\times \left[1 + \frac{8p}{5\alpha} + \frac{4}{5\alpha^2} \left(p^2 - \frac{2}{3}g^2 \right) - \frac{4pg^2}{5\alpha^3} + \frac{4}{15} \frac{g^4}{\alpha^4} \right] \right\}, \qquad (6.23)$$

where n is the concentration of the paramagnetic impurity centers, $p = R_{mn}/a_{\text{ex}}$, $g = \alpha (G/k_0 T)^{1/2}$, and $\Phi(x)$ is the probability function. In Fig. 6.2, calculated according to formula (6.23), the dependence of $\mathcal{I}_{\text{eff}}$ on the distance between the PCs is presented with the following parameters: $\alpha = \beta = 1/2$, $G = 10$ meV, $a_{\text{ex}} = 60$ Å, $\mathcal{I} = 10^{-14}$ erg, $n = 10^{16}$ cm^{-3}, $n_{\text{ex}} = 10^{17}$ cm^{-3}, and $T = 4.2$ K. From (6.23) and Fig. 6.2, one sees that the indirect exchange interaction of the PC through the field of excitons depends on the temperature, and at the given temperature it slows down monotonously with increasing distance between the paramagnetic ions. The effective exchange integral $\mathcal{I}_{\text{eff}}$ at $R_{mn} \lesssim 100$ Å appears to be of the order of the constant of the exchange interaction of the electron (hole) with the PC for the selected values of the parameters.

6.5 The Effect of Giant Spin Splitting of the Exciton Band in Diluted Magnetic Semiconductors

Zeeman splittings of the energy levels of excitons in pure semiconductors are interpreted in terms of the effective g-factors that depend on the symmetry of the wave functions of the electron, hole, and the function of their relative motion, as well as on the effective masses of the carriers [113]. An investigation of the Zeeman effect of excitons in undoped CdTe crystals showed that the crystal terms Γ_3, Γ_4, and Γ_5 [456] correspond to the lowest states of the relative motion of the $1S$ type of electron and hole in the exciton. In a tetrahedral crystal field the triplet Γ_5 is optically allowed, but the doublet Γ_3 and triplet Γ_4 may be observed only in strong magnetic fields, and their center of gravity is chipped off from the level Γ_5 on the magnitude $\delta E_{J_{\text{eh}}}$ because of the electron–hole exchange interaction (according to the data of [456], the magnitude of $\delta E_{J_{\text{eh}}}$ is 0.6 meV). This agrees with the accepted suggestion that in cubic direct-band A_2B_6 crystals the holes from the upper (split-off by spin–orbital interaction) valence subband Γ_8 and the electrons from the conduction band Γ_6 are coupled into excitons.

In semiconductors that are strongly doped by paramagnetic impurities we should expect that the splitting of the exciton spectrum will not correspond to the Zeeman effect for excitons in pure (and weakly doped) crystals, if we suppose that the doping does not lead to the localization of the exciton, and the free exciton (its hole and the electron apart) participates in the exchange interaction with $3d$ or $4f$ electrons of the impurity center. This is what was revealed for the first time during the investigation of magneto–optical effects in the exciton spectral range of the crystal CdTe:Mn^{2+} [71–73]. The exchange interaction of PCs with carriers coupled into excitons has already been displayed during the concentration of the magnetic impurities $C \sim 0, 1$ mol.%, which is influenced by the large exciton radius and the movement of excitons in crystals. This interaction leads to giant spin splitting of the exciton states in CdTe:Mn^{2+} crystals and solid solutions Cd$_{1-x}$Mn$_x$Te [71–73, 457–460], in ZnTe:Mn^{2+} [461, 462], ZnSe:Mn^{2+}, and ZnSe:Fe^{2+} [463], CdSe:Mn^{2+} [464], and CdS:Mn^{2+} [465]. Thus, the experimental results have shown that effects of giant spin splitting of excitons are a general feature of direct-band (cubic and hexagonal) A_2B_6 semiconductors if they are strongly doped by paramagnetic ions. We can also conclude that these effects must be characteristic to all semiconductors strongly doped by paramagnetic ions in which excitons are observed.

The Hamiltonian of the exciton in the approximation of spherical symmetry in an external magnetic field for a crystal doped by magnetic impurities has the form

$$\mathcal{H} = \mathcal{H}_{\text{ex}}^{\text{orb}}(\mathbf{K}, \mathbf{H}) + g_e\beta\mathbf{H}\mathbf{S}_e + g_h\beta\mathbf{H}\mathbf{S}_h^{\text{eff}} + J_{\text{eh}}\mathbf{S}_e\mathbf{S}_h^{\text{eff}}$$
$$+ \sum_i J_{eM,i}\mathbf{S}_e\mathbf{S}_M + \sum_i J_{hM,i}\mathbf{S}_h^{\text{eff}}\mathbf{S}_M + \sum_{ij} J_{ij}^{(M)}\mathbf{S}_{Mi}\mathbf{S}_{Mj}, \qquad (6.24)$$

where $\mathcal{H}_{\text{ex}}^{\text{orb}}(\mathbf{K}, \mathbf{H})$ is the Hamiltonian of the orbital movement of the exciton and the carriers in the magnetic field $\mathbf{H}$ that enter its structure ($\mathbf{K}$ is the wave vector of the exciton); S_e, S_h^{eff}, and S_M are the electron spin, the effective spin of the hole ($S_h^{\text{eff}} = 3/2$), and the spin of the PC, g_e and g_h are the g-factors of the electron and hole, J is the exchange constant for different interactions, and i and j denote the paramagnetic ions.

According to [456] the diamagnetic shifts of $1S$ exciton levels for $A_2 B_6$ crystals in magnetic fields $\lesssim 40$ kOe are very small; therefore, we shall ignore the first item from (6.24). Further, the last term of the Hamiltonian appears not to be essential in (6.24), because in the investigated crystals the EPR spectra of the same paramagnetic ions in magneto-concentrated and magneto-diluted crystals are observed at the same magnetic field (for the determination we shall consider the ions of Mn^{2+}).

The free exciton interacts simultaneously with large numbers of Mn^{2+} ions, so in the terms containing the exchange interaction of the electron and hole with the Mn^{2+} ions we may use the model of the "molecular field." Then in the indicated terms, instead of the summation on the Mn^{2+} ions with their individual spins, the average value of the Mn^{2+} ions' spin may be introduced. The term describing the Zeeman interaction of Mn^{2+} ions with the external magnetic field leads to the Brillouin dependence of the magnetization of Mn^{2+} ions, and the exchange terms e–M and h–M determine the magneto–optical effects in the exciton band. In addition, the effective magnetic field of the exchange origin will act on the electron spin, the magnitude of which, as it appears from [71–73, 457–463], greatly exceeds the values of the magnetic fields usually used in magneto–optical experiments. In this case the Hamiltonian (6.24) can be approximately rewritten in the form

$$\mathcal{H}_{\text{ex}}^{S}(\mathbf{H}) = \mathbf{G}_e \mathbf{S}_e + \mathbf{G}_h \mathbf{S}_h^{\text{eff}} + J_{\text{eh}} \mathbf{S}_e \mathbf{S}_h^{\text{eff}}, \qquad (6.25)$$

where

$$\mathbf{G}_{e(h)} = g_{e(h)} \beta \mathbf{H} + I_{e(h)} \langle \mathbf{S}_M \rangle,$$

and for the components of the vector $\langle \mathbf{S}_M \rangle$ in the molecular field approximation we have

$$\langle \mathbf{S}_{XM} \rangle = \langle \mathbf{S}_{YM} \rangle = 0, \qquad \langle \mathbf{S}_{ZM} \rangle = -\frac{5}{2} B_{5/2} \left(\frac{g_M \beta H S_M}{k_0 T} \right). \qquad (6.26)$$

Here $B_S(y)$ is the Brillouin function. The constants $I_{e(h)}$ that characterize the carrier–impurity exchange interaction and depend on the concentration of the PC (for example, for ZnSe:Mn^{2+} the exchange constant $I_e = 9 \cdot 10^{-5}$ eV at $C_M = 6 \cdot 10^{18}$ cm^{-3} and $I_e = 1.8 \cdot 10^{-3}$ eV at $C_M = 2 \cdot 10^{20}$ cm^{-3} [463]), are subject to experimental determination.

We note that the exchange interaction of the carriers with the PCs is a necessary but not sufficient condition of the appearance of the giant spin splitting of excitons in semiconductors doped by magnetic impurities. There should still exist an external magnetic field that is necessary for polarization of PCs and to satisfy the criterion of molecular field theory (formula (6.26)).

The external magnetic field practically does not influence the excitons; their energy spectra are determined by a strong exchange interaction of the carriers that form these excitons with the system of partially spin polarized PCs.

6.6   Giant Magneto–Optical Effects in Diluted Magnetic Semiconductors

In [71–73] a sharp amplification of the magneto–optical effects (Faraday rotation of the plane of polarization of light, magnetic circular dichroism of the reflection) was revealed upon doping of CdTe crystals by Mn^{2+} ions. It was established that the effect causes splitting of the exciton band $1S$ into six components (two central components are weakly separated) under the action of both the internal magnetic field of an exchange nature and the external magnetic field. The splitting is proportional to the concentration and spin polarization of the system of impurity ions. The effective field in which the splitting of the exciton band takes place achieves some hundreds of kilooersteds in the external field of 30 kOe at manganese ion concentrations of $\sim 8 \cdot 10^{18}$ cm^{-3}. The change of the spin polarization of the impurity system at the saturation of microwave transitions between spin levels of Mn^{2+} ions at a frequency of about 10 GHz leads to a decrease in the magnitude of the splitting. All measurements, the results of which are discussed in [71–73], were performed at the temperature $T = 1.7$ K.

Figure 6.3a shows the dispersion dependence of the effect of Faraday rotation (FR) on the long-wavelength slope of the exciton band in CdTe:Mn^{2+} crystals with different degrees of doping for the case $\mathbf{k} \parallel \mathbf{H} \parallel [110]$ ($\mathbf{k}$ is the wave-vector of the light wave). As we approach the center of the exciton line, the FR increases. With increasing concentration C_M of Mn^{2+} ions, a strong increase of FR takes place: $V_1 = 2.6 \cdot 10^{-3}$ degree/cm · Oe for $C_M = 0$ and $V_2 = 3.6$ degree/cm · Oe at

Fig. 6.3 (a) Dispersion curves of FR, obtained on CdTe:Mn^{2+} samples with different degrees of doping: *1*—$C \approx 6 \cdot 10^{18}$ cm^{-3}, *2*—$C \approx 7.2 \cdot 10^{18}$ cm^{-3}. **(b)** OMDR spectra, recorded on different sections of the dispersion curve of FR in CdTe:Mn^{2+} ($C \sim 6 \cdot 10^{18}$ cm^{-3}): *1*—$h\nu_1 = 1.500$ eV, *2*—$h\nu_2 = 1.568$ eV

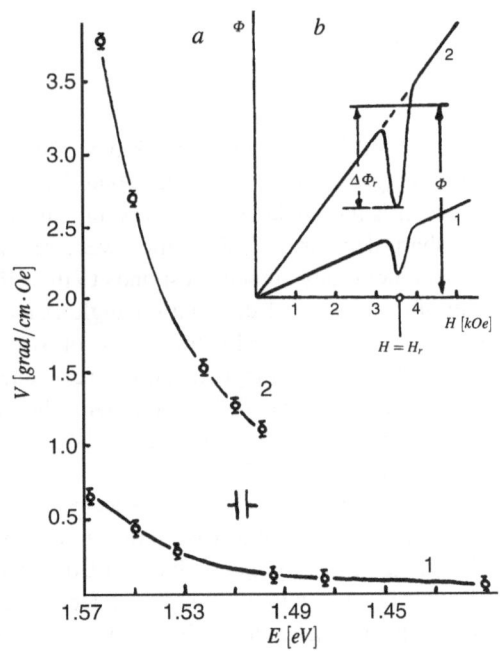

Fig. 6.4 Reflection in the region of the $1S$ exciton band of CdTe:Mn^{2+} (I_{ref}) and MCDR spectra obtained with a crystal having a medium Mn^{2+} concentration ($C \sim 1.2 \cdot 10^{18}$ cm^{-3}) in magnetic fields 3.65 kOe (1), 8 kOe (2), and 18 kOe (3)

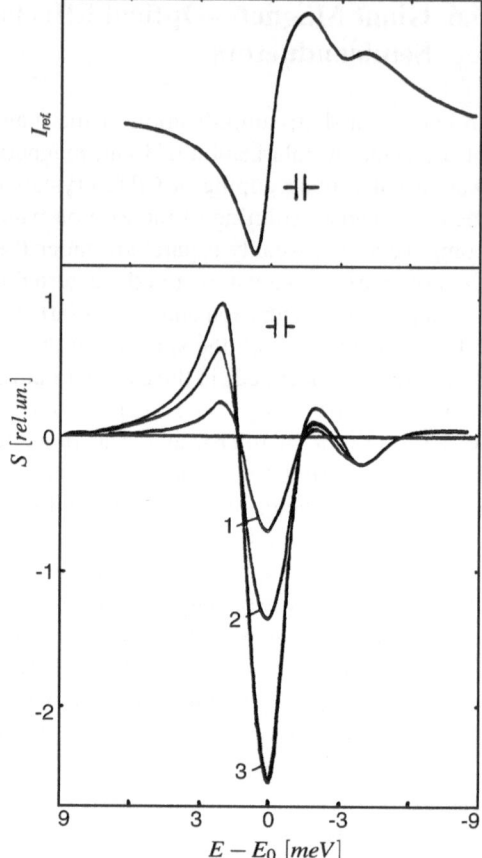

$C_M = 7.2 \cdot 10^{18}$ cm^{-3}; i.e., the Faraday effect in the doped crystal is three orders of magnitude greater than in the pure one. Here $V = \Phi/Hd$, Φ is the angle of FR in degrees, d is the thickness of the crystal, and $H = 3.65$ kOe.

A sharp decrease of FR at microwave pumping of the samples takes place in a magnetic field range that corresponds to the EPR of Mn^{2+} in CdTe (Fig. 6.3b). The depth of the change of the rotation angle $\Delta\Phi_r$ achieves maximum value during the complete saturation of EPR transitions of Mn^{2+} ions. The EPR was detected by its optical effect (optico-magnetic double resonance, OMDR). The relative depth of the resonance dip of $\Delta\Phi_r/\Phi$ increases with the growth of the impurity concentration of Mn^{2+}. In the strongly doped crystals, $\Delta\Phi_r/\Phi \approx 70$ % and does not depend on the frequency of the incident light.

In the doped CdTe crystals the spectra of the magnetic circular dichroism of the reflection (MCDR) at $\mathbf{k} \parallel \mathbf{H}$ were registered. In Fig. 6.4 the spectra of the reflection in the range of the $1S$ exciton band of CdTe:Mn^{2+} (I_{ref}) and the MCDR (the signal $S(E)$) for crystals with an average concentration of Mn^{2+} ($C_M \sim 1.2 \cdot 10^{18}$ cm^{-3}) at different magnetic fields H are presented. With increasing H the intensity of

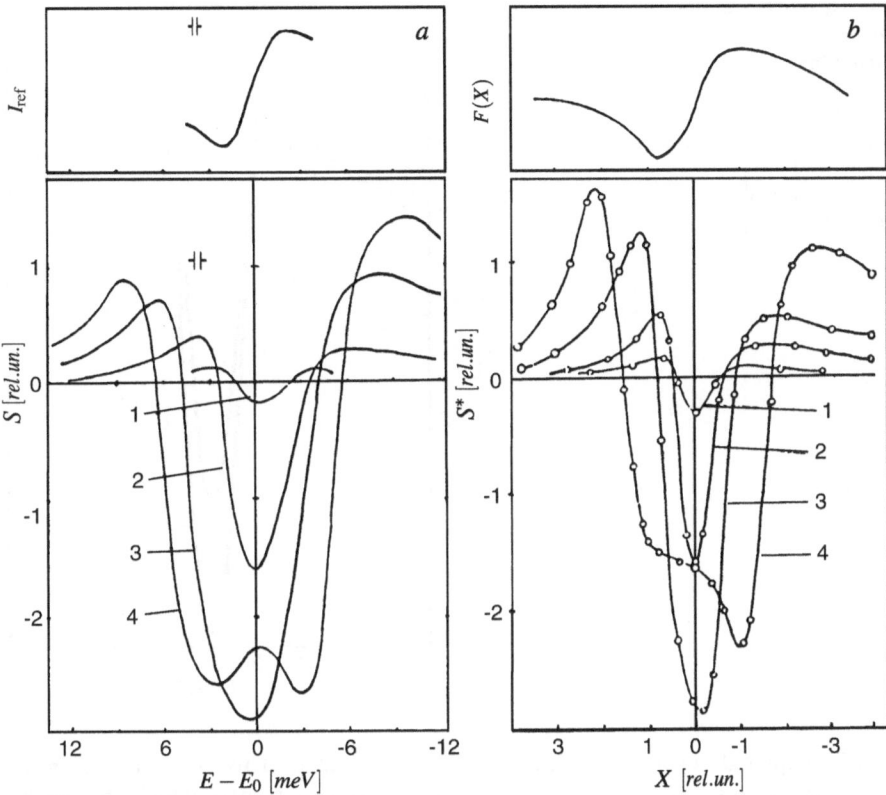

Fig. 6.5 (a) Reflection in the region of the $1S$ exciton band of CdTe:Mn^{2+} (I_{ref}) and MCDR spectra obtained with a sample having a large concentration ($C \approx 8 \cdot 10^{18}$ cm^{-3}) at the following values of the magnetic field: 1—0.6; 2—4.2; 3—10; 4—40 kOe. (b) Contour of tabulated reflection lines $F(x)$ and the calculated dependencies of the quantity $S^*(X) = F(X + \Delta X/2) - F(X - \Delta X/2)$ at the following values of ΔX: 1—0.05; 2—0.2; 3—0.6; 4—1.5

the MCDR line increases. In the strongly doped sample ($C_M \sim 8 \cdot 10^{18}$ cm^{-3}) in magnetic fields of nearly 20 kOe and higher, the manifestation in the MCDR of an anomalous great splitting of the exciton band (Fig. 6.5a) was detected. For comparison, we note that this magnitude of splitting ($\Delta E = 9$ meV) should be expected in a magnetic field $H \approx 760$ kOe. In the spectra of MCDR in Fig. 6.5a (curve 4) two lines are clearly seen, shifted relative to each other in frequency and observed at the two different circular polarizations.

During the microwave pumping of EPR transitions in Mn^{2+} ions the MCDR signal changes similarly to FR: the value of MCDR decreases strongly upon saturation of the EPR of the ground state of the ion Mn^{2+} (Fig. 6.6). The degree of decrease of MCDR during the saturation of EPR transitions of Mn^{2+} ions increases with increasing concentration of the magnetic impurities.

In order to find values of splittings of the exciton band $\Delta E(H)$ from MCDR spectra of the type represented in Fig. 6.5a, the contour of the reflection line

Fig. 6.6 MCDR spectra
obtained with the crystal
CdTe:Mn^{2+}
($C \approx 7 \cdot 10^{18}$ cm^{-3}):
1—without microwave
pumping, 2—with microwave
pumping of the EPR
transition of Mn^{2+};
$H = 3.65$ kOe

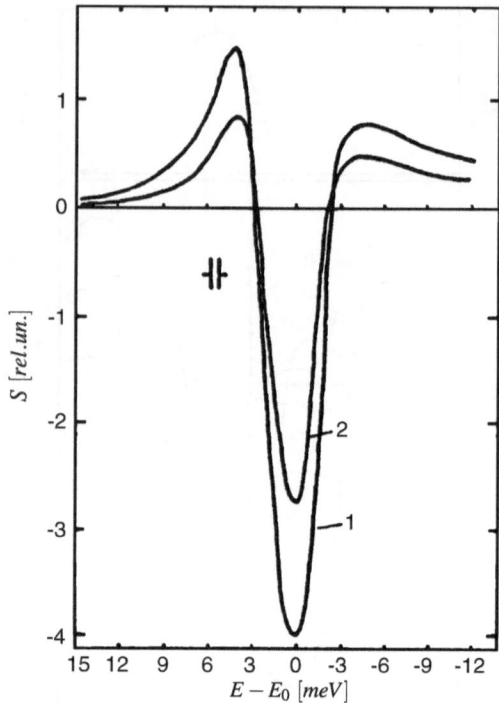

$I_{ref}(E)$ depicted separately in Fig. 6.5a was tabulated. This contour was introduced in the form of a function $F(X)$ in the computer, after which the curves $S^*(X) = F(X + \Delta X/2) - F(X - \Delta X/2)$ were calculated for the consecutive series of the values of the shift ΔX. Some of the calculated curves $S^*(X)$ are presented in Fig. 6.5b. The comparison of the dependencies $S^*(X)$ and $S(E)$ allowed us to establish the character of the dependence of the splitting ΔE from H given in Fig. 6.7. The obtained dependence $\Delta E(H)$ coincides well enough with the Brillouin curve $B_S(y)$ for magnetization of the paramagnetic system with spin $S = 5/2$. This peculiarity is evidently connected with the presence in the crystals of Mn^{2+} ions, possessing in the $^6S_{5/2}$ ground state spin $S_M = 5/2$. In a similar magnetic field H the splitting ΔE for different crystals are greater, the greater the degree of their doping by Mn^{2+} ions. These and the preceding experimental results indicate that the splitting of the exciton band is determined, not by the external magnetic field as such, but by the magnetization of the spin system of the paramagnetic Mn^{2+} ions in CdTe that appears in the external magnetic field H.

Finally, the splitting of the $1S$ exciton reflection band in π and σ polarizations during the propagation of light in the direction $\mathbf{k} \perp \mathbf{H}$ (Fig. 6.8) was registered spectroscopically in the strongly doped crystal CdTe:Mn^{2+} ($C_M \sim 8 \cdot 10^{18}$ cm^{-3}) in a magnetic field of 30 kOe. As shown in Fig. 6.8, five equidistant components of the $1S$ exciton band are observed: two in π polarization ($\mathbf{E} \parallel \mathbf{H}$, $\mathbf{E}$ is the electric field vector of the light wave), separated by $\Delta E \approx 5 \pm 0.5$ meV, and three in σ polarization ($\mathbf{E} \perp \mathbf{H}$)—two separated by $\Delta E \approx 10 \pm 0.5$ meV, and a weak one at the

Fig. 6.7 *Points*—dependence of the splitting $\Delta E(H)$ for the σ^+ and σ^- components of the MCDR spectrum (Fig. 3.3a) obtained by comparison with calculation (Fig. 3.3b). The *solid line* is a plot of the Brillouin dependence $B_S(Y)$ of the magnetization on the field H at $S = 5/2$, obtained at $Y = g\beta HS/kT$, $g = 2$, $T = 1.7$ K

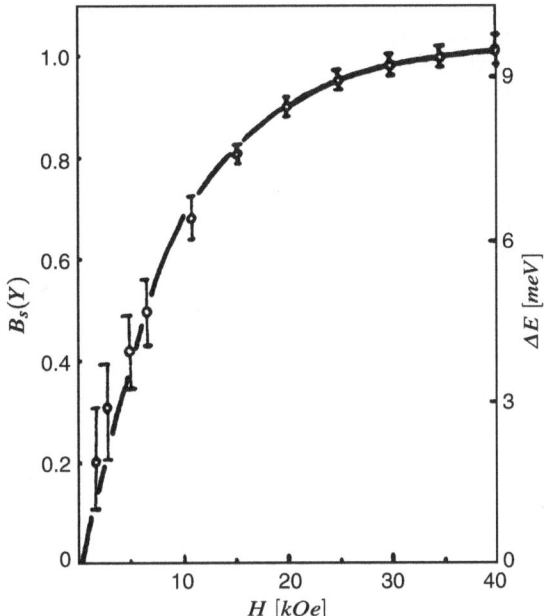

center near E_0. More precise measurements showed that the central line is weakly split into two, each of which is active in one of the σ polarizations. This behavior of the exciton band, at which the giant spin splitting leads to the six-component spectrum of reflection (absorption), in the center of which there are two lines active in two σ polarizations, was detected later in a series of cubic semiconductors A_2B_6 [457–463]. This polarisation dependence of the exiton absorption (reflection) spectrum components can be easily interpreted taking into account the spin–orbital interaction in the valence band (leading to formation of the uppen valence subband Γ_8) and the antiferromagnetic interaction of valence electrons with paramagnetic ions. It was found experimentally that in cubic crystals the giant spin splitting of the exciton band does not depend on the orientation of the magnetic field **H** relative to the crystallographic axis.

The giant magneto–optical effects of the $1S$ exiton band were investigated for a many diluted magnetic semiconductors [457–469].

The change in the MCDR signal at microwave pumping of the spin transitions of Mn^{2+} ions (Fig. 6.6) allows us to realize the optical detection of the EPR of magnetic impurities (OMDR). The degree of spin polarization of the impurity system determines the value of the effective exchange field, and causes the giant spin splitting of the exciton band. Therefore, the change of this degree of spin polarization must change the spectral position of the lines, corresponding to the transitions from the ground state of the crystal to an exciton state. With increasing microwave pumping of EPR transitions of Mn^{2+} ions, the components of the split exciton line of absorption (reflection) of the light, the extremes of which are separated by a giant spin splitting with a value of $\sim$9 meV (in the crystal ZnTe this splitting is

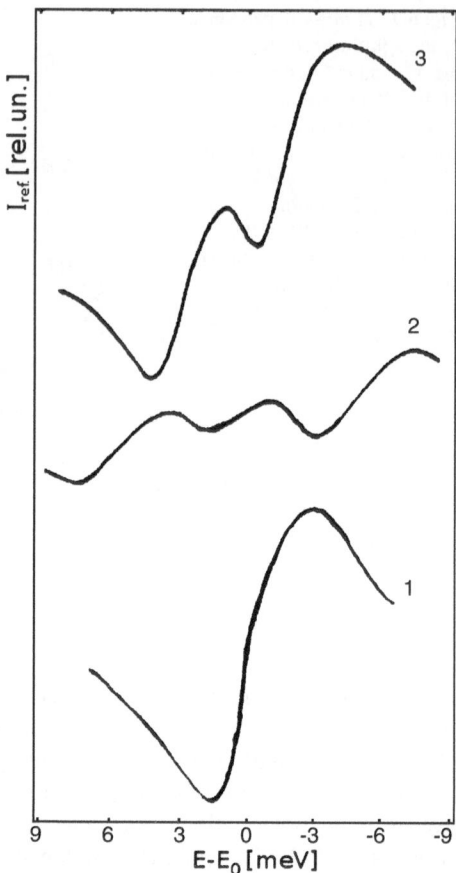

Fig. 6.8 Spectra of $1S$ exciton reflection band obtained with strongly doped CdTe:Mn^{2+} ($C \approx 8 \cdot 10^{18}$ cm^{-3}) in measurements across the field ($\mathbf{H} \perp \mathbf{k}$) at the following magnetic field values: 1—$H = 0$; 2—$H = 30$ kOe, σ polarization ($\mathbf{E} \perp \mathbf{H}$); 3—$H = 30$ kOe; π polarization ($\mathbf{E} \parallel \mathbf{H}$)

even greater [461, 462]), begin to approach each other. Therefore, it follows that by changing the level of microwave pumping of the PCs with which the exchange interaction of carriers forming the excitons takes place, we can smoothly change the spectral position of the exciton lines in a relatively broad interval (~ 0.1 eV).

Thus, the discussed magneto–optical effects are completely explained by the exchange interaction between electrons and holes, coupled into free excitons, and the isovalent impurity Mn^{2+}. The diamagnetic shifts of the exciton bands are determined only by the direct action of the external magnetic field on the electrons and holes. Therefore, such shifts must not be essential in comparison with the spin splittings, caused by the effective exchange field. The diamagnetic shifts may occur on the background of the giant spin splitting effects for excited states of the excitons, as occurs, for example, for $2S$ excitons in ZnTe:Mn^{2+} [462].

The splitting of the exciton levels in the exchange field of the magnetic impurities has the same order of magnitude as the exciton binding energy (for $2S$ excitons in ZnTe:Mn^{2+} this splitting is even greater than the binding energy of the exciton state [462]). However, this does not lead to ionization of the exciton, as the effective exchange fields acts only on the electron spin and hole angular momentum and is not included in the vector potential which determines the movement of the carriers in the magnetic field.

The dependence of the values of carrier–impurity exchange interaction constants on the type of wave function of the relative electron–hole motion in the exciton is established in [462]. The experimentally determined correlation $|I_e|_{2S} > |I_e|_{1S}$ may be due to a decrease in the shielding of the exchange interaction of each carrier with magnetic impurities at the transition of the exciton from the ground to the excited state. This is an additional proof of the fact that the carriers participating in the exchange interaction with the PCs are the components of the excitons.

The results of the more recent investigations in the field of diluted magnetic semi-conductors as well as an introduction to the physics of these magnetic materials are given in Kossut and Gaj's monograph [470], which was published not long ago.

Chapter 7
Effects of Deep Saturation

In the presence of a strong electromagnetic (or hypersonic) field, which is in reso-nance with a pair of exciton levels (bands) and has an intensity at which the exciton states become unsteady, the energy spectrum of excitons is restructured and trans-formed into a quasi-energy spectrum. The concept of quasi-energy was first intro-duced by Zel'dovich [471] and Ritus [472]. Later this representation was used, in particular, to investigate the influence of the resonance laser radiation on semicon-ductors.

In a strong electromagnetic field where the frequency exceeds the energy gap, as a result of the law of conservation of energy and impulse, the electric dipole moment operator binds only two states (one from the valence band and one from the conduction band) [473]. In this case the task is similar to the one for a two-level system in an external variable field [29, 413, 474, 475] and is precisely solved. This solution is obtained by using a canonical transformation, as a result of which the quasi-particles are determined with a new law of dispersion.

The characteristic feature of the energy spectrum of a semiconductor in a strong laser field, causing interband transitions at $\mathbf{k} = 0$, is the presence of an energy gap. The energy gap also appears when the frequency of the light is in resonance at some wave vector $\mathbf{k}_0$ with two conduction bands, the carriers of which have masses with opposite signs [476]. In the works [172, 476] the energy spectrum for the elementary excitations of a crystal was studied when the frequency of the laser radiation was resonancing with respect to the transitions between the degenerate exciton levels. However, neither these nor other studies discuss time-reversal symmetry with regard to quasi-energy states. The influence of strong coherent hypersound on excitons and the behavior of Zeeman systems with an arbitrary spin in a strong alternative magnetic field of resonance frequency, with the application of the convenient quasi-energy formalism, also remain uninvestigated.

In Sect. 7.1, based on the analysis of the first Brillouin zone in the space of quasi-energies, we consider time-reversal symmetry for quantum mechanical systems with quasi-energy spectra.

In Sect. 7.2 the quasi-energy spectrum of excitons in the field of a hypersonic wave of high intensity is obtained by taking into account the interband exciton–

I. Geru, D. Suter, *Resonance Effects of Excitons and Electrons*,
Lecture Notes in Physics 869, DOI 10.1007/978-3-642-35807-4_7,
© Springer-Verlag Berlin Heidelberg 2013

phonon transitions. The intraband scattering of low-density excitons on hypersonic phonons of high density is considered in Sect. 7.3.

In Sect. 7.4 the quasi-energy spectrum for systems with a finite number of degrees of freedom having an equidistant energy spectrum in the presence of a strong electromagnetic field of resonance frequency is obtained by using the multiboson representation of the angular momentum.

The main results of this chapter are published in the works [170, 340, 477–480].

7.1 Unsteady States of Quantum Systems

It is convenient to use the quasi-energy concept [471, 472] to describe unsteady states of particles (quasi-particles) in an external periodical field of great intensity. This description is possible because, in the presence of a periodic perturbation with frequency ω, the Hamiltonian of the system possesses the property

$$H(t + \tau) = H(t), \quad \tau = 2\pi/\omega.$$

The solution of the time-dependent Schrödinger temporal equation may be represented in the form

$$\Psi(\mathbf{r}, t) = \varphi(\mathbf{r}, t)e^{-\frac{i}{\hbar}\mathcal{E}t},$$

where

$$\varphi(\mathbf{r}, t) = \varphi(\mathbf{r}, t + \tau).$$

Here $\mathcal{E}$ is the quasi-energy, and $\mathbf{r}$ includes both the coordinate and the spin variables. The periodic time-dependent wave function $\varphi(\mathbf{r}, t)$ is squarely integrable and satisfies Schrödinger's equation

$$\left[H(t) - i\hbar\frac{\partial}{\partial t} \right] \varphi(\mathbf{r}, t) = \mathcal{E}\varphi(\mathbf{r}, t). \tag{7.1}$$

The time-translation operators $T(m, \tau)$ $(m = 0, \pm1, \pm2, \ldots)$ commute with the operator $H(t) - i\hbar\frac{\partial}{\partial t}$ and form an Abelian symmetry group, the irreducible representations of which must be numbered by means of quasi-energy values $\mathcal{E}$ (τ is fixed and is given by an external source) which satisfy (7.1):

$$T(m\tau)\Psi(\mathbf{r}, t) = e^{\frac{i}{\hbar}m\mathcal{E}\tau}\Psi(\mathbf{r}, t).$$

In the general case, a countable set of functions corresponds to the given value of quasi-energy $\mathcal{E}$:

$$\varphi_{lm}(\mathbf{r}, t) \equiv f_l(\mathbf{r})e^{im\omega t},$$

$$(l = 1, 2, \ldots; \ m = 0, \pm1, \pm2, \ldots).$$

The function $\varphi_{lm}(\mathbf{r}, t)$ may be considered as state vectors in the generalized Hilbert space $\mathcal{R} + \mathcal{T}$ (functions $f_l(\mathbf{r})$ are determined in the space $\mathcal{R}$ and $\exp(im\omega t)$ in the space $\mathcal{T}$) [481]. Any linear Hermitian operator in the space $\mathcal{R}$ (or $\mathcal{T}$) remains linear and Hermitian in $\mathcal{R} + \mathcal{T}$. In the space $\mathcal{R} + \mathcal{T}$ the operator $-i\hbar\frac{\partial}{\partial t}$ is Hermitian, and therefore "Hamiltonian" $\mathcal{H} = H(t) - i\hbar\frac{\partial}{\partial t}$ is also Hermitian in this space. Consequently, every value of the quasi-energy is a real number, and the functions $\varphi_{lm}(\mathbf{r}, t)$ belonging to different values of the quasi-energy are orthogonal.

It is useful to perform an analogy with Brillouin's zones in crystals. If $\mathcal{E}$ is the eigenvalue (the quasi-energy) of the operator $\mathcal{H}$ and $\varphi(\mathbf{r}, t)$ is its eigenfunction, then

$$\mathcal{E}' = \mathcal{E} + m\hbar\omega, \qquad \varphi'(\mathbf{r}, t) = \varphi(\mathbf{r}, t)e^{im\omega t} \tag{7.2}$$

are also, respectively, the eigenvalue and eigenfunction of the operator $\mathcal{H}$ for any integer m. Besides, the full wave function of the system does not change:

$$\varphi(\mathbf{r}, t)e^{-\frac{i}{\hbar}\mathcal{E}t} = \varphi'(\mathbf{r}, t)e^{-\frac{i}{\hbar}\mathcal{E}'t},$$

as a result of which all the solutions (7.2) are physically equivalent at any integer m.

The existence of the equivalent states of the quasi-energy allows us to bring any value of the quasi-energy to the point in the first Brillouin zone of the quasi-energies space (similarly to the first Brillouin zone for an unidimensional periodic lattice). This is realized by determining the real number E for which

$$E - \frac{1}{2}\hbar\omega < \mathcal{E} \leq E + \frac{1}{2}\hbar\omega. \tag{7.3}$$

The selection of the zone, that is, the selection of E, is arbitrary.

Let us clear up how the functions $\varphi(\mathbf{r}, t)$ are transformed under action of the time-reversal operator. Equation (7.1) is not invariant relative to the usual time-reversal operator, since Wigner's operator $\mathcal{K} = \mathcal{U}\mathcal{K}_0$ [398] ($\mathcal{U}$ is the unitary operator and $\mathcal{K}_0$ is the operator of complex conjugation) does not commute with the operator $\mathcal{H}$. However, it is possible to determine an operator $\mathcal{K}_t$, relative to which the "Hamiltonian" $\mathcal{H}$ is invariant [477, 478, 480]:

$$\mathcal{K}_t^{-1}\mathcal{H}\mathcal{K}_t = \mathcal{H}.$$

We shall determine the operator I_t for this by analogy with the operator of inversion in the coordinate space $I_\mathbf{r}$ ($I_\mathbf{r}\mathbf{r} = -\mathbf{r}$):

$$I_t t = -t, \quad I_t^2 = 1.$$

The operator I_t transforms the function $F(t)$, obviously depending on time, into the function $F(-t)$

$$I_t F(t) = F\left(I_t^{-1}t\right) = F(I_t t) = F(-t). \tag{7.4}$$

As the states $\varphi_{lm}(\mathbf{r}, t) = f_l(\mathbf{r}) \exp(im\omega t)$ and $\varphi^*_{\mathbf{r}, t} = \mathcal{U} f^*_l(\mathbf{r}) \exp(-im\omega t)$ belong to the same value of the quasi-energy $\mathcal{E}$, then from (7.4) the time-reversal operator for systems with a quasi-energy spectrum has the form

$$\mathcal{K}_t = \mathcal{U} \mathcal{K}_0 I_t. \tag{7.5}$$

According to perturbation theory, for systems with a quasi-energy spectrum [481],

$$\left[\mathcal{H}^{(0)} + \lambda V(t) - \mathcal{E}(\lambda)\right] \varphi(\mathbf{r}, t, \lambda) = 0, \tag{7.6}$$

where $V(t)$ is the perturbation operator, λ is a small real parameter of the decom-position, and $\varphi(\mathbf{r}, t, \lambda)$ is the continuous function of the parameter λ in the space $\mathcal{R} + \mathcal{T}$. The function $\varphi(\mathbf{r}, t, \lambda)$ is a periodic time-dependent function,

$$\varphi(\mathbf{r}, t, \lambda) = \varphi(\mathbf{r}, t + \tau, \lambda),$$

it is normalized,

$$\langle\langle \varphi(\mathbf{r}, t, \lambda) | \varphi(\mathbf{r}, t, \lambda)\rangle\rangle = 1,$$

and its phase multiplier is fixed by the condition

$$\langle\langle \varphi(\mathbf{r}, t, 0) | \varphi(\mathbf{r}, t, \lambda)\rangle\rangle = \langle\langle \varphi(\mathbf{r}, t, \lambda) | \varphi(\mathbf{r}, t, 0)\rangle\rangle.$$

Here

$$\langle\langle \varphi | \Phi \rangle\rangle = \frac{1}{\tau} \int_{-\tau/2}^{\tau/2} \varphi^* \Phi \, d\mathbf{r} \, dt.$$

For $\lambda = 0$, from (7.6) it follows that

$$\mathcal{H}^{(0)} \varphi(\mathbf{r}, t, 0) = \mathcal{E}(0) \varphi(\mathbf{r}, t, 0). \tag{7.7}$$

Let E_n and $f_n(\mathbf{r})$ be, respectively, the eigenvalue and eigenfunction of the oper-ator $H^{(0)}$:

$$H^{(0)} f_n(\mathbf{r}) = E_n f_n(\mathbf{r}), \tag{7.8}$$

where the Hamiltonian of zeroth approximation $H^{(0)}$ is defined in the $\mathcal{R}$ space. Then the solution of (7.8) will have the form

$$\mathcal{E}(0) = E_n + m\hbar\omega, \qquad \varphi(\mathbf{r}, t, 0) = f_n(\mathbf{r}) e^{im\omega t},$$

where m is an integer number. The selection of the first Brillouin zone (7.3) for the unperturbable solution $\{\mathcal{E}(0), \varphi(\mathbf{r}, t, 0)\}$ fixes the same phase multiplier $\exp(im\omega t)$ for all λ and, in this way, the physically equivalent solutions of (7.5) have the form $\{\mathcal{E}(\lambda), \varphi(\mathbf{r}, t, \lambda)\}$ and $\{\mathcal{E}(\lambda) + m\hbar\omega, \varphi(\mathbf{r}, t, \lambda) e^{im\omega t}\}$ independent of λ.

If E_k is the eigenvalue of the operator $H^{(0)}$ having a set of eigenvalues $E_\varkappa, E_{\varkappa'}, \ldots$ which satisfy the conditions $E_k = E_\varkappa + m\hbar\omega$, $E_k = E_{\varkappa'} + m'\hbar\omega$, $E_k = E_{\varkappa''} + m''\hbar\omega, \ldots$ for some integers $m, m', m'', \ldots$, then the functions $f_k(\mathbf{r})$,

$f_{\varkappa}(\mathbf{r})\exp(im\omega t)$, $f_{\varkappa'}(\mathbf{r})\exp(im'\omega t)$, $f_{\varkappa''}(\mathbf{r})\exp(im''\omega t)$, ... are the eigenfunctions of the operator $\mathcal{H}^{(0)}$ and belong to eigenvalue E_k. This shows that even if the initial eigenvalue E_κ of operator $H^{(0)}$ is not degenerate in the space $\mathcal{R}$, then the eigenvalue of the operator $\mathcal{H}^{(0)}$ in the space $\mathcal{R}+\mathcal{T}$ may become degenerate. From the linearity of the operator $\mathcal{H}^{(0)}$, the eigenfunction $\varphi_c(\mathbf{r},t,0)=c_k f_k(\mathbf{r})+c_\varkappa f_\varkappa(\mathbf{r})\exp(im\omega t)+c_{\varkappa'}f_{\varkappa'}(\mathbf{r})\exp(im'\omega t)+c_{\varkappa''}f_{\varkappa''}(\mathbf{r})\exp(im''\omega t)+\cdots$ is also the solution of (7.7) and belongs to the eigenvalue $\mathcal{E}(0)=E_k$ (c_p are complex numbers).

As an example, we shall consider the case when the operator $H^{(0)}$ from (7.8) has no degenerate eigenvalue E_0. Let us assume that $H^{(0)}$ does not have discrete eigenvalues in the surroundings E_0, $E_0\pm\hbar\omega$, $E_0\pm 2\hbar\omega$, and $E_0\pm 3\hbar\omega$. In second-order perturbation theory, the wave function which satisfies (7.6) is determined by the expression [481]:

$$\varphi^{(2)}(\mathbf{r},t,\lambda)=\left\{f_0(\mathbf{r})+\lambda\left[f_1^{(1)}(\mathbf{r})e^{i\omega t}+f_{-1}^{(1)}(\mathbf{r})e^{-i\omega t}\right]\right.$$
$$+\lambda^2\left[f_2^{(2)}(\mathbf{r})e^{2i\omega t}+f_{-2}^{(2)}(\mathbf{r})e^{-2i\omega t}+2f_0^{(2)}(\mathbf{r})\right]\right\}$$
$$\times\exp\left\{-2i\lambda\left[E^{(1)}\frac{\sin\omega t}{\omega}+\lambda E^{(2)}\frac{\sin 2\omega t}{2\omega}\right]\right\}, \qquad (7.9)$$

where

$$E^{(1)}=\langle f_0(\mathbf{r})|V^{(1)}|f_0(\mathbf{r})\rangle,$$
$$E^{(2)}=\frac{1}{2}\langle f_0(\mathbf{r})|V^{(1)}|(f_{+1}^{(1)}(\mathbf{r})+f_{-1}^{(1)}(\mathbf{r}))\rangle.$$

Here, $V^{(1)}$ is the time-independent Hermitian operator from (7.6): $V(t)=V^{(1)}\times\cos\omega t$; μ and ν at functions $f_\nu^\mu(\mathbf{r})$ indicate, respectively, the order of the perturbation theory and the value of m. By direct action of the operator $\mathcal{K}_t$ from (7.6) on the wave function $\varphi^{(2)}(\mathbf{r},t,\lambda)$ from (7.9), we can verify that the functions $\varphi^{(2)}(\mathbf{r},t,\lambda)$ and $\mathcal{K}_t\varphi^{(2)}(\mathbf{r},t,\lambda)$ belong to the same value of the quasi-energy $\mathcal{E}(\lambda)$. If $\mathbf{r}$ does not include the spin coordinate, then $\varphi^{(2)}(\mathbf{r},t,\lambda)$ and $\mathcal{K}_t\varphi^{(2)}(\mathbf{r},t,\lambda)$ are linearly dependent, and the additional degeneration connected with the time-reversal symmetry is absent. Accounting for the spin, it is possible to have additional degeneration of the quasi-energy levels affected by the time-reversal symmetry. For free spin systems the matrix of the unitary operator $\mathcal{U}$ has matrix elements on the secondary diagonal equal in turn to $+1$ and -1 and the other matrix elements equal to zero [391, 398]. If the total spin takes an integer value, then the matrix of the operator $\mathcal{U}$ is symmetric and there is no additional degeneration. The additional degeneration of the quasi-energy levels due to time-reversal symmetry exists in the case of half-integer total spin when $\mathcal{U}^2=-1$ and correspondingly $\mathcal{K}_t^2=-1$. In the presence of a crystal field it is necessary to use operator $\mathcal{U}$ in the space of the basis functions of the irre-

ducible representations of the point groups of symmetry, an evident form of which
is given in Appendix B.

7.2 Unsteady States of Excitons at Interband Scattering on High-Density Hypersonic Phonons

The investigation of exciton–phonon interaction in crystals is of interest in the case
when the exciton states in the field of a strong hypersonic wave are unsteady. Experimental data relative to the behavior of excitons in the field of an intensive acoustic
wave at present do not exist. In Chap. 3 it was already noted that there is not a
single experimental work on the investigation of the absorption of hypersound by
excitons in the gaseous phase, even under simple conditions of acoustic pumping
(in the absence of a quasi-energy spectrum). Nevertheless, as will be explained below, there are real possibilities for the experimental investigation of the quasi-energy
spectra of an exciton–phonon system. The intensity of the existing powerful sources
of hypersound is sufficient for observing the effects of lowering the degeneration
in the system "excitons + strong hypersound wave" on account of exciton–phonon
interaction (see the numerical estimation at the end of this section).

Let us consider the case when in conditions of resonance transitions between
exciton bands it is possible to ignore the intraband scattering of excitons on phonons
in comparison with the interband transitions under the influence of hypersound,
and the Hamiltonian of exciton–phonon interaction is determined by the formula
(3.24). This situation for the case of steady exciton states was considered in detail
in Sect. 3.3. The quasi-energy spectrum of the system of excitons and high-density
hypersonic phonons, under the condition that this correlation between diagonal and
nondiagonal elements of the exciton–phonon interaction operator is preserved at
high intensities of the hypersound (in conditions of saturation), was investigated in
the work [340].

The phonon mode will be considered as macrocompleted, that is, always performed under conditions of saturation. Then the temporal dependence of the operator of phonon destruction $b_{\mathbf{q}}$ may be represented in the form

$$b_{\mathbf{q}} = B_{\mathbf{q}} \exp(-i\omega t), \quad B_{\mathbf{q}} = \sqrt{N_{\mathbf{q}}}, \tag{7.10}$$

where $N_{\mathbf{q}}$ is the number of phonons with wave vector $\mathbf{q}$. By means of the unitary
operator

$$U(t) = \exp\left\{-i\omega_{\mathbf{q}}t \sum_{\mathbf{k}} \left[\alpha_{\lambda}(\mathbf{k}) A^{+}_{\lambda,\mathbf{k}} A_{\lambda,\mathbf{k}} \right.\right.$$
$$\left.\left. + \alpha_{\lambda'}(\mathbf{k}+\mathbf{q}) A^{+}_{\lambda',\mathbf{k}+\mathbf{q}} A_{\lambda',\mathbf{k}+\mathbf{q}} \right]\right\},$$

we obtain a representation in which the Hamiltonian of exciton–phonon interaction does not depend on the time:

$$H = U^+(t)\mathcal{H}(t)U(t) - i\hbar U^+(t)\frac{\partial}{\partial t}U(t)$$

$$= \sum_{\mathbf{k}}\left[E_\lambda(\mathbf{k})A^+_{\lambda,\mathbf{k}}A_{\lambda,\mathbf{k}} + E_{\lambda'}(\mathbf{k}+\mathbf{q})A^+_{\lambda',\mathbf{k}+\mathbf{q}}A_{\lambda',\mathbf{k}+\mathbf{q}}\right]$$

$$+ \sum_{\mathbf{k}}\left[\sqrt{N_\mathbf{q}}V_{\lambda'\lambda}(\mathbf{q})A^+_{\lambda',\mathbf{k}+\mathbf{q}}A_{\lambda,\mathbf{k}} + \text{H.c.}\right] + \hbar\omega_\mathbf{q}N_\mathbf{q}, \qquad (7.11)$$

where

$$E_l(\varkappa) = \varepsilon_l(\varkappa) - \alpha_l(\varkappa)\hbar\omega,$$

$$\left(l = \lambda, \lambda';\ \varkappa = \mathbf{k},\ \mathbf{k}+\mathbf{q}\right). \qquad (7.12)$$

Here $\alpha_\lambda(\mathbf{k})$ and $\alpha_{\lambda'}(\mathbf{k}+\mathbf{q})$ are the functions of the wave vectors of the excitons in the bands λ and λ', $V_{\lambda'\lambda}(\mathbf{q})$ is the binding constant of the excitons with the hypersonic phonons, and $\varepsilon_l(\varkappa)$ is the energy of an exciton with wave vector $\varkappa$ in the band l in the absence of the strong interaction with the hypersound.

The Hamiltonian (7.11) does not depend on time if the following condition holds:

$$\alpha_{\lambda'}(\mathbf{k}+\mathbf{q}) - \alpha_\lambda(\mathbf{k}) = 1. \qquad (7.13)$$

By canonical transformation, this Hamiltonian achieves the diagonal form

$$H = \sum_{\nu,\mathbf{k}}\mathcal{E}_\nu(\mathbf{k})\mathcal{A}^+_{\nu,\mathbf{k}}\mathcal{A}_{\nu,\mathbf{k}}. \qquad (7.14)$$

The new operators $\mathcal{A}^+_{\nu,\mathbf{k}}$ and $\mathcal{A}_{\nu,\mathbf{k}}$ satisfy the commutation relations for bosons and are expressed through the exciton creation and destruction operators in the following way:

$$\mathcal{A}_{\nu,\mathbf{k}} = \sum_{n=-\infty}^{\infty}\sum_{i=\lambda,\lambda'} C^i_{n,\nu}A_{i,\mathbf{k}+n\mathbf{q}},$$

$$\mathcal{A}^+_{\nu,\mathbf{k}} = (\mathcal{A}_{\nu,\mathbf{k}})^+.$$

The system of an infinite number of equations for the coefficients $C^i_{n,\nu}$ splits into pairs of independent equations for $C^{\lambda'}_{n,\nu}$ and $C^\lambda_{n-1,\nu}$:

$$\left[E_{\lambda'}(\mathbf{k}+n\mathbf{q}) - \mathcal{E}_\nu(\mathbf{k})\right]C^{\lambda'}_{n,\nu} + V_{\lambda'\lambda}(\mathbf{q})\sqrt{N_\mathbf{q}}C^\lambda_{n-1,\nu} = 0,$$

$$V_{\lambda'\lambda}(\mathbf{q})\sqrt{N_\mathbf{q}}C^{\lambda'}_{n,\nu} + \left[E_\lambda(\mathbf{k}+(n-1)\mathbf{q}) - \mathcal{E}_\nu(\mathbf{k})\right]C^\lambda_{n-1,\nu} = 0. \qquad (7.15)$$

The renormalized spectrum of the system of excitons and high-density hypersonic phonons is found by solving the system of (7.15). In the case of precise resonance

$$\hbar\omega_{\mathbf{q}} = \varepsilon_{\lambda'}(\mathbf{k} + n\mathbf{q}) - \varepsilon_{\lambda}(\mathbf{k} + (n-1)\mathbf{q}), \tag{7.16}$$

from (7.15) on account of (7.12) we get

$$\mathcal{E}_{1,2}(\mathbf{k}) = \frac{1}{2}\left\{\varepsilon_{\lambda'}(\mathbf{k} + n\mathbf{q}) - \varepsilon_{\lambda}(\mathbf{k} + (n-1)\mathbf{q}) - \left[1 - 2\alpha_{\lambda'}(\mathbf{k} + n\mathbf{q})\right]\hbar\omega_{\mathbf{q}}\right\}$$

$$\pm \frac{1}{2}\left\{\left[\varepsilon_{\lambda'}(\mathbf{k} + n\mathbf{q}) - \varepsilon_{\lambda}(\mathbf{k} + (n-1)\mathbf{q}) - \hbar\omega_{\mathbf{q}}\right]^2 + 4N_{\mathbf{q}}\left|V_{\lambda'\lambda}(\mathbf{q})\right|^2\right\}^{1/2}, \tag{7.17}$$

where $\alpha_{\lambda'}(\mathbf{k} + n\mathbf{q})$ satisfies equation (7.13) and is determined by the selection of the initial data.

Equation (7.16) represents by itself the condition of the degeneration in a compound exciton–phonon system, which is composed of excitons in the bands λ and λ' and hypersonic phonons. The functions $\mathcal{E}_1(\mathbf{k})$ and $\mathcal{E}_2(\mathbf{k})$ from (7.17) are the branches of the renormalized energy of the excitons from the bands λ' and λ with the wave vectors $\mathbf{k} + n\mathbf{q}$ and $\mathbf{k} + (n-1)\mathbf{q}$ due to the exciton–phonon interaction. These branches appear due to the lowering of the degeneration by exciton–phonon interaction that we have indicated above.

Special interest is represented by the case of crystals in which the splitting $\Delta_{\lambda'\lambda}$ of exciton bands in the crystal field exceeds the kinetic energy of the excitons moving with the velocity of the hypersound. In this case, as was shown in Sect. 3.3, the excitons from the low band with small wave vectors ($|\mathbf{k}| < k_c = b_{\lambda\lambda'}$, where $b_{\lambda\lambda'}$ is determined by formula (3.7)) do not participate in the processes of exciton–phonon interaction. Therefore, the energy spectrum in the range of wave vectors corresponding to these excitons will not change with increasing hypersonic wave amplitude and upon reaching the condition of strong saturation, unlike the remaining part of the energy spectrum of excitons, which leads to the gap in the quasi-energy spectrum. This gap appears in the low exciton band between $-\mathbf{k}_c$ and $\mathbf{k}_c$. In the interval $(-\mathbf{k}_c, \mathbf{k}_c)$ the energy of the excitons in the low band remains the same as that before the interaction with the strong hypersonic wave (Fig. 7.1).

The presence of breaches in the quasi-energy spectrum of the exciton–phonon system may be detected in experiments on photon–phonon double resonance. In particular, the line of absorption of the "weak" light from the ground state of the crystal to the exciton band $2P_0$ in the presence of an intensive acoustic pumping at the frequency $\omega_{2P_0,2S}$ must be split into two components, the distance between which depends linearly on the intensity of the hypersonic wave.

Let us estimate the value of acoustic pumping necessary for appearance of the quasi-energy branch $\mathcal{E}_1^{2P_0}$ and $\mathcal{E}_2^{2P_0}$ in the case $\alpha_{\lambda'}(\mathbf{k} + n\mathbf{q}) = 0$, which corresponds to the selection of the stationary exciton band $2P_0$ as the beginning of the counting-out of the quasi-energy. According to (7.17) the distance between the branches of quasi-energy in conditions of precise resonance is equal to $2|V_{2S,2P_0}(\mathbf{q})|\sqrt{N_{\mathbf{q}}}$.

Fig. 7.1 The gap in the quasi-energy spectrum of the system of excitons and resonant hypersonic phonons at $\Delta_{\lambda\lambda'} > \frac{1}{2}m_{ex}v_s^2$

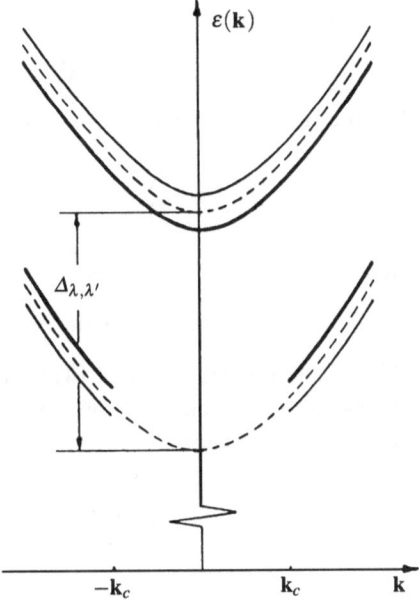

Using the parameters of the crystal CdS given in Sect. 3.3 and taking into account that excitons with wave vectors $\mathbf{k} < \mathbf{k}_c$ are not involved in the processes of exciton–phonon interaction, we find that this distance reaches the value $0.1\Delta_{2P_0,2S}$ at a concentration of hypersonic phonons of $n_\mathbf{q} \approx 10^{15}$ cm^{-3}.

7.3   Unsteady States of Excitons at Intraband Scattering on High-Density Hypersonic Phonons

Let us consider the case when the intraband scattering of excitons on hypersonic phonons is not ignored, but can be omitted in the following Hamiltonian, the terms of which are responsible for the absorption of hypersound with transitions of the excitons from one band to another:

$$\mathcal{H} = \hbar\omega_\mathbf{q} N_\mathbf{q} + \sum_\mathbf{k} \varepsilon(\mathbf{k}) A_\mathbf{k}^+ A_\mathbf{k} + \sum_\mathbf{k} \left[\sqrt{N_\mathbf{q}} V(\mathbf{q}) A_{\mathbf{k}+\mathbf{q}}^+ A_\mathbf{k} \exp(-i\omega_\mathbf{q} t) + \text{H.c.} \right].$$

(7.18)

Here, for the phonon operators, the representation (7.10) was used, corresponding to the approximation to the macrocompleted mode.

The unitary operator $U(t)$, which in this case performs the transition to a representation in which the Hamiltonian (7.18) does not depend on time, has the form

$$U(t) = \exp\left\{ -i\omega_\mathbf{q} t \sum_\mathbf{k} \alpha(\mathbf{k}) A_\mathbf{k}^+ A_\mathbf{k} \right\}$$

under the condition

$$\alpha(\mathbf{k} + \mathbf{q}) - \alpha(\mathbf{k}) = 1. \tag{7.19}$$

Then the Hamiltonian which does not depend on time,

$$H = \hbar \omega_q N_q + \sum_{\mathbf{k}} E(\mathbf{k}) A_{\mathbf{k}}^+ A_{\mathbf{k}} + \sum_{\mathbf{k}} \left[\sqrt{N_q} V(\mathbf{q}) A_{\mathbf{k}+q}^+ A_{\mathbf{k}} + \text{H.c.} \right],$$

where

$$E(\mathbf{k}) = \varepsilon(\mathbf{k}) - \alpha(\mathbf{k}) \hbar \omega_q, \tag{7.20}$$

achieves the diagonal form (7.14) by means of the canonical transformation. The new operators $\mathcal{A}_{v,\mathbf{k}}^+$ and $\mathcal{A}_{v,\mathbf{k}}$ are determined in the form of linear superpositions of exciton Bose operators

$$\mathcal{A}_{v,\mathbf{k}} = \sum_{n=-\infty}^{\infty} C_{n,v} A_{\mathbf{k}+nq}, \quad \mathcal{A}_{v,\mathbf{k}}^+ = (\mathcal{A}_{v,\mathbf{k}})^+. \tag{7.21}$$

However, here, in contradiction to the case considered in the preceding section, the system of the infinite number of equations for the coefficients $C_{n,v}$ is not divided into a pair of independent equations, and the quasi-energy spectrum may be calculated only approximately. The approximate unhooking of the system of equations for $C_{n,v}$ can be easily made if the linewidth of intraexiton band absorption of a "weak" hypersound (in the absence of quasi-energy spectrum) does not exceed the value $2q$.

In the case of a monochromatic fascicle of the phonons and considering in (7.21) only the terms with $n = 0, \pm 1$ (three-level model), the branch of the quasi-energy spectrum is found by solving the cubic equation

$$\mathcal{E}^3(\mathbf{k}) - \left[E(\mathbf{k} - \mathbf{q}) + E(\mathbf{k}) + E(\mathbf{k} + \mathbf{q}) \right] \mathcal{E}^2(\mathbf{k})$$

$$+ \left\{ E(\mathbf{k}) \left[E(\mathbf{k} - \mathbf{q}) + E(\mathbf{k} + \mathbf{q}) \right] + E(\mathbf{k} - \mathbf{q}) E(\mathbf{k} + \mathbf{q}) - 2N_q |V(\mathbf{q})|^2 \right\}$$

$$\times \mathcal{E}(\mathbf{k}) + N_q |V(\mathbf{q})|^2 \left[E(\mathbf{k} - \mathbf{q}) + E(\mathbf{k} + \mathbf{q}) \right] - E(\mathbf{k} - \mathbf{q}) E(\mathbf{k}) E(\mathbf{k} + \mathbf{q}) = 0,$$

where $E(\mathbf{k})$ is determined by formula (7.20). Under the additional condition

$$\frac{E(\mathbf{k} - \mathbf{q}) E(\mathbf{k}) E(\mathbf{k} + \mathbf{q})}{E(\mathbf{k} - \mathbf{q}) + E(\mathbf{k} + \mathbf{q})} = N_q |V(\mathbf{q})|^2,$$

we have

$$\mathcal{E}_{1,2}(\mathbf{k}) = \frac{1}{2} \left[\xi_q(\mathbf{k}) \pm \sqrt{\xi_q^2(\mathbf{k}) + \eta_q^2(\mathbf{k})} \right],$$

$$\mathcal{E}_3(\mathbf{k}) = 0, \tag{7.22}$$

where

$$\xi_{\mathbf{q}} = \varepsilon(\mathbf{k} - \mathbf{q}) + \varepsilon(\mathbf{k}) + \varepsilon(\mathbf{k} + \mathbf{q})$$
$$+ \left[\alpha(\mathbf{k} - \mathbf{q}) + \alpha(\mathbf{k}) + \alpha(\mathbf{k} + \mathbf{q})\right]\hbar\omega_{\mathbf{q}},$$

$$\eta_{\mathbf{q}}^2(\mathbf{k}) = 8N_{\mathbf{q}}|V(\mathbf{q})|^2 - 4\{[\varepsilon(\mathbf{k}) - \alpha(\mathbf{k})\hbar\omega_{\mathbf{q}}]$$
$$\times \left[\varepsilon(\mathbf{k} - \mathbf{q}) + \varepsilon(\mathbf{k} + \mathbf{q})\right] + \left[\alpha(\mathbf{k} - \mathbf{q}) + \alpha(\mathbf{k} + \mathbf{q})\right]\hbar\omega_{\mathbf{q}}$$
$$+ \left[\varepsilon(\mathbf{k} - \mathbf{q}) - \alpha(\mathbf{k} - \mathbf{q})\hbar\omega_{\mathbf{q}}\right]\left[\varepsilon(\mathbf{k} + \mathbf{q}) - \alpha(\mathbf{k} + \mathbf{q})\hbar\omega_{\mathbf{q}}\right]\}.$$

If the wave vector of the phonon $\mathbf{q}$ satisfies the equation

$$\frac{1}{4}\left(\frac{\hbar}{m_{\text{ex}}}\right)^3 q^3 + \frac{1}{2}\left(\frac{\hbar}{m_{\text{ex}}}\right)^2 v_s \left[\frac{3}{2}(\alpha(\mathbf{q}) + \alpha(-\mathbf{q})) + \alpha(0)\right]q^2$$
$$+ \frac{\hbar}{m_{\text{ex}}} v_s^2 \left\{\alpha(\mathbf{q})\alpha(-\mathbf{q}) + (\alpha(\mathbf{q}) + \alpha(-\mathbf{q}))\left[\alpha(0)\right.\right.$$
$$\left.\left. + \frac{1}{2}(\alpha(\mathbf{q}) + \alpha(-\mathbf{q}))\right]\right\}q + v_s^3\left[\alpha(0)(\alpha(\mathbf{q}) + \alpha(-\mathbf{q}))^2\right.$$
$$\left. + \alpha(\mathbf{q})\alpha(-\mathbf{q})(\alpha(\mathbf{q}) + \alpha(-\mathbf{q}) - 2\alpha(0))\right] = 0$$

and the excitons are found in the state with $\mathbf{k} = 0$, then from (7.22) we obtain

$$\mathcal{E}_1(0) = \frac{\hbar^2 q^2}{m_{\text{ex}}} + \left[\alpha(-\mathbf{q}) + \alpha(0) + \alpha(\mathbf{q})\right]\hbar\omega_{\mathbf{q}}, \tag{7.23}$$
$$\mathcal{E}_2(0) = \mathcal{E}_3(0) = 0.$$

Here v_s is the velocity of hypersound in the crystal. As is seen from (7.23), the levels of the quasi-energy of the exciton–phonon system may be degenerate in spite of the fact that the initial exciton band was not degenerate.

We notice that in both cases of unsteady exciton states, i.e., at their interband and intraband scattering on high-density hypersonic phonons, the formulas (7.13) and (7.19), leading to the time-independent Hamiltonian of the exciton–phonon interaction are similar.

7.4 Quasi-Energy Spectrum for a System with Equidistant Energy Levels

In a typical situation for ENDOR, the degree of saturation of EPR transitions influences only the intensity of spectral lines without changing the position of the spin levels. Despite this, we shall consider the case of strong saturation of the Zeeman electron system, when the spin state of the paramagnetic center becomes unstationary [479]. Besides, due to the strong spin–photon interaction, a "lowering of the degeneration" in the system "PC + coherent microwave" takes place.

It is convenient to describe the quasi-energy spectrum of the paramagnetic center in a strong microwave field on the basis of the multiboson representation of the angular momentum considered in [482]. For simplicity, we shall consider the initial spectrum of Zeeman energy levels to be equidistant.

The three-level system ($S = 1$)   Let us present the operators of the spin $S = 1$ using the Bose operators of creation and destruction $a^+_{M_s}$ and a_{M_s} in the states with spin projections $M_s = 1, 0, -1$:

$$S_z = a^+_1 a_1 - a^+_{-1} a_{-1},$$

$$S_+ = \sqrt{2}\left(a^+_1 a_0 + a^+_0 a_{-1}\right), \qquad S_- = \sqrt{2}\left(a^+_0 a_1 + a^+_{-1} a_0\right).$$

The Hamiltonian of the spin–photon system in the case of macrocompleted photon mode has the form

$$\mathcal{H} = N\hbar\omega + \Delta\left(a^+_1 a_1 - a^+_{-1} a_{-1}\right) + \sqrt{2N}G\left(a^+_1 a_0 \right.$$
$$\left. + a^+_0 a_{-1}\right)e^{-i\omega t} + \sqrt{2N}G^*\left(a^+_0 a_1 + a^+_{-1} a_0\right)e^{i\omega t}, \qquad (7.24)$$

where Δ is the electron Zeeman splitting, G is the constant of spin–photon coupling, and N is the number of photons.

By means of the unitary operator

$$U(t) = \exp\left\{-i\omega t\left(\sum_{M_s=-S}^{S} C_{M_s} a^+_{M_s} a_{M_s}\right)\right\} \qquad (7.25)$$

we obtain a representation in which the Hamiltonian (7.24) does not depend on time:

$$H = N\hbar\omega + (\Delta - C_1\hbar\omega)a^+_1 a_1 - C_0\hbar\omega a^+_0 a_0$$
$$- (\Delta + C_{-1}\hbar\omega)a^+_{-1} a_{-1} + \sqrt{2N}\left[G\left(a^+_1 a_0 + a^+_0 a_{-1}\right)\right.$$
$$\left. + G^*\left(a^+_0 a_1 + a^+_{-1} a_0\right)\right]. \qquad (7.26)$$

The coefficients C_{M_s} satisfy the condition

$$C_{M_s+1} - C_{M_s} = 1. \qquad (7.27)$$

After a further transition to the new Bose operators and diagonalization of the Hamiltonian (7.26), we obtain the following equation:

$$\tilde{\varepsilon}^3 + (C_1 + C_0 + C_{-1})\hbar\omega\tilde{\varepsilon}^2 - \left\{\left[\Delta + (C_0 + C_{-1})\hbar\omega\right]\right.$$
$$\times (\Delta - C_1\hbar\omega) - C_0\hbar\omega(\Delta + C_{-1}\hbar\omega) + 4N|G|^2\right\}\tilde{\varepsilon}$$
$$- C_0\hbar\omega(\Delta + C_{-1}\hbar\omega)(\Delta - C_1\hbar\omega) - 2(C_1 + C_{-1})N\hbar\omega|G|^2 = 0, \qquad (7.28)$$

where $\tilde{\mathcal{E}}$ is the self-value of the operator (7.26) in a system of coordinates rotating with frequency ω. For $C_1 = 1$, $C_0 = 0$, and $C_{-1} = -1$, which correspond to the transition to the rotating coordinate system by means of the unitary operator

$$U(t) = \exp(-i\omega t S_z),$$

equation (7.28) has real solutions

$$\tilde{\mathcal{E}} = 0, \qquad \tilde{\mathcal{E}}_{2,3} = \pm\delta, \qquad \delta = \left[(\varDelta - \hbar\omega)^2 + 4N|G|^2\right]^{1/2}. \tag{7.29}$$

After returning to the laboratory coordinate system, on the basis of (7.29) we obtain the following expression for the quasi-energy levels:

$$\mathcal{E}_0 = 0, \qquad \mathcal{E}_{M_S} = E_{M_S} + M_S\delta, \qquad M_S = 1, -1$$

The four-level system ($S = 3/2$)   In the four Bose-operators representation of the spin operators for $S = 3/2$ we have

$$S_z = \frac{3}{2}\left(a_{3/2}^+ a_{3/2} - a_{-3/2}^+ a_{-3/2}\right) + \frac{1}{2}\left(a_{1/2}^+ a_{1/2} - a_{-1/2}^+ a_{-1/2}\right),$$

$$S_+ = \sqrt{3}\left(a_{3/2}^+ a_{1/2} + a_{-1/2}^+ a_{-3/2}\right) + 2a_{1/2}^+ a_{-1/2},$$

$$S_- = (S_+)^+.$$

The time-dependent Hamiltonian of the spin–photon system using a unitary operator of the type (7.25) is transformed into the form

$$H = N\hbar\omega + \frac{3}{2}(\varDelta - C_{3/2}\hbar\omega)a_{3/2}^+ a_{3/2} + \frac{1}{2}(\varDelta - C_{1/2}\hbar\omega)a_{1/2}^+ a_{1/2}$$

$$- \frac{3}{2}(\varDelta + C_{-3/2}\hbar\omega)a_{-3/2}^+ a_{-3/2} - \frac{1}{2}(\varDelta + C_{-1/2}\hbar\omega)a_{-1/2}^+ a_{-1/2}$$

$$+ \sqrt{N}\{G[\sqrt{3}(a_{3/2}^+ a_{1/2} + a_{-1/2}^+ a_{-3/2}) + 2a_{1/2}^+ a_{-1/2}]$$

$$+ G^*[\sqrt{3}(a_{1/2}^+ a_{3/2} + a_{-3/2}^+ a_{-1/2}) + 2a_{-1/2}^+ a_{1/2}]\},$$

where the coefficients C_{M_S} satisfy condition (7.27). For $C_{M_S} = M_S$ the secular determinant reduces to the equation

$$\tilde{\mathcal{E}}_4^4 - 5\left[\frac{1}{2}(\varDelta - \hbar\omega)^2 + 2N|G|^2\right]\tilde{\mathcal{E}}^2 + \frac{9}{16}(\varDelta - \hbar\omega)^4$$

$$+ 9N|G|^2\left[\frac{1}{2}(\varDelta - \hbar\omega)^2 + N|G|^2\right] = 0,$$

and the quasi-energy levels are determined by the expressions

$$\mathcal{E}_{M_S} = E_{M_S} + M_S\delta, \qquad M_S = \pm\frac{3}{2}, \pm\frac{1}{2}.$$

The five-level system ($S = 2$)   The operators of the spin projections for $S = 2$ in the five Bose-operators representation have the forms:

$$S_z = 2\left(a_2^+ a_2 - a_{-2}^+ a_{-2}\right) + \left(a_1^+ a_1 - a_{-1}^+ a_{-1}\right),$$

$$S_+ = 2\left(a_2^+ a_1 + a_{-1}^+ a_{-2}\right) + \sqrt{6}\left(a_0^+ a_{-1} + a_1^+ a_0\right),$$

$$S_- = (S_+)^+.$$

With $C_{M_S} = M_S$ ($M_S = 2, \ldots, -2$) the secular determinant has one trivial solution $\tilde{\varepsilon} = \varepsilon = 0$, and the other four solutions satisfy the bisquare equation

$$\tilde{\varepsilon}^4 - 5\left[(\Delta - \hbar\omega) + 4N|G|^2\right]\tilde{\varepsilon}^2 + 4\left[(\Delta - \hbar\omega)^2 + 4N|G|^2\right]^2 = 0.$$

The quasi-energy levels are determined by the expressions

$$\varepsilon_0 = 0, \qquad \varepsilon_{M_S} = E_{M_S} + M_S\delta, \qquad M_S = \pm 2, \pm 1.$$

$(2S + 1)$-level system   The equations for $\tilde{\varepsilon}$ for $S = 1/2, 1, 3/2, 2, \ldots$ have a similar structure, on the basis of which we can obtain the following secular determinant in the quasi-diagonal form for the case of arbitrary spin S:

$$\begin{vmatrix} P_S & Q_S^* & & & & & & & & \\ Q_S & P_{S-1} & Q_{S-1}^* & & & & & & & \\ & Q_{S-1} & P_{S-2} & Q_{S-2}^* & & & & & & \\ & & Q_{S-2} & P_{S-3} & Q_{S-3}^* & & & & & \\ & & & Q_{S-3} & \ddots & \ddots & & & & \\ & & & & \ddots & P_{3-S} & Q_{3-S}^* & & & \\ & & & & & Q_{3-S} & P_{2-S} & Q_{2-S}^* & & \\ & & & & & & Q_{2-S} & P_{1-S} & Q_{1-S}^* & \\ & & & & & & & Q_{1-S} & P_{-S} & \end{vmatrix} = 0,$$

(7.30)

where $P_S = P(m)|_{m=S}$ and $Q_S = Q(m)|_{m=S}$. The functions $P(m)$ and $Q(m)$ are:

$$P(m) = m\Delta - C_m\hbar\omega - \tilde{\varepsilon},$$

$$Q(m) = G\sqrt{N(S + m)(S - m + 1)},$$

$$S_- = (S_+)^+.$$

All elements of the secular determinant that are not indicated in (7.30) are equal to zero.

The solutions of the secular equation (7.30) are

$$\varepsilon_{M_S} = E_{M_S} + M_S\delta,$$

(7.31)

where $M_S = S, S - 1, S - 2, \ldots, 2 - S, 1 - S, -S$ and δ is determined by (7.29).

The quasi-energy levels determined by using (7.27) are situated not far from energy levels being above (below) from them if $M_S > 0$ ($M_S < 0$). Using the formula $C_{M_S} - C_{M_S+1} = 1$ instead of (7.27), we found the quasi-energy levels situated above (below) the energy levels with $M_S < 0$ ($M_S > 0$).

Thus, nearly of each stationary energy level E_{M_S} (except in the case E_0) there are two symmetrically situated quasi-energy levels. The distance between these levels is equal to $2M_S\delta$.

Chapter 8
Basics of Quantum Information Processing

Over the last decades, science as well as economy have been driven by enormous improvements in information processing, which is not only growing in speed and affordability, but is also becoming continuously more efficient and ubiquitous. The fundamentals of this development are the computational paradigms developed by Turing, von Neumann, and others in the first half of the twentieth century. At the lowest levels, all current commercial devices rely on Boolean logic gate operations. Therefore, it came as a big surprise when it was discovered that alternative types of logic may be possible and that they may provide a computational paradigm that is qualitatively more powerful than what had become so successful over the last decades. This alternative procedure stores information in quantum mechanical states and uses unitary transformations of these states for processing the information. Feynman and others showed that universal computation on the basis of quantum mechanical systems is possible and that such devices can, for certain problems, be exponentially faster than any classical computer.

A physical realization of a quantum computer requires well-defined systems that evolve according to the laws of quantum mechanics. Nuclear and electronic spins have almost ideal properties for this purpose: a well-defined finite Hilbert space that can be used to store the information in the form of quantum bits (qubits). Logical operations can be implemented by applying static and resonant magnetic fields. In this chapter, we discuss the basic requirements for implementing quantum information processing and harnessing the potential that this radically new approach holds.

In Sect. 8.1, we discuss the physical principles that allow one to store, process, and distribute information and the limits to which physical devices are subject—these are different from the mathematical limitations of computability. Some of these limitations are specific to quantum computers, others for classical computers, and some aspects concern both approaches to information processing. In Sect. 8.2, we introduce the basic ideas and the historical developments that allow one to use quantum systems for information processing. Section 8.3 describes how information can be stored in quantum mechanical systems, and Sect. 8.4 shows the most popular approach for processing this information. In Sect. 8.5, we introduce quantum gates as the elementary operations of information processing. In addition to the

I. Geru, D. Suter, *Resonance Effects of Excitons and Electrons*,
Lecture Notes in Physics 869, DOI 10.1007/978-3-642-35807-4_8,
© Springer-Verlag Berlin Heidelberg 2013

fundamental operations on single qubits, operations on two qubits are required for universal quantum computation; these are introduced in Sect. 8.6. Section 8.7 shows how quantum gate operations can be made robust, i.e., implemented so that they work well even if the physically realized control fields deviate from the ideal ones required by the algorithm. While the quantum gate operations are unitary transformations, any algorithm also requires nonunitary (and thus irreversible) operations, in particular for initialization and readout. This aspect is discussed in Sect. 8.8. Section 8.9 introduces the decay of quantum information, which is possibly the biggest obstacle for the realization of quantum computers that are large enough to unleash the potential that the approach holds in principle. Section 8.10 briefly mentions one of the other potential applications of quantum mechanical systems in everyday technology: using single photons for communication allows one to transmit messages in such a way that the contents cannot be read by an eavesdropper.

The standard approach to quantum information processing breaks the processed information down into effective spins 1/2 (qubits). In Sect. 8.11, we introduce another possible scheme, which is based on Bose operators.

8.1 Information and Physics

Information—its exchange, storage, and processing—has always been one of the most important human activities. Originally, information existed in a number of different physical forms, and exchange of information always required a human intermediary. With the evolution of civilization and industry, new tools were developed to store and spread information, such as writing, printing, and telecommunication. The introduction of programmable digital computers opened the possibility (and, to some degree, the necessity) to encode virtually every type of information electronically. Some results of this paradigm shift are that the exchange between different physical forms as well as between different locations has become largely automated and the amount of information that is available almost instantaneously increases exponentially. Processing is mostly achieved by microprocessors built from semiconducting materials, and information processing has developed an independent and mature status, both as a science and as an industry.

The progress achieved in the efficiency and capacity of information processing devices (including transmission and storage) has been extraordinary [483]. This progress was always associated with investigations and improvements of the physical properties of the materials used [484], such as the mobility of charge carriers in semiconductors and the dielectric constants of insulator materials in microprocessors. Many ingenious engineers continue to find ways around these obstacles; one example is speed limits that may be overcome by adjustments to the manufacturing process, e.g., switching from silicon to silicon–germanium or using new types of insulator materials.

However, there are also physical limitations to the power of computational devices that are less obvious but will be much harder to overcome than these materials-related issues [485, 486]. While most of these limits are still quite remote, some are

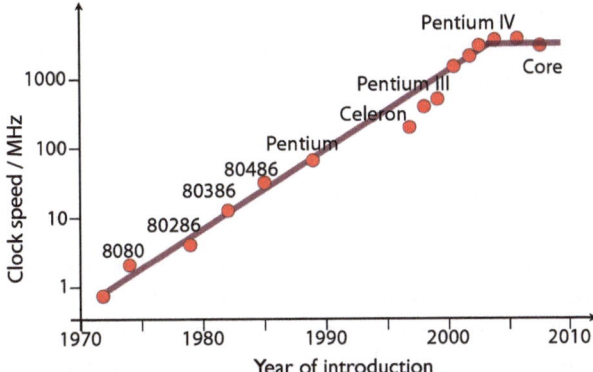

Fig. 8.1 Evolution of the clock speed of Intel microprocessors over the last 40 years

already uncomfortably close. They are starting to make it ever more difficult to continue the exponential improvements in the performance of semiconductor microelectronics that we have become used to [487]. Examples include the minimal amount of energy dissipated per logical operation or the limitations on the size of electronic circuits by the finite size of atoms [488]. These limitations cannot be overcome by simply switching the material basis of the electronic circuit technology.

Figure 8.1 illustrates the consequences of such a limitation. It shows the evolution of the clock rate of Intel microprocessors over the last decades. For many years, the evolution followed an exponential increase of the clock rate, with a doubling time of ≈2.5 years. This was considered a typical example of Moore's law. However, this evolution suddenly came to a stop around the year 2004, when the clock speed stopped increasing. The basic reason was that the processors were running too hot. While the heat dissipation of current microprocessors is still significantly higher than the fundamental physical limit, it is clearly becoming harder to further increase the energy efficiency and thereby reduce the heat dissipation.

Taking the physical basis of information into account [489, 490] can even challenge some fundamentals of computer science. An important example is the famous Church–Turing hypothesis [491, 492], which asserts that most computers are equivalent with respect to computability (not with respect to speed). It allows one to decide if a given problem can be solved on a computer, without having to consider the details of the information processing device used in the actual implementation. However, the strong form of the Church–Turing hypothesis, which states that any problem that can be solved *efficiently* on one computer can be solved efficiently on any other computer, appears to be wrong: some problems have been established to be solvable efficiently if the computer operates according to the laws of quantum mechanics, but not on classical computers.

Quantum mechanics is, as far as we can judge today, the fundamental physical theory. In particular, we need quantum mechanics to explain, e.g., the differences between metals, insulators, and semiconductors, as well as the effect of doping on the conductivity of semiconductor materials. In this sense, quantum mechanics has always been the basis for electronic computers and thus for most of today's information processing. However, while quantum mechanics is necessary for understanding and designing different semiconductor materials, the engineers who build the computers from these materials need no knowledge of quantum mechanics for building ever more powerful devices.

8.2 Quantum Information

Computer science classifies computational problems in terms of their scalability: problems are classified as computationally hard if no efficient solution is known. Examples include optimization problems or the factorization of large numbers. For some of these examples, no algorithms are known that provide an efficient solution that can be implemented on a classical computer, but efficient algorithms have been worked out for quantum mechanical devices. These quantum algorithms require that the information be represented by the quantum state of a physical system whose evolution is governed by Schrödinger's equation [493–496]. This possibility was first suggested by Feynman [497, 498], who hypothesized that quantum mechanical systems might be a more powerful basis for computational devices than classical ones, and by Benioff [499] who showed that quantum systems can be used as universal computers. The fact that they can be qualitatively more powerful than classical computers was proven in 1993 by Bernstein and Vazirani [500], and shortly thereafter, the first algorithms were devised that operate exponentially faster on quantum computers than on classical computers [501, 502]. These algorithms earned much publicity in the media. However, for physicists, Lloyd's proof that quantum computers can also efficiently simulate arbitrary quantum mechanical systems may be even more important. This result may be considered as an indication that Feynman's 1993 suggestion to use quantum systems for simulating quantum systems is actually being realized and may eventually open interesting new possibilities for almost every field of physics [496].

These developments created an enormous interest and a rapidly expanding field of research. On the theoretical side, this research effort [503] is directed towards developing algorithms that work more efficiently on quantum computers than on classical computers. At the same time, efforts are under way to realize physical systems that can implement these quantum algorithms. These systems have to fulfill a number of conditions [504]. In particular, they must support a sufficiently large number of well-controlled quantum mechanical two-level systems, in which the information can be stored and processed and finally read out. The first [505, 506] and still the most successful implementations of quantum information processing were based on nuclear spins in liquids. Since then, nuclear magnetic resonance (NMR) has been used as a testbed for implementing quantum algorithms, testing new concepts, and developing techniques for efficiently controlling quantum systems [507].

The fact that nuclear spins were the first physical system in which quantum information processing was realized, should not be considered accidental but can be linked to some simple physical facts that make them particularly suitable for this purpose.

- Spins 1/2 are the only quantum systems whose Hilbert space is two dimensional and thus exactly realizes a qubit.
- Nuclear spins are quite well isolated from the environment, thus preserving the quantum information for a long time.
- Precise control of the evolution of nuclear spins by resonant radio-frequency fields is well established and can be implemented in commercially available spectrometers.

On the other hand, some of these properties also make it difficult to realize large-scale quantum computers. In particular, the weak coupling between nuclear spins and the environment also implies that it is very hard to detect individual spins. In most cases, this problem can only be overcome by working, not with individual spins, but with ensembles of identically prepared spins, which together generate a signal that can be detected easily enough. Here, we will consider mostly electronic spins, which share some of the special properties of nuclear spins, but interact much more strongly with their environment. In many situations, this reduces their lifetime as quantum objects, but it also makes it much easier to detect individual electron spins.

8.3 Quantum Bits

Storing and processing of information are the main tasks of computers. All information that we consider in this context is represented in binary digital form. The smallest units of information are thus bits, i.e., memory cells that can assume two distinct values, which are usually labeled as 0 and 1. These units of information can be stored in different physical systems that can assume two distinct states, e.g., a capacitor with different amounts of charge or the magnetization of a spot on a magnetic disc, which can be oriented in different directions. Processing this information implies that a string of bits is changed into a different string, according to a finite set of instructions called the algorithm. In a quantum computer, the system that represents the information is a quantum mechanical system, and the logical operations correspond to the temporal evolution of this system [493, 494]. Information processing corresponds to the manipulation of strings of bits, applying rules or algorithms that take one string as input, perform logical operations on it, and finally output a modified string.

In quantum information processing [493, 494, 508], the information is also represented in digital form. However, in this case, the individual bits are stored in quantum systems that have two orthogonal states. Conventionally, these states are labeled $|0\rangle$ and $|1\rangle$. Each such two-level system is described by a two-dimensional Hilbert space. To distinguish these from the classical bits, one usually refers to them as quantum bits or qubits (see Fig. 8.2). The only physical system that has *exactly* two orthogonal states is a spin 1/2. If a different physical system is chosen to represent the quantum bit, a two-dimensional subspace of the full state space is chosen. Since any quantum mechanical two-level system is equivalent to a spin-1/2 system,

Fig. 8.2 The quantum bit (qubit) is the basic unit of information in a quantum computer

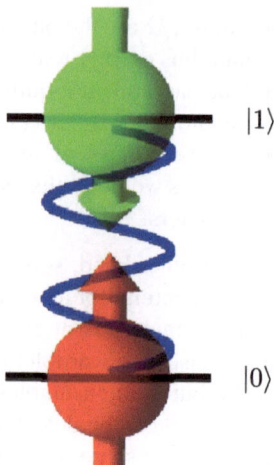

$|1\rangle$

$|0\rangle$

these systems are typically termed pseudospin systems. For simplicity, and because in many cases actual spins are used, we will usually call them qubits or spins.

In principle, any orthogonal pair of states from this two-dimensional Hilbert space can be chosen as the computational basis states $|0\rangle$ and $|1\rangle$. However, in most cases, one chooses the eigenstates of the Hamiltonian for this purpose. For spins in a magnetic field, these are the states $|\uparrow\rangle$ and $|\downarrow\rangle$, where the spin is oriented parallel or antiparallel to the direction of an external magnetic field.

The main difference between the classical and the quantum computer, which results in the qualitatively higher capabilities of the quantum computer, is that the quantum information does not only have to be in the computational basis states $|0\rangle$ and $|1\rangle$. Instead, the superposition principle of quantum mechanics states that the system can also be in a superposition of these states,

$$|\Psi\rangle = \alpha|0\rangle + \beta|1\rangle.$$

This applies not only to the individual qubits, but equally to the array of qubits that stores the complete information: while a classical register can hold the values $000\ldots0, 000\ldots1, \ldots, 111\ldots1$, the quantum register can be in an arbitrary superposition of these states:

$$|\psi_N\rangle = \sum_{x=00\ldots0}^{11\ldots1} c_x|x\rangle,$$

where the c_x are probability amplitudes with

$$\sum_{x=00\ldots0}^{11\ldots1} |c_x|^2 = 1.$$

Fig. 8.3 The network model of quantum information processing

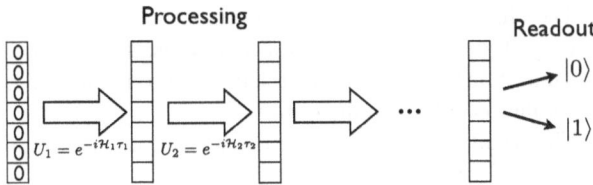

The quantum information is then contained in a state that lives in the Hilbert space spanned by the 2^N computational basis states from $|00\ldots0\rangle$ to $|11\ldots1\rangle$. Accordingly, 2^N complex coefficients are required to identify the state.

Since the Schrödinger equation is linear, an algorithm that works on two different states, also works on any superposition of these states. A quantum information processor that operates on such a superposition state may be considered a highly parallel computer: a suitable input represents the superposition of 2^N orthogonal input states, and the final state the superposition of 2^N calculations performed on all these input states. Algorithms taking advantage of this possibility can therefore increase the speed of computation dramatically—in some cases, the speed increases exponentially with the number of qubits of the processor. This possibility of processing superpositions of basis states is often referred to as "quantum parallelism."

8.4 The Network Model of Quantum Information Processing

The conditions that must be fulfilled by a physical implementation of a quantum information processor have been laid out by DiVincenzo [504]. The exact means of control and the type of inter-qubit interactions can vary and are specific to various proposed architectures for quantum computing [494, 509].

The model that has been most thoroughly investigated so far is called the circuit or network model [493, 494], which is closely related to the processing of classical information. As shown in Fig. 8.3, this model stores the quantum information in a string of qubits called the quantum register. Before the computation can start, this register has to be initialized into a well-defined state containing the input and, possibly, additional qubits required by the algorithm. In the figure, it is assumed that all bits of the quantum register are set to 0. Logical operations are then implemented as unitary transformations U_i acting on the qubits. After the last operation, the result must be read out by measuring suitable observables of some (possibly all) qubits. While the information processing by the unitary operations U_i is clearly reversible, the initial and final steps (initialization and readout) are qualitatively different: they collapse each bit into an eigenstate and are therefore irreversible. In most quantum algorithms, the readout process must act like an ideal quantum measurement, where the state collapses onto an eigenstate of the observable. Typically, these states correspond to the computational basis states $|0\rangle$ and $|1\rangle$, but for some quantum algorithms, the readout projects onto superposition states, such as $\frac{1}{\sqrt{2}}(|0\rangle + |1\rangle)$.

The implementation of quantum information processing therefore requires three types of operations that must be executed on the quantum register.

- The bits of the quantum register must be initialized into a well-defined state.
- Logical operations are performed on the quantum register, according to the algorithm being implemented. Each operation can be represented as a unitary transformation acting on the qubits.
- The result of the processing is read out; i.e., the final state is converted into classical information.

Apart from the network model sketched here and used throughout the rest of this book, different computational models (also called quantum computing architectures) have been introduced. The most important ones are the adiabatic model [510, 511] and measurement-based quantum computation [512].

8.5 Quantum Gates for Single Qubits

All logical operations on information encoded in quantum states can be decomposed into operations on single qubits (single-qubit operations) and operations acting on two qubits (two-qubit operations). The single-qubit operations correspond to rotations of the (pseudo)spins. They can be described in terms of operators acting on the states. The operators can be written as 2×2 complex matrices and expanded in terms of a basis of four orthogonal operators. Suitable operators are, e.g., the unit operator $\mathbf{1}$ together with the three spin operators I_x, I_y, and I_z:

$$\mathbf{1} = \begin{pmatrix} 1 & 0 \\ 0 & 1 \end{pmatrix}, \qquad I_x = \frac{1}{2} \begin{pmatrix} 0 & 1 \\ 1 & 0 \end{pmatrix}, \tag{8.1}$$

$$I_y = \frac{1}{2} \begin{pmatrix} 0 & -i \\ i & 0 \end{pmatrix}, \qquad I_z = \frac{1}{2} \begin{pmatrix} 1 & 0 \\ 0 & -1 \end{pmatrix}. \tag{8.2}$$

An equivalent alternative consists of the set $\mathbf{1}$, $I_{\pm} = I_x \pm iI_y$, and I_z.

The gate operator U_i, which corresponds to a time evolution, must be generated by a suitable Hamiltonian,

$$U_i = e^{-i\mathcal{H}_i \tau_i},$$

where the index i labels the computational step, $\mathcal{H}_i$ the Hamiltonian driving the operation, and τ_i the duration of this step. The experimental apparatus controlling the quantum information processor must be capable of generating these Hamiltonians by suitable control fields. If the qubits are implemented as spins, the control fields are usually resonantly time-dependent magnetic fields that couple to the transitions between the eigenstates of the Hamiltonian. In most cases, it is not possible to generate the Hamiltonian $\mathcal{H}_i$ directly, but it is always possible to decompose it into a sequence of operations that can be generated by suitable sequences of radiofrequency pulses applied to the spins.

For the simplest example, consider z-rotations: they require the Zeeman interaction with a static magnetic field. This interaction is represented by the Hamiltonian

$$\mathcal{H}_Z = \Omega_z I_z,$$

where the Larmor frequency $\Omega_z = -\gamma B_0$ is proportional to the strength of the magnetic field B_0 and the gyromagnetic ratio γ of the spin. Rotations around axes in the xy-plane are generated by pulsed radio-frequency (rf) fields. These operations are best described in what is called the rotating frame, which rotates around the direction of the static magnetic field with a frequency equal to the radio frequency of the applied field. In this reference frame, the interaction with the rf field becomes static and can be written as

$$\mathcal{H}_{rf} = \omega_1 I_\eta,$$

where ω_1 is the strength of the rf field and η is a direction in the xy-plane that can be chosen by the phase of the radio frequency.

Using these interactions, it is possible to perform arbitrary logical operations on individual qubits. Every such operation corresponds to a rotation of the spins. Mathematically, it is described by a unitary transformation

$$U(\delta, \alpha, \beta, \theta) = e^{i\delta} \begin{pmatrix} e^{i(\alpha+\beta)/2} \cos\frac{\theta}{2} & e^{i(\alpha-\beta)/2} \sin\frac{\theta}{2} \\ e^{-i(\alpha-\beta)/2} \sin\frac{\theta}{2} & e^{-i(\alpha+\beta)/2} \cos\frac{\theta}{2} \end{pmatrix}$$

$$= e^{i\delta} R_z(\alpha) R_y(\theta) R_z(\beta).$$

Here, δ is an overall phase factor, while $R_\eta(\alpha)$ is a rotation by an angle α around the η-axis.

8.6 Two-Qubit Operations

Any nontrivial algorithm includes conditional operations. In standard programming languages they are represented by operations like if, while, or for. Clearly, such operations operate on more than one (qu)bit: at least one that controls the operation (the control qubit) and at least one on which the operation is performed (the target qubit). It is possible to show that all multi-qubit operations can be decomposed into one- and two-qubit operations [493, 494]. Therefore, we discuss here only two-qubit operations.

Perhaps the most important multi-qubit operation is the controlled-NOT (CNOT) operation: a NOT operation on one qubit (the target qubit) that is only executed if the second qubit (the control qubit) is 1. If the control qubit is zero, the CNOT operation becomes a unity ($=$ NOP) operation. In the 2-qubit basis $\{|00\rangle, |01\rangle, |10\rangle, |11\rangle\}$, this operation can be expressed in matrix form as

$$\mathrm{CNOT} = \begin{pmatrix} 1 & 0 & 0 & 0 \\ 0 & 1 & 0 & 0 \\ 0 & 0 & 0 & 1 \\ 0 & 0 & 1 & 0 \end{pmatrix}.$$

The first qubit is the control bit. It is never changed by the CNOT operation. The target qubit is inverted if the control bit has the value 1 and left unchanged if the control qubit is 0. The overall effect is therefore an exchange of the two states $|10\rangle$ and $|11\rangle$. This operation can be implemented by single-qubit operations and the two-qubit operation $R_{zizj}(\phi)$ as

$$\mathrm{CNOT} = \mathrm{e}^{-\mathrm{i}\delta}\, R^1_{-y}\left(\frac{\pi}{2}\right) R_{zizj}(\pi) R^1_{-x}\left(\frac{\pi}{2}\right)$$

$$\times R^2_{-z}\left(\frac{\pi}{2}\right) R^2_{-y}\left(\frac{\pi}{2}\right). \tag{8.3}$$

Each rotation R^k_η acts only on a single qubit, indicated by the superscript k. The subscript gives the rotation axis; the argument in parentheses is the rotation angle.

To implement controlled operations in a physical device, information must be exchanged between the qubits participating in the operation, which requires an interaction between them. In the case of spins, this interaction is a spin–spin coupling. In the simplest case (to which we restrict ourselves here), this interaction is represented by the Hamiltonian

$$\mathcal{H}_J = 2 J_{ij} I^i_z I^j_z. \tag{8.4}$$

The interaction can be turned into an operation by letting the system evolve under this Hamiltonian. In reality, the coupling operator (8.4) is only part of the complete system Hamiltonian. For two spins, a typical system Hamiltonian takes the form

$$\mathcal{H}_{\mathrm{NMR}} = \Omega^1_z I^1_z + \Omega^2_z I^2_z + \mathcal{H}_J. \tag{8.5}$$

Evolution under this Hamiltonian will therefore result in a different logical operation than that under the pure coupling operator. To turn this into the desired evolution

$$R_{zizj}(\phi) = \mathrm{e}^{-\mathrm{i}\phi I^i_z I^j_z},$$

we have to refocus the Zeeman terms using π-pulses (rotations by an angle π around an axis in the xy-plane) in the middle and at the end of the evolution period. The π rotation of both spins around the x-axis corresponds to the operation

$$\mathrm{e}^{-\mathrm{i}\pi(I^i_x+I^j_x)},$$

and the free evolution of duration τ to

$$\mathrm{e}^{-\mathrm{i}(2 J_{ij} I^i_z I^j_z + \Omega^i_z I^i_z + \Omega^j_z I^j_z)\tau}.$$

The sequence of a free evolution period, a π rotation, another free evolution period, and another π rotation thus results in the following evolution:

$$\mathrm{e}^{-\mathrm{i}\pi(I^i_x+I^j_x)} \mathrm{e}^{-\mathrm{i}\tau(2 J_{ij} I^i_z I^j_z + \Omega^i_z I^i_z + \Omega^j_z I^j_z)}$$

$$\times \mathrm{e}^{-\mathrm{i}\pi(I^i_x+I^j_x)} \mathrm{e}^{-\mathrm{i}\tau(2 J_{ij} I^i_z I^j_z + \Omega^i_z I^i_z + \Omega^j_z I^j_z)}$$

$$= e^{-i\tau(2J_{ij}I_z^iI_z^j - \Omega_z^iI_z^i - \Omega_z^jI_z^j)}e^{-i\tau(2J_{ij}I_z^iI_z^j + \Omega_z^iI_z^i + \Omega_z^jI_z^j)}$$

$$= e^{-i4J_{ij}\tau I_z^iI_z^j}, \tag{8.6}$$

as required. Depending on the state of the system, the second π-pulse can be omitted.

8.7 Robust Gate Operations

The ability to perform a general quantum gate is one of the five important criteria (DiVincenzo criteria [504]) for the physical realization of quantum information processing. A quantum gate can be described as a target unitary operator U_T, which must be constructed using the control parameters available for the specific system. In general, the gate should be independent of the input state, which may not be known.

Any experimental implementation of such a unitary transformation suffers from imperfections in the control fields, which can only be generated with a finite precision. As a result, the actual gate operation U_{exp} differs from the target operation U_T. As a measure of the deviation, one usually uses the fidelity

$$F = \left| \text{Tr}\{U_T^\dagger U_{exp}\}/N \right|^2, \tag{8.7}$$

where N is the dimension of the operators: if the experimental operation is identical to the target operation, the operator product becomes the unity operator and the fidelity reaches the limiting value of 1.

In any single experiment, one initializes the system in a specific input state $|\psi_{in}\rangle$ and applies the control sequence, which converts the system to the final state $|\psi_{exp}\rangle$, which typically differs from the ideal output state $|\psi_T\rangle$. The difference between the two states can be measured and represents a measure of the quality of the operation; however, for quantum information processing, the operation must perform well for all possible input states. The fidelity measure (8.7), which does not depend on a state, is therefore more appropriate. To measure the fidelity, it is necessary to eliminate the dependence on the specific input state by averaging the deviation over all possible input states. This is achieved by process tomography [513, 514]. Since the fidelity of individual quantum logical gate operations must be very high (of the order of >0.999), it is important to design them in such a way that the fidelity remains close to unity even if the experimental parameters deviate from the ideal values. These types of gate operations are called robust, and their design is the field of coherent control. Methods to achieve such robust gate operations were developed in the field of optimal control [515] and introduced to magnetic resonance [516] and quantum information processing [517].

8.8 Initialization and Readout

Before starting a calculation, the system must be initialized into a well-defined state. In many algorithms, it is assumed that the initial state is the state $|00 \ldots 0\rangle$, which corresponds to the ground state of the spin system. As the last step of the quantum computation, it is necessary to read out the result from the quantum state by performing an appropriate set of measurements. For most algorithms, it is sufficient to measure the polarization of individual qubits independently. However, to improve the understanding of the relevant processes and to check individual computational steps, it is sometimes useful to determine the complete density operator. This can be achieved via *quantum state tomography*: converting those parts of the density operator that are not directly observable into observable magnetization and subsequently measuring the NMR signal allows one to tomographically reconstruct the density operator [505]. For a two-qubit system, quantum state tomography can be achieved with the help of four independent readout pulses [518]; in the case of three qubits, the same result can be achieved with seven readout pulses [519]. For larger systems, the number of readout pulses required to reconstruct the complete density operator increases exponentially with the number of qubits, making this approach unsuitable for large quantum registers. Possible alternatives include a two-dimensional Fourier transform technique [520], but again this approach is not scalable to very large systems.

8.9 Decoherence

Possibly the biggest obstacle for the construction of large quantum information processors is the loss of quantum information before the end of the computation. While loss of information must also be avoided in classical computers, it is much more difficult to preserve quantum information than it is to preserve classical information. The additional difficulty present in quantum mechanics can be tracked to two closely related fundamental properties.

The first is Heisenberg's uncertainty principle: a quantum mechanical measurement cannot be performed without perturbing the variable that is conjugate to the measurement variable. Thus it is not possible to perform measurements on the quantum register while a computation is ongoing. This makes it difficult to detect errors and correct them.

The second problem is the impossibility of exactly duplicating quantum systems, which is known as the no-cloning theorem [521, 522]. In classical computers, the simplest error detection schemes duplicate some information and detect differences between the multiple copies. Since exact duplication is not possible in quantum systems, this procedure is not directly applicable.

The loss of information can be traced to two different mechanisms.

1. The quantum register is never completely isolated from its environment. Couplings to the environment dissipate quantum information. In NMR this effect is

Fig. 8.4 Principle of secure
data transmission between
two partners, Alice and Bob

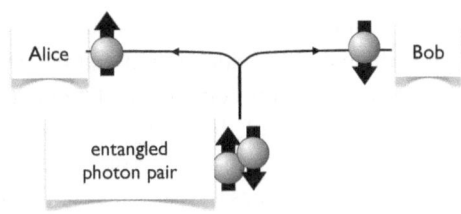

entangled
photon pair

known as relaxation, in quantum information processing it is usually referred to
as decoherence.

2. Experimental imperfections of gate operations and fluctuations in the control
   fields cause the system evolution to differ from the intended evolution.

As discussed in Sect. 8.7 on coherent control, optimizing the control operations
can go a long way towards improving the quality of the gate operations. However,
even if fidelities of individual gate operations can reach values of >0.999, this will
not be sufficient for completing many potentially useful calculations, which may
involve thousands of gate operations acting on hundreds or thousands of qubits. It
will therefore be necessary to develop techniques that detect and correct errors in
quantum mechanical systems.

8.10 Quantum Communication

Quantum communication [523, 524] is among the most advanced subfields of quantum information science. Its main goal is the inherently secure exchange of information between two partners, typically referred to as Alice and Bob, who share entangled pairs of photons (see Fig. 8.4). Suitable measurements on the two spatially
separated particles are completely correlated. This implies that a measurement that
Alice performs on her particle provides her not only the information about the state
of the particle on which she performed the measurement, but also on what the result of a measurement is that Bob performs on his particle. While it is in principle
possible for an eavesdropper to intercept one of the photons and therefore obtain the
information that was meant to be transmitted to one of the partners, the use of single
photons for the information transfer means that the measurement performed by the
eavesdropper cannot remain undetected. Even if the eavesdropper creates another
photon and transmits it to Bob, the no-cloning theorem [521, 522] prevents him from
creating a photon that is in the identical quantum state as the one that he intercepted.
If Alice and Bob do not decide which property of the photon they measure until it is
safely in their laboratory, they will notice the interference of the eavesdropper from
a statistical analysis of a representative subset of their data. This scheme was developed in 1984 by Charles Bennet and Gilles Brassard [525]. It has been implemented
many times, initially in laboratory settings, but then also using standard telecommunication fibers. Several companies now sell equipment for quantum communication,
including id Quantique, MagiQ Technologies, and QuintessenceLabs.

The main difficulty for worldwide introduction of secure quantum communication is the inevitable attenuation that occurs in any medium through which photons are transmitted. This corresponds to an exponential loss of information that (i) limits the bandwidth of the channel and (ii) compromises the security of the transmission, since the interference due to an eavesdropper is hard to distinguish from the natural loss processes. To increase the length over which secure transmission is possible, the concept of a quantum repeater was introduced [526]. It can be compared to the regeneration of classical optical signals in an optical amplifier, except that it re-creates the complete quantum state of a single photon. The regeneration is typically achieved by generating pairs of entangled photons and performing joint measurements. For this (and many other) purposes, it is essential that the photons be available at the required instant, which may be significantly later than the time at which the photon has been emitted. This can only be achieved if photons can be stored in some other form, typically in a material excitation. More details of this will be discussed in the following chapter.

8.11 Quantum Computing with Bose Operators

The Hilbert space of a single spin-$1/2$ particle (qubit) is spanned by the two mutually orthogonal basis states [494]

$$\begin{pmatrix} 1 \\ 0 \end{pmatrix} = |\uparrow\rangle = |0\rangle \quad \text{and} \quad \begin{pmatrix} 0 \\ 1 \end{pmatrix} = |\downarrow\rangle = |1\rangle.$$

To obtain the basic states for a system of two qubits it is necessary to calculate the following Kronecker products of ket vectors: $|00\rangle = |0\rangle \otimes |0\rangle$, $|01\rangle = |0\rangle \otimes |1\rangle$, $|10\rangle = |1\rangle \otimes |0\rangle$, and $|11\rangle = |1\rangle \otimes |1\rangle$. It is easy to show that these four Kronecker products can be mapped to the standard spinor basis of the spin $S = 3/2$. Indeed, the Kronecker product $\otimes$ of an $m \times n$ matrix $\mathbf{A}$ with an $m' \times n'$ matrix $\mathbf{B}$ by definition is an $mm' \times nn'$ matrix given by

$$\mathbf{A} \otimes \mathbf{B} = \begin{pmatrix} a_{11}\mathbf{B} & \dots & a_{1n}\mathbf{B} \\ \dots & \dots & \dots \\ a_{m1}\mathbf{B} & \dots & a_{mn}\mathbf{B} \end{pmatrix},$$

where

$$\mathbf{A} = \begin{pmatrix} a_{11} & \dots & a_{1n} \\ \dots & \dots & \dots \\ a_{m1} & \dots & a_{mn} \end{pmatrix}.$$

Table 8.1   Values of the effective spin S for different numbers N of qubits

N	1	2	3	4	5	6	7	8	9	10	...	N
S	$\frac{1}{2}$	$\frac{3}{2}$	$\frac{7}{2}$	$\frac{15}{2}$	$\frac{31}{2}$	$\frac{63}{2}$	$\frac{127}{2}$	$\frac{255}{2}$	$\frac{511}{2}$	$\frac{1023}{2}$	...	$2^{N-1}-\frac{1}{2}$

According to this definition, the basis vectors for a system of two qubits are given by

$$|00\rangle = \begin{pmatrix} 1 \\ 0 \end{pmatrix} \otimes \begin{pmatrix} 1 \\ 0 \end{pmatrix} = \begin{pmatrix} 1 \\ 0 \\ 0 \\ 0 \end{pmatrix}, \qquad |01\rangle = \begin{pmatrix} 1 \\ 0 \end{pmatrix} \otimes \begin{pmatrix} 0 \\ 1 \end{pmatrix} = \begin{pmatrix} 0 \\ 1 \\ 0 \\ 0 \end{pmatrix},$$

$$|10\rangle = \begin{pmatrix} 0 \\ 1 \end{pmatrix} \otimes \begin{pmatrix} 1 \\ 0 \end{pmatrix} = \begin{pmatrix} 0 \\ 0 \\ 1 \\ 0 \end{pmatrix}, \qquad |11\rangle = \begin{pmatrix} 0 \\ 1 \end{pmatrix} \otimes \begin{pmatrix} 0 \\ 1 \end{pmatrix} = \begin{pmatrix} 0 \\ 0 \\ 0 \\ 1 \end{pmatrix}.$$

For a system of three qubits there are eight basis states $|0\rangle \otimes |00\rangle$, $|0\rangle \otimes |01\rangle$, $|0\rangle \otimes |10\rangle$, $|0\rangle \otimes |11\rangle$, $|1\rangle \otimes |00\rangle$, $|1\rangle \otimes |01\rangle$, $|1\rangle \otimes |10\rangle$, and $|1\rangle \otimes |11\rangle$, which correspond to the basis states of a spin $S = 7/2$. For N qubits the basis vectors correspond to the eigenstates of the spin projection operator S_Z, with spin $S = 2^{N-1} - \frac{1}{2}$.

We notice that making a set of Kronecker products of spinors $\begin{pmatrix} 1 \\ 0 \end{pmatrix}$ and $\begin{pmatrix} 0 \\ 1 \end{pmatrix}$ belonging to different qubits, the spinor basis for the spin $S = 2^{N-1} - \frac{1}{2}$ can be obtained. However, it is not necessary to calculate the Kronecker products for each number N of qubits. It is sufficient to use the known standard basis vectors for the spin S at a given number N of qubits. The values of S at different N are presented in Table 8.1. Thus, in the case of N qubits there are 2^N spinor basis vectors corresponding to the spin $S = 2^{N-1} - \frac{1}{2}$. With an increasing number of qubits the dimension of the S_X, S_Y, and S_Z matrices quickly increases (for example, at $N = 10$ the dimension of spin matrices is 1024×1024, $S = 1023/2$).

It is possible to show that all multi-qubit operations can be decomposed into one- and two-qubit operations (see Sect. 8.6). However, there are possibilities for quantum computing without this decomposition. This approach is based on the theorem about full reduction of the operator loads for the operators of angular momentum projections in the two-Bose operator representation (Appendix H). As a consequence of this theorem, in the two-Bose operator representation the spin operators are invariant relative to orthogonal addition or orthogonal reduction of the spin space basis. Therefore, in this representation the operators $S_\pm = S_X \pm iS_Y$ and S_Z have the form

$$S_+ = a_1^+ a_2, \qquad S_- = (S_+)^+ = a_2^+ a_1, \qquad S_Z = \frac{1}{2}\left(a_1^+ a_1 - a_2^+ a_2\right) \qquad (8.8)$$

independent of the spin value S. Here a_1^+, a_1 and a_2^+, a_2 are the operators of creation and annihilation of two different types of Bose particles. These operators satisfy the commutation relations

$$\left[a_1, a_1^+\right] = \left[a_2, a_2^+\right] = 1, \qquad \left[a_1, a_2\right] = \left[a_1^+, a_2^+\right] = 0$$

and, what is more, satisfy the condition for the operator of the total number of bosons[1]

$$n = n_1 + n_2 = a_1^+ a_1 + a_2^+ a_2 = 2S, \tag{8.9}$$

which limits the number of Bose particle states by means of which the spin wave functions of a system with spin S are determined.

Since in the two-Bose operator representation the form of the S_+, S_-, and S_Z operators does not depend on the spin S of the multi-qubit, specific features on the N-qubit system are determined by the corresponding spin wave functions. For a system with spin $S = 2^{N-1} - \frac{1}{2}$ these spin wave functions in the two-Bose operator representation are

$$\begin{aligned}
&\left|2^N - 1\right\rangle_1 |0\rangle_2, \quad \left|2^N - 2\right\rangle_1 |1\rangle_2, \quad \left|2^N - 3\right\rangle_1 |2\rangle_2, \quad \ldots, \quad |2\rangle_1 \left|2^N - 3\right\rangle_2, \\
&|1\rangle_1 \left|2^N - 2\right\rangle_2, \quad |0\rangle_1 \left|2^N - 1\right\rangle_2,
\end{aligned} \tag{8.10}$$

where $|2^N - k\rangle_i$ is the wave function of the $(2^N - k)$th excited boson state containing $2^N - k$ bosons of the ith type $(i = 1, 2)$ and $|0\rangle_i$ is the vacuum state of the ith Bose field $(i = 1, 2)$. On the basis of a set of pair products of different boson wave functions $|k\rangle_1$ and $|l\rangle_2$ $(k + l = 2^N - 1)$ from (8.10) the basis functions for the systems of one, two, etc., qubits can be easily obtained. The basis functions from (8.10) were obtained as a particular case of the general formula for spin wave functions in the two-Bose operator representation [406]

$$|S, M\rangle = \frac{1}{\sqrt{(S + M)!(S - M)!}} \left(a_1^+\right)^{S+M} \left(a_2^+\right)^{S-M} |0\rangle, \tag{8.11}$$

taking into account the kinematic condition (8.9) for the spin $S = 2^{N-1} - \frac{1}{2}$. In (8.11) M is the eigenvalue of the operator S_Z.

Now it is clear that for quantum computing using the creation operators (a_1^+, a_2^+) and annihilation operators (a_1, a_2) of two Bose fields, it is necessary:

- To transfer the spin operators S_X, S_Y, and S_Z into two-Bose operator representation.
- To use the spin wave functions (8.10) in the two-Bose operator representation.

As an example, let us consider the Hadamard gate. For one qubit the Hadamard gate $\mathbf{H}$ [494] is defined by

$$\mathbf{H}_1 = \frac{1}{\sqrt{2}} (\mathbf{X} + \mathbf{Z}), \tag{8.12}$$

---

[1]$n = 2S$ is one of $2S$ solutions of (H.7) from Appendix H, where J must be changed by S. The other solution of this equation in the case of integer spin is $n = -1$. The remaining solutions of this equation at any values of spin are couples of complex conjugated solutions [527, 528].

where

$$\mathbf{X} = 2S_X, \qquad \mathbf{Z} = 2S_Z. \tag{8.13}$$

Taking into account (8.8) and (8.13), the Hadamard gate from (8.12) in the two-Bose operator representation is

$$\mathbf{H}_1 = \frac{1}{\sqrt{2}} \left[a_1^+ (a_1 + a_2) + a_2^+ (a_1 - a_2) \right]. \tag{8.14}$$

The matrix of the operator $\mathbf{H}_1$ from (8.14), which was found by means of basis functions $|1\rangle_1 |0\rangle_2$ and $|0\rangle_1 |1\rangle_2$, is given by

$$\mathbf{H}_1 = \frac{1}{\sqrt{2}} \begin{pmatrix} 1 & 1 \\ 1 & -1 \end{pmatrix},$$

which coincides with the result obtained using the spinor formalism.

For two qubits the Hadamard gate is

$$\mathbf{H}_2 = \mathbf{H}_1 \otimes \mathbf{H}_1 = \frac{1}{2} (\mathbf{X} + \mathbf{Z}) \otimes (\mathbf{X} + \mathbf{Z}). \tag{8.15}$$

In the two-Bose operator representation the operator $\mathbf{H}_2$, taking into account (8.14) and (8.15), has the form

$$\mathbf{H}_2 = \frac{1}{2} \left[a_1^+ (a_1 + a_2) + a_2^+ (a_1 - a_2) \right] \otimes \left[a_1^+ (a_1 + a_2) + a_2^+ (a_1 - a_2) \right]. \tag{8.16}$$

The matrix $\mathbf{H}_2$ defined in the spinor basis $\{ |3/2\rangle, |1/2\rangle, |-1/2\rangle, |-3/2\rangle \}$, which is equivalent to the two-Bose operator basis $\{ |3\rangle_1 |0\rangle_2, |2\rangle_1 |1\rangle_2, |1\rangle_1 |2\rangle_2, |0\rangle_1 |3\rangle_2 \}$, is

$$\mathbf{H}_2 = \frac{1}{\sqrt{2}} \begin{pmatrix} 1 & 1 & 1 & 1 \\ 1 & -1 & 1 & -1 \\ 1 & 1 & -1 & -1 \\ 1 & -1 & -1 & 1 \end{pmatrix}.$$

In the case of three qubits we have

$$\mathbf{H}_3 = \mathbf{H}_2 \otimes \mathbf{H}_1 = \mathbf{H}_1 \otimes \mathbf{H}_1 \otimes \mathbf{H}_1.$$

In the spinor basis, the $\mathbf{H}_3$ matrix is

$$\mathbf{H}_3 = \frac{1}{2\sqrt{2}} (\mathbf{X} + \mathbf{Z}) \otimes (\mathbf{X} + \mathbf{Z}) \otimes (\mathbf{X} + \mathbf{Z}),$$

while in the two-Bose operator representation it is

$$\mathbf{H}_3 = \frac{1}{2\sqrt{2}} \left[a_1^+ (a_1 + a_2) + a_2^+ (a_1 - a_2) \right] \otimes \left[a_1^+ (a_1 + a_2) + a_2^+ (a_1 - a_2) \right]$$
$$\otimes \left[a_1^+ (a_1 + a_2) + a_2^+ (a_1 - a_2) \right]. \tag{8.17}$$

The matrix $\mathbf{H}_3$ defined in the spinor basis $\{|7/2\rangle, |5/2\rangle, |3/2\rangle, |1/2\rangle, |-1/2\rangle, |-3/2\rangle, |-5/2\rangle, |-7/2\rangle\}$ or in the equivalent two-Bose operator basis $\{|7\rangle_1|0\rangle_2, |6\rangle_1|1\rangle_2, |5\rangle_1|2\rangle_2, |4\rangle_1|3\rangle_2, |3\rangle_1|4\rangle_2, |2\rangle_1|5\rangle_2, |1\rangle_1|6\rangle_2, |0\rangle_1|7\rangle_2\}$ is

$$\mathbf{H}_3 = \frac{1}{\sqrt{2}} \begin{pmatrix} 1 & 1 & 1 & 1 & 1 & 1 & 1 & 1 \\ 1 & -1 & 1 & -1 & 1 & -1 & 1 & -1 \\ 1 & 1 & -1 & -1 & 1 & 1 & -1 & -1 \\ 1 & -1 & -1 & 1 & 1 & -1 & -1 & 1 \\ 1 & 1 & 1 & 1 & -1 & -1 & -1 & -1 \\ 1 & -1 & 1 & -1 & -1 & 1 & -1 & 1 \\ 1 & 1 & -1 & -1 & -1 & -1 & 1 & 1 \\ 1 & -1 & -1 & 1 & -1 & 1 & 1 & -1 \end{pmatrix}.$$

For any number N of qubits the Hadamard operator $\mathbf{H}_N$ is

$$\mathbf{H}_N = \mathbf{H}_1 \otimes \mathbf{H}_1 \otimes \mathbf{H}_1 \otimes \cdots \otimes \mathbf{H}_1,$$

where the operator $\mathbf{H}_1$ occurs in this Kronecker product N times. On the other hand, the operator $\mathbf{H}_N$ is given by the recursive formula

$$\mathbf{H}_N = \frac{1}{\sqrt{2}} \begin{pmatrix} \mathbf{H}_{N-1} & \mathbf{H}_{N-1} \\ \mathbf{H}_{N-1} & -\mathbf{H}_{N-1} \end{pmatrix}. \tag{8.18}$$

In the spinor basis the operator $\mathbf{H}_N$ is

$$\mathbf{H}_N = \frac{1}{2^{N/2}} (\mathbf{X} + \mathbf{Z}) \otimes (\mathbf{X} + \mathbf{Z}) \otimes (\mathbf{X} + \mathbf{Z}) \otimes \cdots \otimes (\mathbf{X} + \mathbf{Z}),$$

while in the two-Bose operator representation it is

$$\mathbf{H}_N = \frac{1}{2^{N/2}} \left[a_1^+ (a_1 + a_2) + a_2^+ (a_1 - a_2) \right] \otimes \cdots \otimes \left[a_1^+ (a_1 + a_2) + a_2^+ (a_1 - a_2) \right]. \tag{8.19}$$

All calculations using the operators from (8.14), (8.16), (8.17), and (8.19) must be performed taking into account the kinematic condition (8.9).

The matrix $\mathbf{H}_N$ is defined in the spinor basis $\{|2^{N-1} - \frac{1}{2}\rangle, |2^{N-1} - \frac{3}{2}\rangle, \ldots, |-2^{N-1} + \frac{3}{2}\rangle, |-2^{N-1} + \frac{1}{2}\rangle\}$ or in the equivalent two-Bose operator basis from (8.10). Using the formula (8.18) it is easy to show that $\mathbf{H}_N^2$ is the identity operator $(\mathbf{H}_N^2 = \mathbf{1})$.

In this way, we can find the transformation properties of the spin basis functions in the two-Bose operator representation for the spin $S = 2^{N-1} - \frac{1}{2}$ under the action of the Hadamard operator for any number N of qubits. In particular, if $N = 77$ then the total number of bosons of first and second types that is necessary to realize the two-Bose operator representation in the case of spin $S = 2^{76} - \frac{1}{2} \sim 2^{76}$, is $4S \sim 3 \cdot 10^{23}$. For such a large number of qubits the kinematic condition (8.9) does not play such an important role as it does at low values of the spin S and, respectively, a small number of qubits. In this case the number of spin states is so large that it can be considered infinite, as is the number of degrees of freedom of the boson field.

The transition from spinor representation to two-Bose operator representation must be done for spin wave functions, spin operators, Zeeman operators, evolution operators, and spin-dependent density operators, etc.

Quantum information processing using the two-Bose operator representation of angular momentum is an effective type of processing for a high number N of qubits, because in this case it is possible to use Wick's theorem [529], which simplifies the calculations.

Chapter 9
Test Systems for Quantum Information Processing

Here, we discuss specific examples of systems that have been used for demonstrating important concepts of quantum information processing (QIP). This includes very different systems like semiconductor quantum dots, dielectric solids containing rare-earth ions, or defects, such as the nitrogen–vacancy center in diamond.

After introducing the basic principles of quantum information processing in Chap. 8, this chapter describes some of the systems that have been proposed for realizing a quantum information processor. Some of these systems have actually been tested in this respect, at least as small-scale versions able to store and process a few quantum bits.

In Sect. 9.1, we summarize the basic requirements that such systems must fulfill. Section 9.2 introduces semiconductor quantum dots as engineered traps for electronic qubits. The nitrogen–vacancy defect of diamond, introduced in Sect. 9.3, is an example of another type of qubit: here, an atomic defect (which occurs in natural diamond, but can also be engineered in industrial diamond) is an atomic trap for an electron spin qubit, which also couples to nearby nuclear spins. The ^{31}P donor in silicon, discussed in Sect. 9.4, is an intermediate example: it is an atomic defect that traps an electron and can be used as a nuclear spin qubit, but it can be controlled through nanoscale electrodes placed close to the center. As discussed in Sect. 9.5, ^{31}P can be embedded in silicon, but it can also be trapped in endohedral fullerenes, like C_{60}; the same holds for ^{14}N and ^{15}N. While all these types of qubits can be used as individual nanoscale information carriers, it is sometimes advantageous to distribute the information in excitations of ensembles of atomic systems. A good example is that of rare-earth atomic ions embedded in dielectric crystals. This approach is discussed in Sect. 9.6. Section 9.7 finally introduces "molecular magnets," consisting of clusters of strongly coupled atoms that together have a high spins.

9.1 Requirements

The basic requirements that a system must have for successful implementation of quantum information processing were stated by DiVincenzo [504]:

I. Geru, D. Suter, *Resonance Effects of Excitons and Electrons*,
Lecture Notes in Physics 869, DOI 10.1007/978-3-642-35807-4_9,
© Springer-Verlag Berlin Heidelberg 2013

1. **Well-defined qubits.**

    A useful qubit must consist of at least two orthogonal quantum states. For scalable quantum computation, it must not only be possible to produce many of them, but it must also be possible to identify and address them without perturbing the others. These basic computational states must be chosen in such a way that the system can be switched between them and that arbitrary superposition states of the computational basis states can be generated by external control operations (see also criterion 4 below).

2. **Initialization into a well-defined state.**

    Every algorithm requires a well-defined initial state, which must be encoded into the quantum register. This initialization process is an irreversible process, since it must work independently of the previous state of the qubits. For solid state systems at low temperature, this initialization process often proceeds naturally, by cooling the sample into its ground state. However, this is not a feasible scheme for scalable operation, since large-scale quantum information processing requires a steady supply of freshly initialized qubits for quantum error correction [530]. However, this problem can be solved if a reliable readout of the corresponding qubits is available, which is also a projective, irreversible operation. If necessary, they can be inverted after the readout.

3. **Long decoherence times.**

    The lifetime of the states, including superposition states, must be long enough to finish all the required computational steps. It is thus not possible to express this in numbers, but the required lifetime depends on the duration of the individual gate operations and the number of gate operations dictated by the algorithm.

4. **A universal set of quantum gates.**

    As discussed in Chap. 8, a universal set of quantum gate operations is required to implement arbitrary algorithms. Such a universal set of gates can be built from at least two rotations of each individual qubit around different axes and a set of two-qubit operations. In most systems, the single-qubit operations are relatively easy to implement, typically by resonant excitation with alternating electric or magnetic fields. The two-qubit operations require interactions with external fields, and also couplings between the qubits. Ideally, these couplings should be switchable, but techniques have been developed to also work with natural "always-on" couplings [493, 507].

5. **A qubit-selective readout.**

    Depending on the algorithm, the result of the computation may be encoded in a superposition state of one or several qubits. The readout process converts this superposition state into classical information. The readout is again a nonunitary process. Typical readout processes in solid state materials use electric readout, e.g., through single-electron transistors or Coulomb blockade measurements [531].

9.2   Semiconductor Quantum Dots

Semiconductor materials offer a range of additional possibilities for defining qubits, since the techniques for making very small structures with very precise system parameters have been developed over the last decades to a perfection that has not been reached for any other material system. For quantum computing applications, quantum dots appear most promising: they trap conduction band electrons and holes in the valence band by spatially confined effective potentials. Enclosing the electron in an effective potential fulfills several important requirements for using them in quantum information processing devices: it localizes the electrons so that gates can address them selectively. If the confinement is smaller than the coherence length of the electron, it leads to discrete quantum states, which is necessary for defining a qubit. In addition, the confinement makes the states more stable against environmental perturbations, since they cannot be scattered by infinitesimal excitations. Compared to free excitons, the confinement increases the binding energy of an electron–hole pair.

Confinement of the electrons can be achieved by spontaneous formation of quantum dots. This occurs typically during deposition of semiconductor layers from the vapor phase, e.g., by molecular beam epitaxy (MBE). If the material being deposited has a different lattice constant than the substrate material, the strain energy induced by the lattice mismatch can be reduced if the freshly deposited atoms do not form continuous layers but instead assemble into islands [532]. This may occur from the first layer or after an initial wetting layer. An alternative approach confines electrons in a layer by appropriate deposition techniques and then uses electrodes on top of the structure to confine the electrons to small islands within this layer [533].

The capacitance of a quantum dot drops with its dimensions, and the energy required for adding an additional electron to it therefore increases. As a result, the charge state of a quantum dot can be controlled by an external potential and electrons can be added "one by one." This makes it possible to encode a quantum bit by assigning, e.g., the logical 0 to a state with n electrons on the quantum dot and the logical state 1 to $n + 1$ electrons on the same dot.

Independent of the type of confinement used, the qubit can be associated with different degrees of freedom. The exciton can be used as a qubit, with the ground state (no exciton) representing $|0\rangle$ and the excited state (exciton present) representing $|1\rangle$. An early demonstration was given by Zrenner et al. [534], who used a laser to drive coherent oscillations between the ground state of the quantum dots and the state with a single exciton. The presence or absence of the exciton was measured by the electric current through the device. The Rabi oscillations can also be observed optically, as shown by Borri et al. [535].

It may then be advantageous not to encode a qubit in the presence or absence of an exciton, but rather in the position of electron and hole in two coupled quantum dots. Figure 9.1 shows a possible encoding scheme: the first qubit is associated with the position of the electron, the second with that of the hole. If the particle is in the first quantum dot, it encodes a logical 0; if it resides in the second quantum dot, it encodes the value 1.

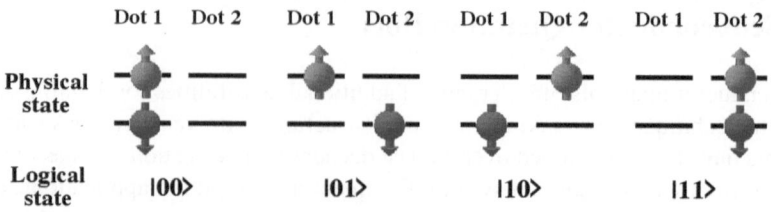

Fig. 9.1 Representation of a logical qubit by two quantum dots

While the electron charge can be used as a qubit, the associated decay times are quite short. Longer coherence times can be obtained, e.g., by using the spin degrees of freedom. While the group IV materials Si and Ge were mostly used in implementations on the basis of impurity spins, III/V materials like GaAs are preferred for most of the other approaches. Using III/V materials is particularly important for implementations that use optical excitation or readout, which requires direct bandgap materials.

The disadvantage of the III/V materials for spin-based qubits is that the natural abundance materials all have nuclear spins with which the electron spin interacts via the hyperfine interaction

$$\mathcal{H}_{\mathrm{hf}} = S_z \sum_k A^k I_z^k,$$

where we have assumed that the electron spin is quantized in a strong magnetic field $\parallel z$. The sum runs over all nuclear spins I^k, and the hyperfine coupling constant A^k is proportional to the electron density at the location of the corresponding nucleus. While the interaction of the individual nuclear spin with the electron is relatively weak, the number of interacting nuclei is very large. As a result, the combined interaction of the nuclear spins within the envelope of the electron wave function generates an effective magnetic field $B_N \approx \langle \sum_k A^k I_z^k \rangle$. This "nuclear field" adds to the Larmor precession of the electron spin with frequencies in the gigahertz range. Since the orientation of the nuclear spins is not constant in time, this effective field fluctuates and leads to a loss of coherence. This is a much smaller problem in Si, where the most abundant species does not have a nuclear spin and therefore does not interact with the electron spin. Readout of electron spin qubits can be performed by converting the spin state to a charge state and using single electron techniques for readout or optically, e.g., by Kerr rotation measurements [536].

Coupled quantum dots, called "quantum dot molecules," can be used to encode multiple qubits [537, 538]. They can be grown by different techniques, including overgrowth of cleaved edges of semiconductor quantum well structures [539] and confinement of the electrons of a two-dimensional electron gas by electrostatic potentials [540].

Charged excitons (trions) are a useful tool for manipulating electron spin qubits [541]. As shown in Fig. 9.2, the spin of the excess electron imposes a selection rule on the generation of an exciton. The spin of the electron has a much

Fig. 9.2 An excess electron in a quantum dot allows generation of an exciton only if the excited electron has antiparallel spin

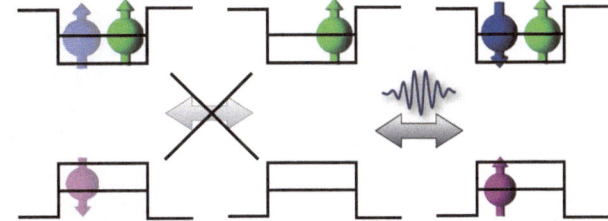

longer lifetime than the exciton, but the exciton can be controlled on a picosecond time scale by laser pulses. The interaction between the three constituents of the trion ensures control over the excess electron. Similarly, the exciton can be used to read out the spin state of the excess electron [542].

In contrast to silicon-based systems, where isotopically enriched ^{28}Si material is free of nuclear spins, GaAs consists of three nuclear isotopes, all of which have a nuclear spin $I = 3/2$. Therefore, electron spins are always subject to hyperfine interaction with the nuclei over which the electron wave function extends, and this interaction yields a significant contribution to the dephasing of electron spins in GaAs [543, 544].

9.3 The Diamond Nitrogen–Vacancy Center

Diamond has a number of well-characterized defects, of which the most prominent one is the nitrogen–vacancy (NV) center [545, 546]. It consists of a nitrogen at a carbon site and an adjacent vacancy, i.e., a missing carbon. The electrons of the defect combine to an $S = 1$ total spin. The attractive properties of this center include the long coherence times at room temperature, and the special optical properties: the photostability is very high, allowing experiments on single centers for months.

For the purpose of this section, we will not discuss bulk experiments on NV centers, but only experiments with single centers. Each of the bright spots in the right-hand part of Fig. 9.3 represents a single NV center. While it is not possible to determine this from the image alone, which was taken by scanning confocal microscopy with a resolution of $\approx$300 nm, it is possible to estimate it from the observed count rate. However, a much cleaner signature is obtained by measuring the correlation function of the arrival times of the photons on the detector. If we measure the delays τ between the arrival times of individual photons, we find that the probability of detecting a second photon immediately after the first drops to zero for short times [547]. This is easy to understand by considering that after the emission of a photon, the center is in the ground state and cannot emit another photon until it has absorbed one.

Figure 9.4 shows an example: the emission rate drops almost to zero for short delays, and it takes $\approx$15 ns for the emission probability to rise again to its average value. This rise time decreases with increasing laser intensity.

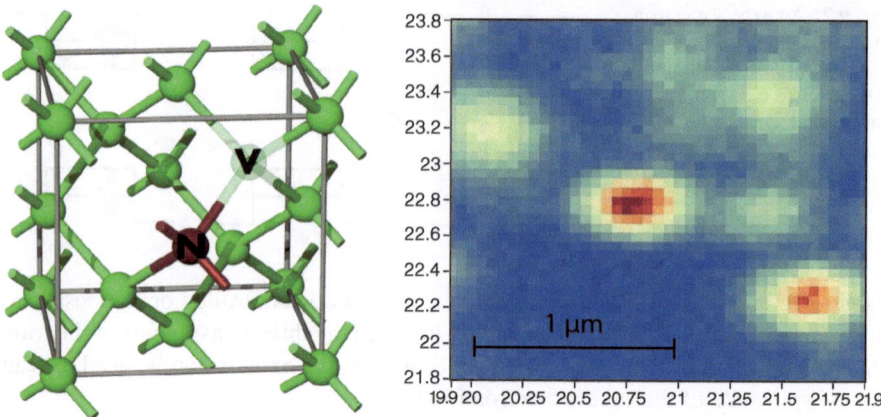

Fig. 9.3 Structure of the N/V center in diamond. The right-hand side represents an image of a diamond surface, recorded by a scanning confocal microscope. Each *bright spot* represents a single NV center

Fig. 9.4 Photon correlation function for a single NV center

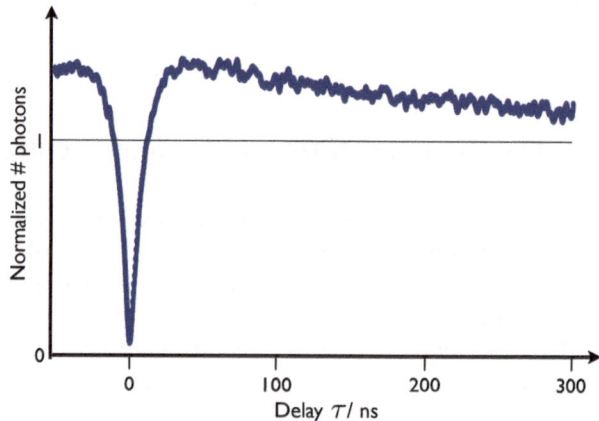

Initialization as well as readout rely on absorption–emission cycles between the $^3A_2$ electronic ground state and the ^3E electronically excited state, whose zero-phonon line has a wavelength of $\lambda_0 = 637$ nm. The phonon sidebands can be excited by green laser light (e.g., at $\lambda = 532$ nm). Between these two electronic triplet states are two singlet states, which can be populated by intersystem crossing processes. These processes are spin dependent. Pumping the system for ≈ 0.5 µs with 1 mW of green laser light leaves it with high probability in the $m_S = 0$ spin state. When the system is in the $m_S = 0$ state, the scattering rate for unpolarized green light is about twice that of the $m_S = \pm 1$ states, which allows a relatively straightforward detection of the individual spin states.

In the absence of a magnetic field, the $m_S = \pm 1$ spin states are degenerate, but separated from the $m_S = 0$ state by a zero-field splitting of $D = 2.87$ GHz. A magnetic field lifts the degeneracy of the $m_S = \pm 1$ states. In addition, the electron spin

Fig. 9.5 Spectrum from a single NV center showing resolved hyperfine couplings to the ^{14}N and one ^{13}C nuclear spin

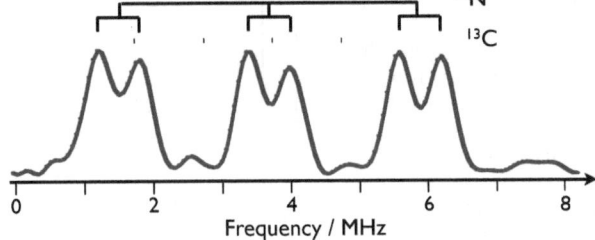

is coupled to the nitrogen nuclear spin (usually ^{14}N, $I = 1$) and to those carbon sites that are occupied by a ^{13}C isotope ($I = 1/2$) with a hyperfine coupling constant that starts at 130 MHz for the carbon sites adjacent to the vacancy and decreases with the distance [548]. Thus, the most important terms in the ground-state Hamiltonian of the NV defect are

$$\mathcal{H} = DS_z^2 + \mu_0 g \mathbf{BS} + A_N \mathbf{SI}_N + \sum_k A_C^k \mathbf{SI}_C^k,$$

where the sum runs over all sites occupied by ^{13}C isotopes.

Figure 9.5 shows a typical spectrum: the $m_S = 0 \rightarrow m_S = -1$ transition of the electron spin is split by the hyperfine interaction with the ^{14}N nuclear spin ($A_N = 2.17$ MHz) and one ^{13}C nuclear spin ($A_C = 0.58$ MHz). Many additional nuclear spins couple to the electron spin with hyperfine coupling constants ≤ 0.3 MHz, which do not lead to resolved splittings, but to a broadening of the resonance line.

The decay of electron spin coherence by the hyperfine interaction with the ^{13}C nuclear spins can be refocused by the usual spin echo experiments. As shown in Fig. 9.6, a single refocusing pulse, corresponding to the Hahn echo, can generate echoes for delays of up to 10 µs. For longer times, the refocusing does not work, because fluctuations in the environment make the refocusing inefficient. As in molecular diffusion, it then becomes necessary to apply multiple refocusing pulses with shorter delays between them [549, 550]. As shown by the other curves in Fig. 9.6, sequences of refocusing pulses can extend the coherence time up to about 1 ms.

The experimental data of Fig. 9.6 show that the decay is not exponential. The curves drawn through the experimental points were obtained by fitting a "stretched exponential" $e^{-(t/T_2)^\beta}$ to the experimental data. For a small number of refocusing pulses, the exponent is close to 1, but it becomes smaller for larger numbers of pulses, indicating a complex dynamics in the environment. One contribution to this comes from the Larmor precession of the ^{13}C nuclear spins, which is synchronized by the microwave pulses applied to the electron spins. At the start of the experiment, the laser pulse initializes the electron spin into the $m_S = 0$ state. In this state, the secular part $S_z I_z$ of the hyperfine interaction vanishes, and the nuclear spin interacts only with the external magnetic field. The $\pi/2$ microwave pulse then puts the system into a superposition of the $m_S = 0$ and $m_S = 1$ state. If the electron is in the $m_S = 1$ state, the nuclear spins interact not only with the external magnetic field, but also experience an effective field from the electron spin, which is oriented along

Fig. 9.6 Refocusing of
electron spin coherence by
spin echo experiments. The
curves in the lower panel
show the decay of the echo
amplitude as a function of the
total measurement time for
different experiments with
increasing number of
refocusing pulses

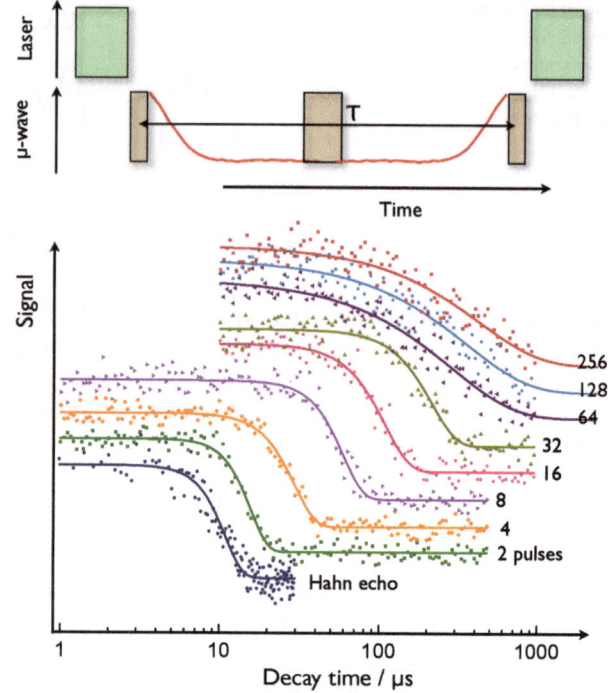

the symmetry axis of the NV center. If this axis does not coincide with the direction of the external magnetic field, the two states have different quantization axes for the nuclear spin, and the microwave pulse creates not only a superposition of electron spins, but also a superposition of the nuclear spins, which evolves between the $\pi/2$ and π pulses. The refocusing pulse cannot completely refocus such a time-dependent environment, and the echo amplitude decreases. However, this evolution of the environment is partly coherent, since the Larmor frequency is the same for all ^{13}C spins. The environment therefore refocuses after a Larmor period, and if the electron spin refocusing pulse is applied at this particular time $\tau = 2\pi/\Omega_C$ (or a multiple thereof), the echo amplitude recovers [551, 552].

Defects with similar properties have also been identified in SiC [553], although they have not all been as well characterized as the NV defect in diamond.

9.4 ^{31}P in Silicon

A very prominent example of a defect center proposed as a qubit is the ^{31}P donor in silicon—the only $I = 1/2$ shallow (group V) donor in Si. In its neutral state, the center consists of an unpaired electron and an $I = 1/2$ nuclear spin, which are coupled by hyperfine interaction. The Si:^{31}P system was exhaustively studied in the first electron–nuclear double-resonance experiments [554]. At sufficiently low

Fig. 9.7 Electron and nuclear spin states in the ^{31}P:Si donor. The two lower energy states, both with the electron spin ↓, define the qubit

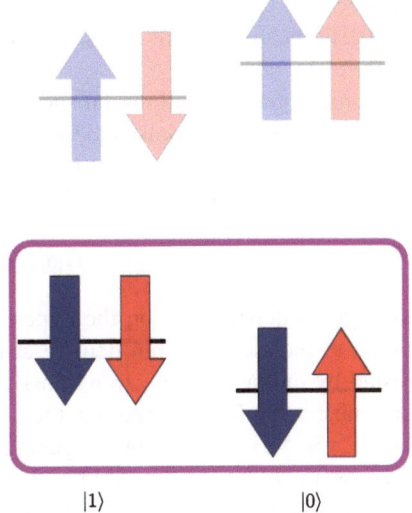

$|1\rangle$ $|0\rangle$

^{31}P concentrations at temperature $T = 1.5$ K, the electron spin relaxation time is thousands of seconds, and the ^{31}P nuclear spin relaxation time exceeds 10 hours. In principle, both spins can be used as qubits, and their long spin lifetimes make them very attractive candidates for spin qubits.

One possible scheme for using this center in quantum information processing was proposed by Kane [555]: he suggested using the nuclear spin of ^{31}P as a qubit. The relevant Hamiltonian for the defect center then consists of the interaction of the electron and nuclear spins with each other and with the external magnetic field,

$$\mathcal{H} = \Omega_S S_z - \Omega_I I_z + A I_z S_z,$$

where S is the electron spin, I is the nuclear spin, Ω_S and Ω_I are the Larmor frequencies of the two spins, and A is the hyperfine coupling constant. The four eigenstates of this system are $|\downarrow\uparrow\rangle$, $|\downarrow\downarrow\rangle$, $|\uparrow\downarrow\rangle$, and $|\uparrow\uparrow\rangle$, where the first position refers to the electron and the second to the nuclear spin state.

Their energies are $(-\Omega_S - \Omega_I)/2 - A/4$, $(-\Omega_S + \Omega_I)/2 + A/4$, $(\Omega_S + \Omega_I)/2 - A/4$, and $(\Omega_S - \Omega_I)/2 + A/4$. As shown in Fig. 9.7, Kane's scheme does not use the electron spin for storing and processing information, but as a control mechanism for the nuclear spin, which stores the qubit. The information is stored in the two lower energy states, corresponding to the electron spin antiparallel to the field. The two qubit states are thus

$$|0\rangle = |\downarrow\uparrow\rangle, \qquad |1\rangle = |\downarrow\downarrow\rangle$$

and their energies $(-\Omega_S - \Omega_I)/2 - A/4$ and $(-\Omega_S + \Omega_I)/2 + A/4$. For initialization, the system can be cooled in a strong magnetic field. After a sufficient delay, the electron spin is in the ↓ state with a high probability. The nuclear spin will generally be in a mixed state, but this can be corrected by exciting the (forbidden) transition between the $|\downarrow\downarrow\rangle$ and the $|\uparrow\uparrow\rangle$ state and letting the electron spin relax again. If this cycle is repeated several times, the system is initialized into the $|0\rangle = |\downarrow\uparrow\rangle$ ground state.

Logical gate operations on the nuclear spin qubits can be generated by resonant radio-frequency pulses [556, 557]. In general, all the ^{31}P nuclei in the sample would respond to such radio-frequency pulses in the same way. However, selective addressing can be implemented by adjusting the transition frequency

$$\Omega_{|0\rangle\leftrightarrow|1\rangle} = \Omega_I + A/2$$

through its dependence on the hyperfine coupling constant A: its value is proportional to the electron spin density at the position of the nucleus. This can be adjusted by moving the electron density through external electric potentials. For the parameters that Kane suggested, the transition frequency $\Omega_{|0\rangle\leftrightarrow|1\rangle}$ could be changed by $\approx 2\pi 50$ MHz, sufficient for generating resonant gate operations. In the readout process, the qubit state would first be transferred from the nuclear spin to the electron spin and from there into the charge degree of freedom. Single electrons can be detected via their effect on the conductivity through nanoscale islands and channels [558].

A significant contribution to the decoherence in this system is given by the hyperfine interaction with ^{29}Si ($I = 1/2$) nuclear spins. This effect can be suppressed by using isotopically pure material consisting only of ^{28}Si (which has no nuclear spin). Gate operations could be performed by resonant radio-frequency pulses. However, this requires that the qubits have unique resonance frequencies, which is not normally the case. Kane therefore suggested to use the hyperfine interaction with the electron to shift the nuclear spin frequencies. The strength of the hyperfine interaction depends on the electron density at the position of the nucleus. This can be adjusted by applying electrostatic potentials that shift the position of the electron wave function. In a model calculation, Kane showed that gate voltages of ≈ 0–1 V should be able to tune the nuclear spin Larmor frequency between 50 and 100 MHz. Similarly, gate operations could shift the position of two neighboring electron spins in such a way that an indirect coupling between the nuclear spins of two adjacent sites is generated. This allows one to apply all the necessary quantum gate operations. Most of the required operations have been demonstrated experimentally, e.g., entanglement between the electron and nuclear spins in an ensemble experiment [559].

One of the main challenges of this approach is the placement of the individual donor atoms, which should occur with atomic precision. The second main challenge is the detection of the individual spins. Recent experiments [560] achieved single-spin readout on the basis of a transfer to a single-electron transistor. They observed a spin lifetime of ≈ 6 seconds at a magnetic field of 1.5 T.

The concept of using donor atoms in silicon can also be modified by using Si/Ge heterostructures [561, 562], rather than bulk Si. An attractive feature of such heterostructures is that the g-factor of the electron spin depends on the material. Using electrodes, the electrons can be pushed into the Si or Ge material, thereby changing their resonance frequency and providing addressability for single-qubit gates.

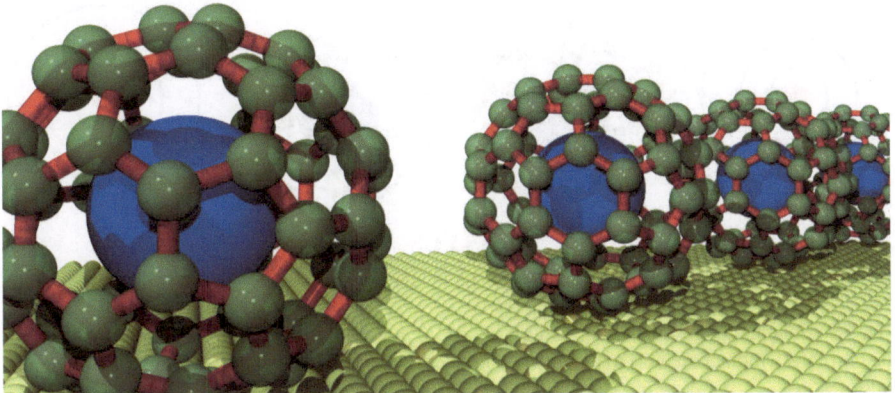

Fig. 9.8 Endohedral fullerenes

9.5 Endohedral Fullerenes

Fullerenes are another carbon variation that has generated immense interest in the scientific community. In the context of quantum computing, the main interest in fullerenes arises from the possibility of using them as room-temperature, nanometer-sized traps for neutral atoms [563]. In particular, nitrogen and phosphorous atoms are attractive candidates, as they are hard to trap with other methods. They both have a half-full p shell, which results in a total electron spin $S = 3/2$. The electron spin is coupled to the nuclear spin by hyperfine interaction. The relevant Hamiltonian of the spin system can be written as

$$\mathcal{H}_S = g\mu_B \mathbf{B}_0 \cdot \mathbf{S} - \gamma_n \mathbf{B}_0 \cdot \mathbf{I} + A\mathbf{S} \cdot \mathbf{I}.$$

Here, $\mathbf{S}$ is the electron spin, $\mathbf{I}$ the nuclear spin, g, μ_B, and γ_n are the electron g-factor, Bohr's magneton, and the nuclear gyromagnetic ratio, $\mathbf{B}_0$ is the magnetic field, and A is the hyperfine coupling constant. For the atoms trapped in a C_{60} molecule, the corresponding values are

Nucleus	Spin/$\hbar$	A/MHz
^{14}N	1	15.88
^{15}N	1/2	22.26
^{31}P	1/2	138.4

Using the electronic as well as the nuclear spin degrees of freedom allows one, in principle, to encode up to three qubits in each molecule.

Figure 9.8 shows a possible use of these molecules as qubits: each C_{60} molecule acts as a trap for a nitrogen or phosphorus atom, whose spins encode the quantum information. The major properties that make this system so attractive for quantum information processing is that (i) the spins have very long lifetimes, with the longitudinal relaxation time T_1 exceeding 1 s at low temperature [564], and (ii) they are

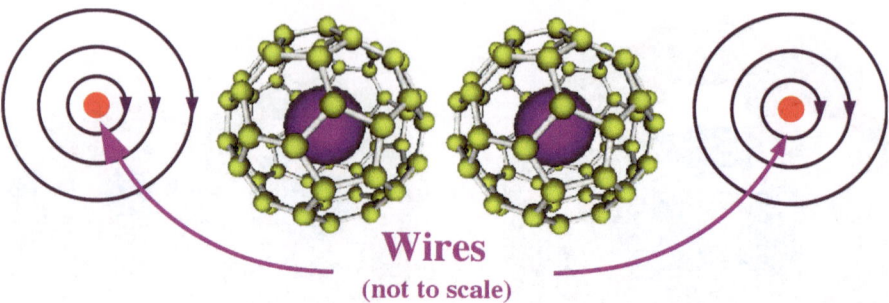

Wires
(not to scale)

Fig. 9.9 Scheme for resonant addressing of N@C$_{60}$ molecular spins: the wires carry copropagating currents, which generate a magnetic field gradient superimposed over a static external magnetic field

easier to manipulate. It would be possible, e.g., to deposit them on the surface of a suitable material, such as silicon, and manipulate them using a scanning tunneling microscope [565].

Gate operations can be performed by resonant microwave pulses applied to the electron spins and radio-frequency pulses applied to the nuclear spin transitions [566]. One can address the individual molecules, e.g., by applying a magnetic field gradient that shifts the resonances of the individual molecules [567]. By depositing copper wires on the Si surface and running currents of the order of 1 A through two parallel wires, one can generate a magnetic field that combines with the homogeneous background magnetic field to form a magnetic field gradient between the two wires, as shown schematically in Fig. 9.9. For a distance between the wires of the order of 1 μm, the resulting gradient would be of the order of 4×10^5 T/m. For two N@C$_{60}$ molecules separated by 1.14 nm (the diameter of the molecules), this results in a frequency splitting of 12.7 MHz, which should allow precise qubit addressing in frequency space. If larger distances are chosen between the molecules, the frequency difference is correspondingly larger.

One major difficulty of the system is that the magnetic dipole couplings between the molecules are static; i.e., they cannot be switched as required by the algorithm. This problem can be solved by using the electron and nuclear spin for encoding a single logical qubit. Figure 9.10 shows the relevant energy level scheme for the ^{15}N@C$_{60}$ or ^{31}P@C$_{60}$ electron–nuclear spin system. The four nuclear spin transitions and the electron spin transitions are split by a hyperfine coupling of 22 or 138 MHz.

Using both degrees of freedom allows one to store the information in the nuclear spin degree of freedom. Since the nuclear spin couples only weakly to other degrees of freedom, the quantum information stored in it has a long lifetime. It is also effectively isolated from the other molecules, since the magnetic dipole–dipole couplings between nuclear spins is $\approx 10^9$ times smaller than that between electron spins. When the algorithm requires an active coupling between two qubits, it can be generated by switching both qubits into the electron spin degrees of freedom, thereby switching the coupling on. The two-qubit operation is then performed on

Fig. 9.10 Energy levels of the $^{15}N@C_{60}$ or $^{31}P@C_{60}$ electron–nuclear spin system. I refers to the nuclear spin, S to the electron spin

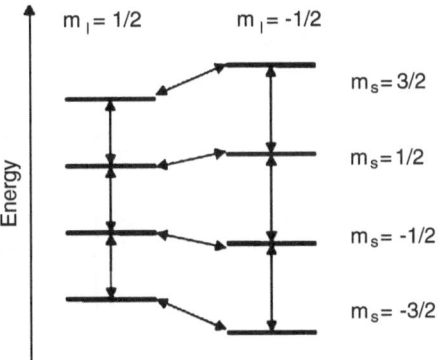

the electron spins, and the qubits are switched back to the nuclear spins when the gate operation is complete [563, 567, 568].

While several elements of this scheme have been tested, the readout of the qubits remains a significant challenge. Experimental evidence [569] shows that it is possible to electrically contact individual magnetic $N@C_{60}$ molecules and measure spin excitations in their electron tunneling spectra. The tunneling spectra allow the identification of the charge and spin states of the molecule. If such measurements can be combined with the other elements, a quantum computer based on endohedral fullerenes appears possible.

9.6 Rare-Earth Ions

The electronic properties of rare-earth ions, i.e., the elements from lanthanum ($Z = 57$) to lutetium ($Z = 71$), distinguish them from almost all other elements. The states that are responsible for these special properties are the partly filled $4f$ orbitals. The relevant transitions that fall into the visible or near-IR range of the spectrum are all forbidden by parity and often also by spin selection rules. This results in long lifetimes and narrow natural linewidths [570]. Furthermore, the states are only weakly affected by crystal field effects, which results also in relatively small inhomogeneous broadening. These properties have fascinated physicists working in atomic spectroscopy as well as physicists and engineers interested in optical data storage [571] or optical data processing [572]. Rare-earth ions were also found to be useful qubits for quantum information processing, either stored in electromagnetic traps [573] or as dopant ions in dielectric crystals [574].

An additional use for rare-earth ions in solid materials came with the search for quantum memories [575]. These devices must store the complete quantum state of a photon in a suitable material for times of microseconds to seconds [576, 577]. For this purpose, it is necessary to convert "flying qubits" into stationary qubits and vice versa. This is achieved when a photon interacts with an optical transition. These processes can proceed directly, and they can also be assisted by different experimental techniques, such as electromagnetically induced transparency (EIT) [578].

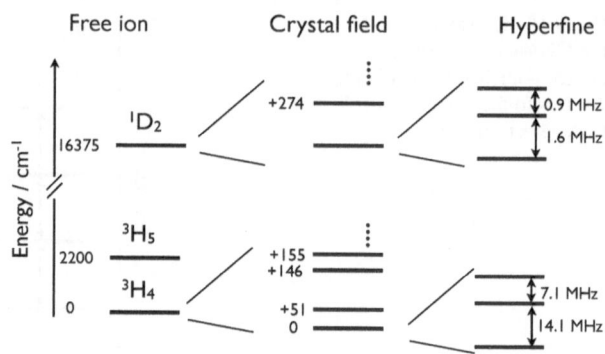

Fig. 9.11 Energy levels of the Pr^{3+} ion substituting for Y^{3+} in $YAlO_3$

Compared to conventional optical storage, quantum memories require storage of the complete quantum state of a photon. This is nontrivial, since it is not possible to convert the quantum state into classical information; this is usually specified in terms of the "no-cloning theorem" [521]. Quantum storage thus requires not only that the populations of the relevant states be conserved, but also the relative phases between them. This requirement is extremely difficult to meet in almost all solid state materials. Crystals containing doped rare-earth ions are the major exception [575, 579, 580]: due to their electronic structure, the optical dephasing times are unusually long.

The materials used for this purpose are mostly based on Pr^{3+} or Eu^{3+} substituting for La^{3+} or Y^{3+}, such as $Pr:La_2(WO_4)_3$, $Pr:YAlO_3$, or $Pr:Y_2SiO_5$. Relevant criteria include the accessibility of suitable transition frequencies by available lasers, the linewidth of these transitions, the lifetimes of the electronic and nuclear spin states, and the transition strengths and absorption depths for a given amount of doping. High levels of doping can generate stress in the crystal and therefore broadening of the resonance lines, particularly if the ionic radii of the host and guest ion differ significantly.

Figure 9.11 shows, as a typical example, the simplified energy level scheme of the Pr^{3+} ion substituting for Y^{3+} in a $YAlO_3$ crystal. Among the many possible optical transitions, that between the $^3H_4$ electronic ground state and the $^1D_2$ electronically excited state is easily accessible by high-resolution ring dye lasers, with a transition energy of $16\,375\ cm^{-1}$, which corresponds to a wavelength of 610.7 nm. The highly degenerate states of the free ion split in the presence of a crystal field. At the same time, the crystal field also quenches the orbital angular momentum of the electrons. On a much smaller energy scale, the states split further due to the interaction of the nuclear spin with external magnetic fields and the nuclear quadrupole moment with the electric field gradient tensor of the crystal. Both interactions are enhanced by the second-order hyperfine interaction.

As illustrated in Fig. 9.12, the simplest optical storage scheme can be realized as a photon echo: absorption creates a superposition state of two electronic states $|g\rangle$ and $|e\rangle$, which contains the information about the quantum state of the absorbed photon. The inhomogeneous dephasing in the material can be reversed by an echo pulse, resulting in the emission of a photon at a later time, in a direction which

Fig. 9.12 Photon echo as a short-time optical memory

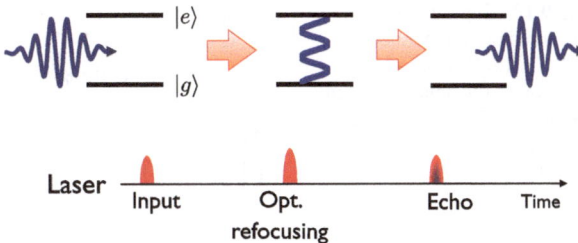

is determined by the directions of the incident pulse and the refocusing pulse. This simple photon echo experiment has several limitations: its efficiency is quite small—only a few percent of the incident light is typically recovered, the rest is lost to absorption, and the storage times are relatively short. Different solutions have been proposed for these problems [575, 581–584], and a number of these improvements have been tested experimentally (see, e.g., [585, 586]). The ultimate goal is to store the state of a single photon with fidelity close to unity for a time of the order of seconds.

Rare-earth ions can be used for more than information storage; once the information has been input into the system, it can also be processed by the usual quantum gate operations. Optical pulses as well as radio-frequency fields can be used to generate the quantum logical gate operations. In the form of trapped atomic ions, rare-earth ions were relatively quickly adopted for quantum computing applications (see, e.g., [587]). Rare-earth ions in solid state materials offer in principle the same potential [588]. Compared to many other solid state systems, they can be operated at relatively "warm" temperatures close to 4.2 K. The optical as well as the magnetic dipole degrees of freedom offer many possibilities for generating gate operations. Several demonstration experiments have verified this potential. Important milestones include the demonstration of optical coherence lifetimes of 4.4 ms in Er^{3+}:Y_2SiO_5[589]. Significantly longer coherence lifetimes can be achieved, if the phase information is transferred from the optical transition to a hyperfine transition, where coherence lifetimes of several seconds can be achieved with suitable techniques [577]. All these experiments involve the use of coherent magneto-optics, discussed in the following section.

The transfer of the information from the electronic degrees of freedom into the nuclear spin state increases the lifetime by several orders of magnitude. Even more important is that it now becomes easier to use established experimental techniques for further extending the lifetime. The main limitation for the decay of quantum information stored in the nuclear spin degrees of freedom of rare-earth ions is the magnetic fluctuations of the environment. Their influence can be suppressed by different techniques, including the application of suitable magnetic fields, which suppress the effect of magnetic field fluctuations on the transition frequency to first order [590], or by sequences of radio-frequency pulses that refocus the dephasing induced by the environment [591].

Figure 9.13 summarizes some results on the storage of optical states. The left-hand panel shows a series of photon echoes, measured with a $\pi/2 - \tau - \pi - \tau$

Fig. 9.13 Experimentally measured photon echoes in $Pr^{3+}:La_2(WO_4)_3$ as a function of the delay between the pulses. The fitted curve corresponds to a dephasing rate of $T_2 = 9.34$ μs

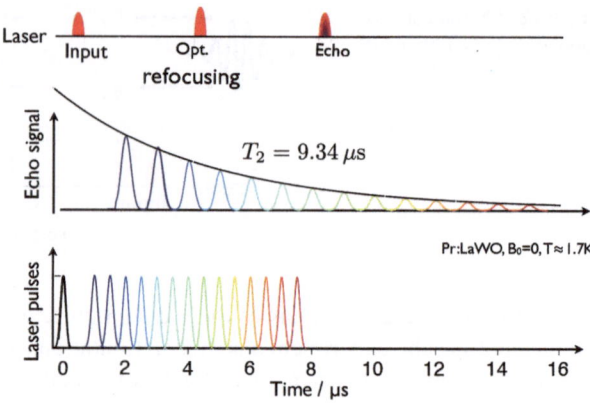

Fig. 9.14 Increase of the lifetime of nuclear spin coherence by several orders of magnitude, using either Hahn echoes or the Carr–Purcell–Meiboom–Gill sequence

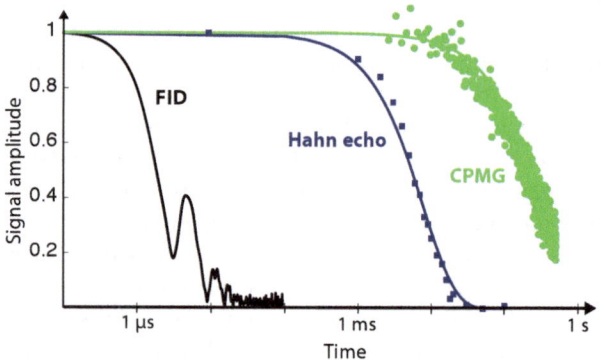

sequence. The decay of the echoes with increasing pulse separation can be fitted as an exponential decay $\propto e^{-4\tau/T_2}$ with a dephasing time $T_2 = 9.34$ μs. This corresponds to the lifetime of quantum states in the electronic degrees of freedom for this material.

We also implemented the transfer from the electronic to the nuclear spin degrees of freedom. The black curve in the right-hand panel of Fig. 9.13, labeled FID, shows the decay after the transfer.

This decay is dominated by magnetic interactions and can be refocused either by a simple refocusing pulse ("Hahn echo" in Fig. 9.13) or, more effectively, by a series of pulses (points labeled CPMG in Fig. 9.14. Clearly, these refocusing techniques extend the lifetime of the coherence in the material by more than five orders of magnitude.

9.7 Molecular Magnets

Molecules containing clusters of transition-metal ions have also been proposed as possible qubit systems [592]. The ions in these "molecular magnets" are strongly

Fig. 9.15 Mn$_{12}$O$_{12}$ cluster, forming the central part of a "molecular magnet" qubit

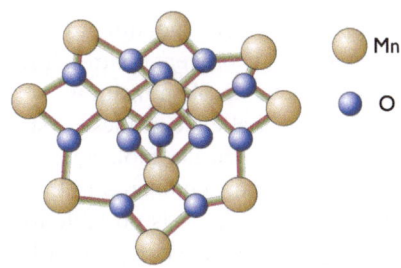

Fig. 9.16 Energy levels of an $S = 10$ system, corresponding to the Hamiltonian (9.1), with $E/D = 0.05$

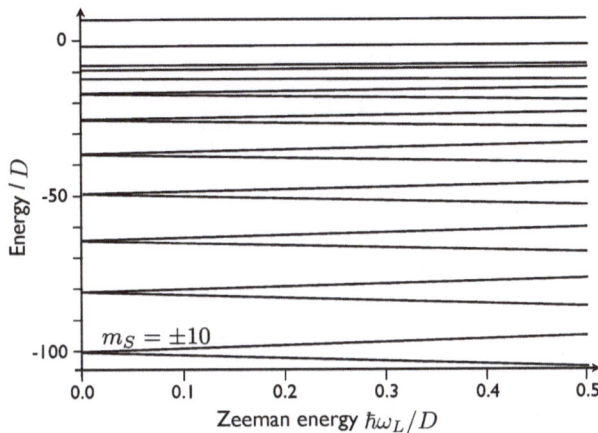

coupled by exchange interaction and have large total spins. Examples include clusters like Mn$_{12}$ [593, 594] shown in Fig. 9.15 and Fe$_8$ [595] with total spin $S = 10$. The spin interacts with its anisotropic environment, resulting in a large zero-field splitting energy

$$\mathcal{H}_{ZF} = -DS_z^2,$$

which stabilizes the ground states with $m_S = \pm S$. In most cases, the environment does not have axial symmetry, and thus the Hamiltonian contains an additional anisotropy term,

$$\mathcal{H}_{ZF2} = -DS_z^2 - E\frac{1}{2}(S_+^2 + S_-^2).$$

If a magnetic field is applied along the z axis, the total Hamiltonian becomes

$$\mathcal{H}_{mm} = -DS_z^2 - E\frac{1}{2}(S_+^2 + S_-^2) - \hbar\omega_L S_z. \tag{9.1}$$

As shown in Fig. 9.16, this completely lifts the degeneracy of the system. The $2S + 1$ energy levels offer a wide range of possible schemes for storing the quantum information in this system. However, since the energies are spread over a range of >100 GHz $\hbar$, it is an enormous challenge to implement coherent control for the complete system. As an alternative, it was proposed to use only the two lowest energy levels, corresponding to $m_S = \pm S$. This does not eliminate the challenge of implementing coherent control, however, since these states are not directly coupled by a magnetic dipole transition.

A major challenge of these systems for quantum information applications is the coupling between the molecules and their environment, which leads to relatively fast decoherence, compared to the more rigid solid state systems discussed above. For some systems, it is possible to deposit the metals as monolayers on a surface (often gold layers, see, e.g., [596]), without significantly changing their magnetic properties.

Chapter 10
Conclusions

The study of resonance effects on excitons and electrons in condensed matter allows us to obtain information about the energetic structure of excitons, exciton molecules (biexcitons), and local electron centers. On the other hand, under resonance conditions excitons manifest interesting properties from a practical point of view, such as laser IR radiation, generation of coherent hypersound and terasound in piezoelectric semiconductors, and the generation of coherent magnons in magnetic semiconductors. Quantum electronic devices, based on these effects, can be used to solve problems in solid state physics, radio astronomy, and cosmic telecommunications.

Microwave methods allow us to detect excitons at extremely low concentrations ($\sim 10^8$ cm^{-3}), which are not accessible by the methods of optical spectroscopy. These methods are very valuable in those cases when exciton states cannot be studied by optical means (optically forbidden transitions, extremely low concentrations of excitons, etc.).

When applying resonance methods, it turns out to be possible to separate the contributions from excitons, biexcitons, free charge carriers, and impurities in the absorption of hypersound by crystals, as well as to distinguish the contributions from excitons, biexcitons, and free charge carriers in the processes of nuclear spin relaxation. A similar situation is also characteristic for the spin relaxation of paramagnetic centers. In this case there is a substantial decrease in the spin relaxation time of paramagnetic centers for the Coulomb binding of charge carriers in excitons in comparison with the spin relaxation time of paramagnetic centers via free charge carriers.

Exciton paramagnetic, paraelectric, and acoustic resonances represent effective methods for studying the individual properties of excitons. In particular, for low concentrations of excitons, isotopic substitution of atoms in the crystal lattice sites leads to a shift of the lines of interserial exciton paraelectric resonance which can be comparable in magnitude to the frequency of resonance transition (the giant isotope shift).

For high levels of optical excitation of crystals, the peculiarities of the resonance absorption of hypersound allow us to distinguish the contributions from excitons and biexcitons, and the anomalous temperature behavior of the exciton Knight

I. Geru, D. Suter, *Resonance Effects of Excitons and Electrons*,
Lecture Notes in Physics 869, DOI 10.1007/978-3-642-35807-4_10,
© Springer-Verlag Berlin Heidelberg 2013

shift of the NMR line represents a supplementary evidence (in comparison with the δ-shaped peak in the optical spectrum of the exciton recombination radiation) of the presence of a phase transition of excitons to the state of Bose–Einstein condensation.

The interaction of paramagnetic centers of relatively high concentrations ($\sim 10^{19}$ cm^{-3}) with electrons and holes, forming a low-density exciton gas, leads to giant spin splittings of exciton states which determine giant magneto–optical effects. These effects have been particularly promising for various practical applications, which led to the development of the physics of diluted (semimagnetic) semiconductors.

The influence on excitons of hypersonic waves of high amplitude leads to the transformation of the energy spectrum of the exciton system to the quasi-energy spectrum of an exciton–phonon system. For a deep saturation of transitions between two exciton bands, when the exciton states in these bands become unsteady, a discontinuity in the quasi-energy spectrum appears. This represents a consequence of the energy and momentum conservation laws during exciton–phonon interaction, and it occurs when the distance between the exciton bands exceeds the kinetic energy of the exciton, which moves with the speed of the hypersonic wave. Similarly, a system of equidistant electron or nuclear Zeeman levels (with arbitrary spin value) in the presence of a resonance electromagnetic wave of high amplitude is transformed into a quasi-energy spectrum of a spin–photon system. For a half-integer value of the spin no quasi-energy level coincides with any energy level, whereas for an integer spin there is one such coincidence for sure.

Nuclei that are transformed into each other during the operation of spatial inversion are physically indistinguishable in electron–nuclear double resonance (ENDOR) spectra, while a hidden acoustical nonequivalence for those nuclei represents a characteristic of acoustical ENDOR spectra. The acoustical nonequivalence of nuclei connected by inversion is "hidden," because the positions of the resonance lines in the acoustical ENDOR spectra, which correspond to these nuclei, are exactly the same. However, acoustical ENDOR lines for these nuclei have different intensities, so the acoustical nonequivalence of nuclei connected by inversion, which is not seen in the acoustical ENDOR spectra, can be detected by the influence of an external constant electric field on the nuclear spin system.

These and other features of the interaction of spin systems with external perturbations allow us to determine the spin states which are suitable for quantum computing. These properties of spins in solids make them very interesting candidates for storing and processing information. As quantum mechanical objects, spins can be put in superposition states, and their evolution is governed by the Schrödinger equation. Accordingly, quantum systems consisting of arrays of coupled spins can form the basic units of quantum computers. These systems are not bound by the same laws as classical computers and therefore promise qualitatively more powerful information processing devices that can solve computational problems that appear intractable in the context of classical computer science.

Appendix A
Irreducible Tensor Operators $Y_M^L(\mathcal{I})$

L	The coefficients of operator equivalents	M	$Y_M^L(\mathcal{I})$
1	**2**	**3**	**4**
1	$i\sqrt{\frac{3}{4\pi}}\cdot\frac{1}{r}$	1	$-\frac{1}{\sqrt{2}}\mathcal{I}_+$
		0	$\mathcal{I}_z$
2	$-\sqrt{\frac{15}{8\pi}}\cdot\frac{1}{r^2}$	2	$\frac{1}{2}\mathcal{I}_+^2$
		1	$-\frac{1}{2}\mathcal{I}_+(2\mathcal{I}_z+1)$
		0	$\frac{1}{\sqrt{6}}(3\mathcal{I}_z^2-\mathcal{I}^2)$
3	$-i\sqrt{\frac{35}{8\pi}}\cdot\frac{1}{r^3}$	3	$-\frac{1}{2\sqrt{2}}\mathcal{I}_+^3$
		2	$\sqrt{\frac{3}{4}}\mathcal{I}_+^2(\mathcal{I}_z+1)$
		1	$-\frac{1}{2}\sqrt{\frac{3}{10}}\mathcal{I}_+(5\mathcal{I}_z^2+5\mathcal{I}_z-\mathcal{I}^2+6)$
		0	$\frac{1}{\sqrt{10}}(5\mathcal{I}_z^3-3\mathcal{I}_z\mathcal{I}^2+\mathcal{I}_z)$
4	$\frac{3}{4}\sqrt{\frac{35}{2\pi}}\cdot\frac{1}{r^4}$	4	$\frac{1}{4}\mathcal{I}_+^4$
		3	$-\frac{1}{2\sqrt{2}}\mathcal{I}_+^3(2\mathcal{I}_z+3)$
		2	$\frac{1}{2\sqrt{7}}\mathcal{I}_+^2(7\mathcal{I}_z^2+14\mathcal{I}_z-\mathcal{I}^2+9)$
		1	$-\frac{1}{\sqrt{56}}\mathcal{I}_+(14\mathcal{I}_z^3+21\mathcal{I}_z^2-6\mathcal{I}^2\mathcal{I}_z+19\mathcal{I}_z+3\mathcal{I}^2+6)$
		0	$\frac{1}{\sqrt{280}}(35\mathcal{I}_z^4-30\mathcal{I}^2\mathcal{I}_z^2+25\mathcal{I}_z^2+3\mathcal{I}^4-6\mathcal{I}^2)$
5	$\frac{21i}{4}\sqrt{\frac{11}{14\pi}}\cdot\frac{1}{r^5}$	5	$-\frac{1}{4\sqrt{2}}\mathcal{I}_+^5$
		4	$\frac{\sqrt{5}}{4}\mathcal{I}_+^4(\mathcal{I}_z+2)$
		3	$-\frac{5}{4\sqrt{90}}\mathcal{I}_+^3(9\mathcal{I}_z^2+27\mathcal{I}_z-\mathcal{I}^2+24)$
		2	$\sqrt{\frac{5}{12}}\mathcal{I}_+^2(3\mathcal{I}_z^3+9\mathcal{I}_z^2-\mathcal{I}^2\mathcal{I}_z+12\mathcal{I}_z-\mathcal{I}^2+6)$

I. Geru, D. Suter, *Resonance Effects of Excitons and Electrons*,
Lecture Notes in Physics 869, DOI 10.1007/978-3-642-35807-4,
© Springer-Verlag Berlin Heidelberg 2013

L	The coefficients of operator equivalents	M	$Y_M^L(\mathcal{I})$
6	$-\frac{45}{8}\sqrt{\frac{13}{\pi}}\cdot\frac{1}{r^6}$	1	$-\sqrt{\frac{5}{336}}\mathcal{I}_+(21\mathcal{I}_z^4+42\mathcal{I}_z^3-14\mathcal{I}^2\mathcal{I}_z^2+63\mathcal{I}_z^2-14\mathcal{I}^2\mathcal{I}_z$ $\quad+42\mathcal{I}_z+\mathcal{I}^4-8\mathcal{I}^2+12)$
		0	$\frac{1}{3\sqrt{56}}(63\mathcal{I}_z^5-70\mathcal{I}^2\mathcal{I}_z^3+105\mathcal{I}_z^3+15\mathcal{I}^4\mathcal{I}_z-50\mathcal{I}^2\mathcal{I}_z+12\mathcal{I}_z)$
		6	$\frac{1}{360}\sqrt{231}\mathcal{I}_+^6$
		5	$-\frac{1}{60}\sqrt{77}\mathcal{I}_+^5(\mathcal{I}_z+\frac{15}{6})$
		4	$\frac{1}{120}\sqrt{14}\mathcal{I}_+^4(11\mathcal{I}_z^2+44\mathcal{I}_z-\mathcal{I}^2+50)$
		3	$-\frac{1}{180}\sqrt{105}\mathcal{I}_+^3(11\mathcal{I}_z^3+99\cdot\frac{1}{2}\mathcal{I}_z^2-3\mathcal{I}^2\mathcal{I}_z+\frac{179}{2}\mathcal{I}_z$ $\quad-\frac{9}{2}\mathcal{I}^2+60)$
		2	$\frac{1}{360}\sqrt{105}\mathcal{I}_+^2(33\mathcal{I}_z^4+132\mathcal{I}_z^3-18\mathcal{I}^2\mathcal{I}_z^2+\frac{811}{3}\mathcal{I}_z^2-\frac{532}{15}\mathcal{I}^2\mathcal{I}_z$ $\quad+\frac{830}{3}\mathcal{I}_z+\mathcal{I}^4-\frac{382}{15}\mathcal{I}^2+\frac{352}{3})$
		1	$-\frac{1}{180}\sqrt{42}\mathcal{I}_+(33\mathcal{I}_z^5+\frac{165}{2}\mathcal{I}_z^4-30\mathcal{I}^2\mathcal{I}_z^3+180\mathcal{I}_z^3-51\mathcal{I}^2\mathcal{I}_z^2$ $\quad+\frac{375}{2}\mathcal{I}_z^2+5\mathcal{I}^4\mathcal{I}_z-55\mathcal{I}^2\mathcal{I}_z+117\mathcal{I}_z+\frac{5}{2}\mathcal{I}^4$ $\quad-20\mathcal{I}^2+30)$
		0	$\frac{1}{180}(231\mathcal{I}_z^6-315\mathcal{I}^2\mathcal{I}_z^4+735\mathcal{I}_z^4+105\mathcal{I}^4\mathcal{I}_z^2-525\mathcal{I}^2\mathcal{I}_z^2$ $\quad+294\mathcal{I}_z^2-5\mathcal{I}^6+40\mathcal{I}^4-60\mathcal{I}^2)$
7	$\frac{21i}{64}\sqrt{\frac{15}{\pi}}\cdot\frac{1}{r^7}$	7	$\frac{\sqrt{429}}{21}\mathcal{I}_+^7$
		6	$-\sqrt{\frac{286}{21}}\mathcal{I}_+^6(\mathcal{I}_z+3)$
		5	$\frac{1}{2}\sqrt{\frac{21}{22}}\mathcal{I}_+^5(13\mathcal{I}_z^2+65\mathcal{I}_z-\mathcal{I}^2+90)$
		4	$-\sqrt{\frac{21}{22}}\mathcal{I}_+^4(13\mathcal{I}_z^3+78\mathcal{I}_z^2-3\mathcal{I}^2\mathcal{I}_z+179\mathcal{I}_z-6\mathcal{I}^2+150)$
		3	$\frac{1}{\sqrt{6}}\mathcal{I}_+^3(143\mathcal{I}_z^4+\frac{6086}{7}\mathcal{I}_z^3-66\mathcal{I}^2\mathcal{I}_z^2+2365\mathcal{I}_z^2-198\mathcal{I}^2\mathcal{I}_z$ $\quad+3234\mathcal{I}_z+3\mathcal{I}^4-186\mathcal{I}^2+1800)$
		2	$-\sqrt{\frac{3}{7}}\mathcal{I}_+^2(143\mathcal{I}_z^5+715\mathcal{I}_z^4-110\mathcal{I}^2\mathcal{I}_z^3+2035\mathcal{I}_z^3-150\mathcal{I}^2\mathcal{I}_z^2$ $\quad+3241\mathcal{I}_z^2+15\mathcal{I}^4\mathcal{I}_z-150\mathcal{I}^2\mathcal{I}_z+2858\mathcal{I}_z+15\mathcal{I}^4$ $\quad-50\mathcal{I}^2+1080)$
		1	$\frac{1}{3\sqrt{14}}\mathcal{I}_+(429\mathcal{I}_z^6+\frac{8870}{7}\mathcal{I}_z^5-495\mathcal{I}^2\mathcal{I}_z^4+\frac{26535}{7}\mathcal{I}_z^4$ $\quad+\frac{38374}{7}\mathcal{I}_z^3+135\mathcal{I}^4\mathcal{I}_z^2+\frac{38312}{7}\mathcal{I}_z^2+135\mathcal{I}^4\mathcal{I}_z$ $\quad+\frac{20980}{7}\mathcal{I}_z-5\mathcal{I}^6+\frac{668}{7}\mathcal{I}^4+\frac{5040}{7})$
		0	$-\frac{2}{21}(429\mathcal{I}_z^7-693\mathcal{I}^2\mathcal{I}_z^5+2310\mathcal{I}_z^5+315\mathcal{I}^4\mathcal{I}_z^3-665\mathcal{I}^2\mathcal{I}_z^3$ $\quad+1785\mathcal{I}_z^3-35\mathcal{I}^6\mathcal{I}_z-385\mathcal{I}^4\mathcal{I}_z-882\mathcal{I}^2\mathcal{I}_z-180\mathcal{I}_z)$

Appendix B
Matrix of Unitary Operator U Defined by Means of Basis Function Operators of the Irreducible Representations of the Symmetry Point Groups

$\mathcal{I}$ is invariant with respect to proper and improper rotations, $\mathcal{I}_{\pm} = \mathcal{I}_x \pm i\mathcal{I}_y$, $u = 3\mathcal{I}_z^2 - \mathcal{I}(\mathcal{I}+1)$, $v = \sqrt{3}(\mathcal{I}_x^2 - \mathcal{I}_y^2)$, $[\mathcal{I}_\alpha \mathcal{I}_\beta] = \frac{1}{2}(\mathcal{I}_\alpha \mathcal{I}_\beta + \mathcal{I}_\beta \mathcal{I}_\alpha)$, where $\alpha, \beta = y, z$; z, x and x, y; $[\mathcal{I}_x \mathcal{I}_y \mathcal{I}_z] = \frac{1}{6}(\mathcal{I}_x \mathcal{I}_y \mathcal{I}_z + \mathcal{I}_y \mathcal{I}_z \mathcal{I}_x + \mathcal{I}_z \mathcal{I}_x \mathcal{I}_y + \mathcal{I}_x \mathcal{I}_z \mathcal{I}_y + \mathcal{I}_z \mathcal{I}_y \mathcal{I}_x + \mathcal{I}_y \mathcal{I}_x \mathcal{I}_z)$, $V_x = [\mathcal{I}_x(\mathcal{I}_y^2 - \mathcal{I}_z^2)]$, $V_y = [\mathcal{I}_y(\mathcal{I}_z^2 - \mathcal{I}_x^2)]$, $V_z = [\mathcal{I}_z(\mathcal{I}_x^2 - \mathcal{I}_y^2)]$.

Symmetry group	Irreducible representations		Bases [44, 127]		U
	Bethe symbols	Mulliken symbols			
1	**2**	**3**	**4**		**5**
$\bar{1}$	Γ_1^+	A_g	$\mathcal{I}_x; \mathcal{I}_y; \mathcal{I}_z$		$-1; +1; -1$
2 and m	Γ_1	A	A'	$\mathcal{I}_z$	-1
	Γ_2	B	A''	$\mathcal{I}_x; \mathcal{I}_y$	$-1; +1;$
222 and $2mm$	Γ_1	A	A	$\mathcal{I}, (222)$	$+1$
	Γ_2	B_2	B_1	$\mathcal{I}_y$	$+1$
	Γ_3	B_1	A_2	$\mathcal{I}_z$	-1
	Γ_4	B_3	B_2	$\mathcal{I}_x$	-1
$2/m$	Γ_1^+	A_g		$\mathcal{I}_z$	-1
	Γ_2^+	B_g		$\mathcal{I}_x; \mathcal{I}_y$	$-1; +1$
4 and $\bar{4}$	Γ_1	A	A	$\mathcal{I}_z$	-1
	Γ_2	B	B	$[\mathcal{I}_x \mathcal{I}_y], (4)$	-1
				$\mathcal{I}_x^2 - \mathcal{I}_y^2, (4)$	$+1$
	Γ_3	$\}E$	$E\{$	$\{-\mathcal{I}_+, \mathcal{I}_-\}$	$\begin{pmatrix} 0 & 1 \\ 1 & 0 \end{pmatrix}$
	Γ_4				
422, $4\,mm$, $\bar{4}\,2\,m$	Γ_1	A	A	$\mathcal{I}$	$+1$
	Γ_2	A_2	A_2	$\mathcal{I}_z$	-1
	Γ_3	B_1	B_1	$\mathcal{I}_x^2 - \mathcal{I}_y^2, (422)$	$+1$
	Γ_4	B_2	B_2	$[\mathcal{I}_x \mathcal{I}_y], (422)$	-1
	Γ_5	E	E	$\{\mathcal{I}_x, \mathcal{I}_y\}$	$\begin{pmatrix} -1 & 0 \\ 0 & 1 \end{pmatrix}$

I. Geru, D. Suter, *Resonance Effects of Excitons and Electrons*,
Lecture Notes in Physics 869, DOI 10.1007/978-3-642-35807-4,
© Springer-Verlag Berlin Heidelberg 2013

Symmetry group	Irreducible representations			Bases	U
	Bethe symbols	Mulliken symbols		[44, 127]	
3	Γ_1	A		$\mathcal{I}; \mathcal{I}_z$	$1; -1$
	Γ_2	$\}\ E$		$\{-\mathcal{I}_+, \mathcal{I}_-\}$	$\begin{pmatrix} 0 & 1 \\ 1 & 0 \end{pmatrix}$
	Γ_3				
32 and $3\,m$	Γ_1	A_1	A_1	$\mathcal{I}$	$+1$
	Γ_2	A_2	A_2	$\mathcal{I}_z$	-1
	Γ_3	E	E	$\{\mathcal{I}_-, -\mathcal{I}_+\}$	$\begin{pmatrix} 0 & 1 \\ 1 & 0 \end{pmatrix}$
6 and $\bar{6}$	Γ_1	A	A'	$\mathcal{I}_z, (6); \mathcal{I}, (6, \bar{6})$	$-1; +1$
	Γ_2	B	B'	$z, (\bar{6})$	$+1$
	Γ_3	$\}E_1$	$E_1'\{$	$\{-zS_+, zS_-\}, (\bar{6})$	$\begin{pmatrix} 0 & 1 \\ 1 & 0 \end{pmatrix}$
	Γ_4				
	Γ_5	$\}E_2$	$E_2'\{$	$\{-\mathcal{I}_+, \mathcal{I}_-\}$	$\begin{pmatrix} 0 & 1 \\ 1 & 0 \end{pmatrix}$
	Γ_6				
$622, 6mm, \bar{6}m2$	Γ_1	A_1	A_1	A_1'	$\mathcal{I}$	1
	Γ_2	A_2	A_2	A_2'	$\mathcal{I}_z$	-1
	Γ_3	B_1	B_2	A_1''	$i(\mathcal{I}_+^3 - \mathcal{I}_-^3), (622)$	$+1$
	Γ_4	B_2	B_1	A_2''	$(\mathcal{I}_+^3 + \mathcal{I}_-^3), (622)$	-1
	Γ_5	E_1	E_1	E''	$\{\mathcal{I}_-, -\mathcal{I}_+\}$	$\begin{pmatrix} 0 & 1 \\ 1 & 0 \end{pmatrix}$
	Γ_6	E_2	E_2	E'	$\{\mathcal{I}_+^2, \mathcal{I}_-^2\}, (622)$	$\begin{pmatrix} 0 & 1 \\ 1 & 0 \end{pmatrix}$
					$\{i\mathcal{I}_-(\mathcal{I}_+^3 - \mathcal{I}_-^3),$ $-i\mathcal{I}_+(\mathcal{I}_+^3 - \mathcal{I}_-^3)\},$ (622)	$\begin{pmatrix} 0 & 1 \\ 1 & 0 \end{pmatrix}$
23	Γ_1	A		$\mathcal{I}; [\mathcal{I}_x\mathcal{I}_y\mathcal{I}_z]$	$+1; +1$
	Γ_2	$\}E$		$\frac{1}{\sqrt{2}}(u - v)$	$\begin{pmatrix} 1 & 0 \\ 0 & 1 \end{pmatrix}$
	Γ_3			$\frac{1}{\sqrt{2}}(u + v)$	
	Γ_4	T		$\{\mathcal{I}_x, \mathcal{I}_y, \mathcal{I}_z\}$	W
432 and $\bar{4}3m$	Γ_1	A_1	A_1	$\mathcal{I}$	1
	Γ_2	A_2	A_2	$[\mathcal{I}_x\mathcal{I}_y\mathcal{I}_z], (432)$	1
	Γ_3	E	E	$\{u, v\}, (432)$	$\begin{pmatrix} 1 & 0 \\ 0 & 1 \end{pmatrix}$
	Γ_4	T_1	T_1	$\{\mathcal{I}_x, \mathcal{I}_y, \mathcal{I}_z\}, (432)$	W
				$\{\mathcal{I}_x^3, \mathcal{I}_y^3, \mathcal{I}_z^3\}, (432)$	W
				$\{\mathcal{I}_x^5, \mathcal{I}_y^5, \mathcal{I}_z^5\}, (432)$	W
				$\{[\mathcal{I}_y\mathcal{I}_z], [\mathcal{I}_z\mathcal{I}_x], [\mathcal{I}_x\mathcal{I}_y]\},$ $(\bar{4}3m)$	W
				$\{V_z, V_y, V_z\}, (\bar{4}3m)$	W
	Γ_5	T_2	T_2	$\{\mathcal{I}_x, \mathcal{I}_y, \mathcal{I}_z\}, (\bar{4}3m)$	W
				$\{\mathcal{I}_x^3, \mathcal{I}_y^3, \mathcal{I}_z^3\}, (\bar{4}3m)$	W
				$\{\mathcal{I}_x^5, \mathcal{I}_y^5, \mathcal{I}_z^5\}, (\bar{4}3m)$	W
				$\{[\mathcal{I}_y\mathcal{I}_z], [\mathcal{I}_z\mathcal{I}_x], [\mathcal{I}_x\mathcal{I}_y]\},$ (432)	W
				$\{V_z, V_y, V_z\}, (432)$	W

$$W = \begin{pmatrix} -1 & 0 & 0 \\ 0 & 1 & 0 \\ 0 & 0 & -1 \end{pmatrix}.$$

$$\begin{pmatrix} 0 & 0 & 1 \\ 0 & 1 & 0 \\ 1 & 0 & 0 \end{pmatrix}$$

Appendix C
Color Symmetry and Time Reversal in Systems with Half-Integer Total Spin [392]

The magnetic symmetry groups built on the basis of expansion of the classical crystallographic group by means of the group $\{\mathcal{K}, \mathcal{K}^2 = e\}$ ($\mathcal{K}$ is the time-reversal operator) [129, 399] refer only to systems with integer total spin. We shall obtain point groups of magnetic symmetry conditioned by a time-reversal transformation, for systems with the Kramers degeneracy.

In the functional $(2S + 1)$-fold space the state of the system with resulting spin S corresponds to the point which under the action of the time-reversal transformation performs an antirotation around the axis passing through the origin of the coordinate system. There is an essential geometric difference between a system of particles with half-integer total spin from that of particles with integer spin. In the first case the antirotation is performed on a $90°$ angle, and the system comes to the initial state with the primary phase at the wave function after four consecutive antirotations in one direction at a similar angle. In the second case the angle is equal to $180°$, and the system comes to the initial state after the second antirotation, which is opposite with respect to the first one. For $S = 1/2, 1,$ and $3/2$, this is shown in Fig. C.1. The operation of anti-identification corresponds to transitions to the complex conjugate wave function. The different geometric behavior of the systems with integer and half-integer S during the time inversion is connected with the Wigner operator $\mathcal{K}$ [392].

On the basis of the system of coordinate axes we shall build a $(2S + 1)$-dimensional cube, inscribed in a $(2S + 1)$-dimensional unit sphere with the center at the origin of the coordinates, where at first the coordinate axes pass through the middle $2S$-dimensional faces. Then we shall turn this cube so that the position selected on the sphere vertex of the $(2S + 1)$-dimensional cube corresponds to the values of $2S + 1$ functions $\psi_\sigma(\mathbf{r}_1, \mathbf{r}_2, \ldots, \mathbf{r}_N)$ that play the role of coefficients in the decomposition of N-particle wave function $\Psi^{(S)}$ on the basic spinors $\xi_\sigma^{(S)}$

$$\Psi^{(S)} = \sum_{\sigma=-S}^{S} \psi_\sigma(\mathbf{r}_1, \mathbf{r}_2, \ldots, \mathbf{r}_N)\xi_\sigma^{(S)} \tag{C.1}$$

I. Geru, D. Suter, *Resonance Effects of Excitons and Electrons*,
Lecture Notes in Physics 869, DOI 10.1007/978-3-642-35807-4,
© Springer-Verlag Berlin Heidelberg 2013

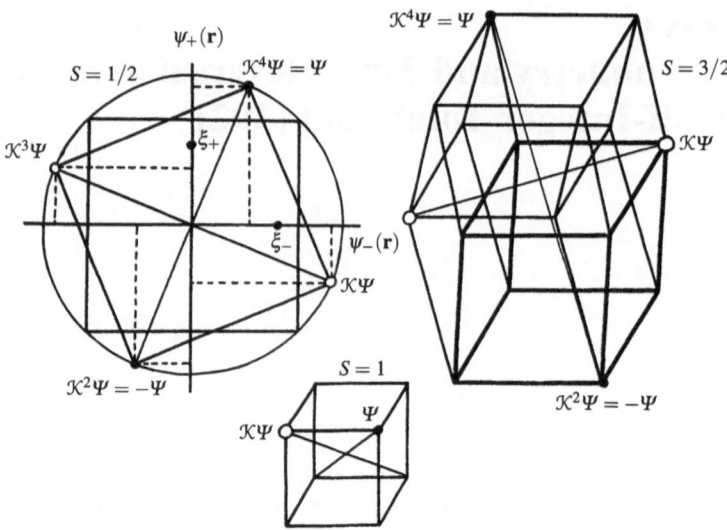

Fig. C.1 The geometric transformation of the wave function Ψ of systems with half-integer spin ($S = 1/2, 3/2$) and integer spin ($S = 1$) at time reversal in the absence of a magnetic field. The wave function is represented as an expansion on spinor basis vectors. The state of the system corresponds to a point at the vertex of a $(2S + 1)$-dimensional cube. •—the state Ψ; ○—the time-reversed state $\mathcal{K}\Psi$

that satisfies the normalization condition

$$\sum_{\sigma=-S}^{S} \int |\psi_\sigma (\mathbf{r}_1, \mathbf{r}_2, \ldots, \mathbf{r}_N)|^2 \, d\mathbf{r}_1 \, d\mathbf{r}_2 \cdots d\mathbf{r}_N = 1. \tag{C.2}$$

During the time-reversal operation there is an antirotation of the "state vector," directed on one of the main cube diagonals with a transition to the other main diagonal. From the Kramers theorem on the additional degeneracy of energy levels conditioned by time-reversal symmetry [398], we have the following geometrical theorem.

Theorem *In an n-dimensional Euclidean cube, mutually orthogonal principal diagonals do not exist, if n is an odd number, and they exist with a precision up to the group of movement of the n-dimensional cube of the single pair of such diagonals at even n.*

For systems with half-integer S an extension of classical groups to groups of magnetic symmetry is necessary by means of a cyclic group of the fourth order $4' \equiv \{\mathcal{K}, \mathcal{K}^2, \mathcal{K}^3, \mathcal{K}^4 = e\}$, since in this case the smaller integer p for which $\mathcal{K}^p = e$ is $p = 4$. At half-integer S the operators $\mathcal{K}^p$ satisfy the relations $\mathcal{K}^2 = -\mathcal{K}^4 = -e$ and $\mathcal{K}^3 = -\mathcal{K}$.

As a result of the higher symmetry of group $4'$ in comparison with $\{\mathcal{K}, e\}$, the number of groups of magnetic symmetry at half-integer S must be less than that for systems with integer S. In [597] groups of generalized symmetry were obtained on the basis of the extension of 32 crystal classes using the same crystal classes. Using the results of this work, we find that during extension using group $4'$ instead of 58 younger point groups of magnetic symmetry [129, 399] corresponding to the systems with integer S, there exist four younger groups of magnetic symmetry at half-integer S. These are Belov's point groups of four-color symmetry, $4^{(4'_z)}$, $\bar{4}^{(4'_z)}$, $4^{(4'_z)}/m^{(1)}$ and $4^{(4'_z)}/m^{(2)}$, where the bar denotes mirror rotation and the prime indicates antirotation. The generators for the first two groups are, respectively, the groups 4 and $\bar{4}$, while the other two groups are generated with the help of group $4/m$.

If along with antirotations forming the cyclic group $4'$ we consider the other elements of generalized symmetry of the square with two identically colored opposite vertices situated in the plane perpendicular to the antirotation axis (the reflections $m_{xy}^{(i)}$ in the diagonals and antireflection $m_x'^{(i)}$ in lines passing through the centers of the opposite sides $i = 1, 2$), then the extension of the classical symmetry point group must be fulfilled by means of the group $4'm'm$. In group $4'm'm$ the time-reversal transformation may be represented in the form of the product of a reflection of type m_{xy} on the respective antireflection of type m_x':

$$m_x'^{(i)}m_{xy}^{(i)} = -m_{xy}^{(i)}m_x'^{(i)} = m_{xy}^{(i)}m_x'^{(j)} = -m_x'^{(j)}m_{xy}^{(i)} = \mathcal{K},$$

$$m_x'^{(i)}m_{xy}^{(i)} \pm m_{xy}^{(i)}m_x'^{(i)} = m_{xy}^{(i)}m_x'^{(j)} \pm m_x'^{(j)}m_{xy}^{(i)} = \eta_{\pm}, \tag{C.3}$$

where $\eta_+ = 0$, $\eta_- = 2\mathcal{K}$, $i, j = 1, 2$ $(i \neq j)$. The straight lines $m_{xy}^{(1)}$ and $m_{xy}^{(2)}$ pass through the black and white verticles of the square, correspondingly. From all 32 point groups of symmetry only the groups 422, $4mm$, and $\bar{4}2m$ allow extension by means of group $4'm'm$. In this more general case there also exist four younger point groups of the generalized symmetry-groups of four-color symmetry $4^{(4'_z)}2^{(m'_x)}2^{(m_{xy})}$, $4^{(4'_z)}m^{(m'_x)}m^{(m_{xy})}$, $\bar{4}^{(4'_z)}2^{(m'_x)}m^{(m_{xy})}$, and $\bar{4}^{(4'_z)}2^{(m_{xy})}m^{(m'_x)}$. The generators for the first two groups are 422 and $4mm$, while for the other two groups the generator is the group $\bar{4}2m$.

Thus, unlike systems with integer spin, for systems with resulting half-integer spin the magnetic symmetry groups conditioned by the time-reversal transformation do not belong to the set of antisymmetry groups, but are groups of four-color symmetry. The sharp decrease in the number of groups of four-color symmetry with half-integer S compared with the number of antisymmetry groups with integer S is connected with the additional degeneracy of the energy levels of the systems with half-integer total spin due to time-reversal symmetry.

The point groups of magnetic symmetry can be applied to describe the types of spin ordering in the magnetic clusters. Among the symmetry elements of the above-mentioned color symmetry point groups there are no rotations around the threefold, fivefold, or sixfold (proper and mirror) axes. Therefore, in particular, trinuclear and pentanuclear clusters cannot be symmetric. In connection with this, we shall note the necessity of further investigating the pentanuclear iron cluster [598], for which the supposition about its trigonal bipyramid geometrical structure is still insufficiently well-grounded. Moreover, for this cluster with degenerated energy levels due to time-reversal symmetry the threefold axes of symmetry are forbidden.

Another high-spin system with an odd number of electrons is the molecule Mn_5 ($S = 25/2$) in the krypton matrix. The penta-atomic molecule Mn_5 was investigated by the EPR method [599], and 25 transitions with the rules of selection $\Delta M_S = \pm 1$ were detected. However, the supposition of these authors that such a molecule must have the form of a plane pentagon requires additional grounding. None of the groups listed above describe the symmetry of the magnetic spin ordering of the atoms at the vertices of pentagonal molecules. Therefore, the assumption from [599] is true only if it is not in contradiction with the structure of four-color symmetry groups.

Finally, the problem of magnetic ordering in systems with high total spin has a direct relation to the consideration in Sects. 6.5 and 6.6 of semiconductors doped by magnetic impurities. In [600] it was shown that the difference of the binding energies of the exciton bonded on the neutral acceptor and free exciton in the crystal $Cd_{1-x}Mn_x Te$ sometimes exceeds that for CdTe. This excess is explained by the supposition about the magnetic ordering of the Mn^{2+} ions in the domain of the localization of the bond excitons, the number of which reaches 15 ($S = 75/2$).

Thus, the four-color symmetry point groups discussed above describe the magnetic symmetry of localized spin systems with a resulting half-integer angular momentum.

Appendix D
Operators ξ and η

$I_p\,(\mathbf{R}_k)$ is the spin projection operator of the nuclei ^{19}F situated in the $\mathbf{R}_k$ node of the crystal lattice, $E(0|\mathbf{R}_k)$ is identity matrix 2×2; $p = x, y, z$.

$$E(0|\mathbf{R}_k; \mathbf{R}_{k'}) = E(0, \mathbf{R}_k) + E(0, \mathbf{R}_{k'}),$$

$$\xi_1 = \frac{1}{\sqrt{8}} \sum_k E(0, \mathbf{R}_k),$$

$$\xi_{4x}^{(1)} = \frac{1}{\sqrt{8}} \sum_k I(x|\mathbf{R}_k), \qquad \xi_{4y}^{(1)} = \frac{1}{\sqrt{8}} \sum_k I(y|\mathbf{R}_k),$$

$$\xi_{4z}^{(1)} = \frac{1}{\sqrt{8}} \sum_k I(z|\mathbf{R}_k),$$

$$\xi_{4x}^{(2)} = \frac{1}{4}\big[I(y+z|111; \bar{1}\bar{1}\bar{1}) + I(z-y|\bar{1}1\bar{1}; 1\bar{1}1)$$
$$+ I(y-z|11\bar{1}; \bar{1}\bar{1}1) - I(y+z|1\bar{1}\bar{1}; \bar{1}11)\big],$$

$$\xi_{4y}^{(2)} = \frac{1}{4}\big[I(x+z|11\bar{1}; \bar{1}\bar{1}1) + I(z-x|\bar{1}1\bar{1}; 1\bar{1}1)$$
$$+ I(x-z|11\bar{1}; \bar{1}\bar{1}1) - I(x+z|1\bar{1}\bar{1}; \bar{1}11)\big],$$

$$\xi_{4z}^{(2)} = \frac{1}{4}\big[I(x+y|111; \bar{1}\bar{1}\bar{1}) + I(x-y|\bar{1}1\bar{1}; 1\bar{1}1)$$
$$- I(x+y|11\bar{1}; \bar{1}\bar{1}1) + I(y-x|1\bar{1}\bar{1}; \bar{1}11)\big],$$

$$\xi_{5x}^{(1)} = \frac{1}{\sqrt{8}}\big[E(0|111; \bar{1}\bar{1}\bar{1}) + E(0|11\bar{1}; \bar{1}\bar{1}1)$$
$$- E(0|\bar{1}1\bar{1}; 1\bar{1}1) - E(0|1\bar{1}\bar{1}; \bar{1}11)\big],$$

I. Geru, D. Suter, *Resonance Effects of Excitons and Electrons*,
Lecture Notes in Physics 869, DOI 10.1007/978-3-642-35807-4,
© Springer-Verlag Berlin Heidelberg 2013

$$\xi_{5y}^{(1)} = \frac{1}{\sqrt{8}}\left[E(0|111;\bar{1}\bar{1}\bar{1}) + E(0|\bar{1}11;1\bar{1}1)\right.$$
$$\left. - E(0|11\bar{1};\bar{1}\bar{1}1) - E(0|1\bar{1}\bar{1};\bar{1}11)\right],$$

$$\xi_{5z}^{(1)} = \frac{1}{\sqrt{8}}\left[E(0|111;\bar{1}\bar{1}\bar{1}) + E(0|11\bar{1};\bar{1}\bar{1}1)\right.$$
$$\left. - E(0|\bar{1}1\bar{1};1\bar{1}1) - E(0|1\bar{1}\bar{1};\bar{1}11)\right],$$

$$I\big(p \pm p'|\mathbf{R}_k;\mathbf{R}_{k'}\big) = I_p(\mathbf{R}_k) + I_p(\mathbf{R}_{k'}) \pm I_{p'}(\mathbf{R}_k) \pm I_{p'}(\mathbf{R}_{k'}),$$

$$\eta_1 = \frac{1}{8}\sqrt{\frac{1}{30}}\big[2\big(35\mathcal{I}_z^4 + 25\mathcal{I}_z^2 - 30\mathcal{I}_z^2\mathcal{I}^2 - 6\mathcal{I}^2\big) + 5\big(\mathcal{I}_+^4 + \mathcal{I}_-^4\big)\big],$$

$$\eta_{4x}^{(1)} = \mathcal{I}_x, \qquad \eta_{4y}^{(1)} = \mathcal{I}_y, \qquad \eta_{4z}^{(1)} = \mathcal{I}_z,$$

$$\eta_{4x}^{(2)} = \frac{1}{4\sqrt{10}}\left[3(\mathcal{I}_+ + \mathcal{I}_-)\left(1 - \frac{1}{2}\mathcal{I}^2 + \frac{5}{2}\mathcal{I}_z^2\right) + \frac{15}{2}(\mathcal{I}_+ - \mathcal{I}_-)\mathcal{I}_z - \frac{5}{2}\big(\mathcal{I}_+^3 + \mathcal{I}_-^3\big)\right],$$

$$\eta_{4y}^{(2)} = \frac{1}{4i\sqrt{10}}\left[3(\mathcal{I}_+ - \mathcal{I}_-)\left(1 - \frac{1}{2}\mathcal{I}^2 + \frac{5}{2}\mathcal{I}_z^2\right) + \frac{15}{2}(\mathcal{I}_+ + \mathcal{I}_-)\mathcal{I}_z + \frac{5}{2}\big(\mathcal{I}_+^3 + \mathcal{I}_-^3\big)\right],$$

$$\eta_{4z}^{(2)} = \frac{1}{4\sqrt{10}}\big(5\mathcal{I}_z^3 - 3\mathcal{I}_z\mathcal{I}^2 + \mathcal{I}_z\big),$$

$$\eta_{5x}^{(1)} = \frac{i}{\sqrt{2}}\left[(\mathcal{I}_- - \mathcal{I}_+)\mathcal{I}_z - \frac{1}{2}(\mathcal{I}_+ + \mathcal{I}_-)\right],$$

$$\eta_{5y}^{(1)} = \frac{1}{\sqrt{2}}\left[(\mathcal{I}_+ + \mathcal{I}_-)\mathcal{I}_z + \frac{1}{2}(\mathcal{I}_+ - \mathcal{I}_-)\right], \qquad \eta_{5z}^{(1)} = \frac{1}{2\sqrt{2}}\big(\mathcal{I}_+^2 - \mathcal{I}_-^2\big),$$

$$\eta_{5x}^{(2)} = \frac{1}{4\sqrt{6}}\left[(\mathcal{I}_+ + \mathcal{I}_-)\left(\frac{3}{2}\mathcal{I}^2 - \frac{15}{2}\mathcal{I}_z^2 - 3\right) - \frac{15}{2}(\mathcal{I}_+ - \mathcal{I}_-)\mathcal{I}_z - \frac{3}{2}\big(\mathcal{I}_+^3 + \mathcal{I}_-^3\big)\right],$$

$$\eta_{5y}^{(2)} = \frac{1}{4i\sqrt{6}}\left[3(\mathcal{I}_+ - \mathcal{I}_-)\left(1 + \frac{5}{2}\mathcal{I}_z^2 - \frac{1}{2}\mathcal{I}^2\right) + \frac{15}{2}(\mathcal{I}_+ + \mathcal{I}_-)\mathcal{I}_z - \frac{3}{2}\big(\mathcal{I}_+^3 + \mathcal{I}_-^3\big)\right],$$

$$\eta_{5z}^{(2)} = \frac{1}{2}\sqrt{\frac{3}{2}}\big[\big(\mathcal{I}_+^2 - \mathcal{I}_-^2\big) + \big(\mathcal{I}_+^2 + \mathcal{I}_-^2\big)\mathcal{I}_z\big],$$

$$\eta_{5x}^{(3)} = \frac{1}{8i}\sqrt{\frac{7}{2}}\Bigg[-3\big(\mathcal{I}_+^3+\mathcal{I}_-^3\big)-2\big(\mathcal{I}_+^3-\mathcal{I}_-^3\big)\mathcal{I}_z-\frac{1}{7}(\mathcal{I}_++\mathcal{I}_-)$$

$$\times\big(-6+3\boldsymbol{\mathcal{I}}^2-21\mathcal{I}_z^2\big)+\frac{1}{7}(\mathcal{I}_+-\mathcal{I}_-)\big(-6\boldsymbol{\mathcal{I}}^2\mathcal{I}_z+19\mathcal{I}_z-21\mathcal{I}_z^2+14\mathcal{I}_z^3\big)\Bigg],$$

$$\eta_{5y}^{(3)} = \frac{1}{8}\sqrt{\frac{7}{2}}\Bigg[3\big(\mathcal{I}_+^3-\mathcal{I}_-^3\big)+2\big(\mathcal{I}_+^3+\mathcal{I}_-^3\big)\mathcal{I}_z+\frac{1}{7}(\mathcal{I}_--\mathcal{I}_+)$$

$$\times\big(-6+3\boldsymbol{\mathcal{I}}^2-21\mathcal{I}_z^2\big)+\frac{1}{7}(\mathcal{I}_--\mathcal{I}_+)\big(-6\boldsymbol{\mathcal{I}}^2\mathcal{I}_z+19\mathcal{I}_z-21\mathcal{I}_z^2+14\mathcal{I}_z^3\big)\Bigg],$$

$$\eta_{5z}^{(3)} = -\frac{\sqrt{7}}{4i}\Big[\mathcal{I}_+^2\big(9-\boldsymbol{\mathcal{I}}^2+7\mathcal{I}_z^2+14\mathcal{I}_z\big)-\mathcal{I}_-^2\big(9-\boldsymbol{\mathcal{I}}^2+7\mathcal{I}_z^2+14\mathcal{I}_z\big)\Big].$$

Appendix E
The Functions f_{kj}

$$f_{10} = \frac{1}{\sqrt{3}}\left[2a_{1,\alpha_1} + \frac{1}{\sqrt{2}}a_{2,\alpha_2}(3\cos^2\theta - 1) + a_{8,\alpha_8}(1 - \sin^2\theta\cos^2\varphi)\right],$$

$$f_{11} = \sqrt{\frac{3}{2}}a_{2,\alpha_2}(1 - 3\cos^2\theta) + \sqrt{\frac{2}{3}}a_{9,\alpha_9}$$
$$\times\left[\sin^2\theta(3\cos^2\varphi - 1) - 1\right] + \sqrt{2}a_{10,\alpha_9}(1 - \sin^2\theta\sin^2\varphi),$$

$$f_{12} = -\sqrt{\frac{3}{2}}a_{2,\alpha}\sin^2\theta\cos 2\varphi - \sqrt{2}(a_{9,\alpha_9} + a_{10,\alpha_9})\sin^2\theta,$$

$$f_{13} = -\frac{1}{\sqrt{2}}(a_0 + 3a_{4,\alpha_4})\sin 2\theta\sin\varphi,$$

$$f_{14} = \frac{1}{\sqrt{2}}\left(\frac{1}{2}a_0\sin 2\theta\cos\varphi + i\tilde{a}\sin\theta\sin\varphi\right),$$

$$f_{15} = \frac{1}{\sqrt{2}}\left(\frac{1}{2}a_0\sin^2\theta\sin 2\varphi - i\tilde{a}\cos\theta\right),$$

$$f_{16} = \sqrt{\frac{2}{3}}a_{9,\alpha_9}(\sin^2\theta\sin^2\varphi - 1),$$

$$f_{17} = 2\left[\sqrt{\frac{2}{3}}a_{9,\alpha_9}(1 - \sin^2\theta\cos^2\varphi) + \frac{1}{\sqrt{2}}a_{10,\alpha_9}(\sin^2\theta\sin^2\varphi - 1)\right],$$

$$f_{18} = b'(1 - \sin^2\theta\cos^2\varphi) + b''(1 + \sin^2\theta\cos^2\varphi),$$

$$f_{19} = b_0\sin^2\theta\sin 2\varphi - i\tilde{b}\cos\theta,$$

I. Geru, D. Suter, *Resonance Effects of Excitons and Electrons*,
Lecture Notes in Physics 869, DOI 10.1007/978-3-642-35807-4,
© Springer-Verlag Berlin Heidelberg 2013

$$f_{1,10} = b_0 \sin^2 \theta \cos \varphi + i\tilde{b} \sin \theta \sin \varphi,$$

$$f_{20} = \frac{1}{\sqrt{3}} \left[2a_{1,\alpha_1} + a_{8,\alpha_8} \left(1 - 3\sin^2 \theta \sin^2 \varphi\right) + \frac{1}{\sqrt{2}} a_{2,\alpha_2} \left(3\cos^2 \theta - 1\right) \right],$$

$$f_{21} = \sqrt{\frac{3}{2}} a_{2,\alpha_2} \left(1 - 3\cos^2 \theta\right) + \sqrt{\frac{2}{3}} a_{9,\alpha_9} \left[\sin^2 \theta \left(3\sin^2 \varphi - 1\right) - 1\right]$$
$$+ \sqrt{2} a_{10,\alpha_9} \left(1 - \sin^2 \theta \cos^2 \varphi\right),$$

$$f_{22} = -\sqrt{\frac{3}{2}} a_{2,\alpha_2} \sin^2 \theta \cos 2\varphi + \sqrt{2}(a_{10,\alpha_9} - a_{9,\alpha_9}) \sin^2 \theta,$$

$$f_{23} = \frac{1}{\sqrt{2}} \left(\frac{1}{2} a_0 \sin 2\theta \sin \varphi - i\tilde{a} \sin \theta \cos \varphi \right),$$

$$f_{24} = -\frac{1}{\sqrt{2}} (a_0 + 3a_{4,\alpha_4}) \sin^2 \theta \sin 2\varphi,$$

$$f_{25} = \frac{1}{\sqrt{2}} \left(\frac{1}{2} a_0 \sin^2 \theta \sin 2\varphi + i\tilde{a} \cos \theta \right),$$

$$f_{26} = 2 \left[\sqrt{\frac{2}{3}} a_{9,\alpha_9} \left(1 - \sin^2 \theta \sin^2 \varphi\right) + \frac{1}{\sqrt{2}} a_{10,\alpha_9} \left(\sin^2 \theta \cos^2 \varphi - 1\right) \right],$$

$$f_{27} = \sqrt{\frac{2}{3}} a_{9,\alpha_9} \left(\sin^2 \theta \cos^2 \varphi - 1\right),$$

$$f_{28} = b_0 \sin^2 \theta \cos 2\varphi + i\tilde{b} \cos \theta,$$

$$f_{29} = b' \left(1 - \sin^2 \theta \sin^2 \varphi\right) + b'' \left(1 + \sin^2 \theta \sin \varphi\right),$$

$$f_{2,10} = b_0 \sin 2\theta \sin \varphi - i\tilde{b} \sin \theta \cos \varphi,$$

$$f_{30} = \frac{1}{\sqrt{3}} \left[2a_{1,\alpha_1} + \left(a_{8,\alpha_8} - \frac{1}{\sqrt{2}} a_{2,\alpha_2} \right) \left(1 - 3\cos^2 \theta\right) \right],$$

$$f_{31} = \sqrt{\frac{3}{2}} a_{2,\alpha_2} \left(1 - 3\cos^2 \theta\right) + \sqrt{2} \left(\frac{1}{\sqrt{3}} a_{9,\alpha_9} - a_{10,\alpha_9} \right) \left(1 + \cos^2 \theta\right),$$

$$f_{32} = -\sqrt{\frac{3}{2}} a_{2,\alpha_2} \sin^2 \theta \cos 2\varphi + 2\sqrt{2} a_{9,\alpha_9} \sin^2 \theta,$$

$$f_{33} = \frac{1}{\sqrt{2}} \left(\frac{1}{2} a_0 \sin 2\theta \sin \varphi + i\tilde{a} \sin \theta \cos \varphi \right),$$

$$f_{34} = \frac{1}{\sqrt{2}} \left(\frac{1}{2} a_0 \sin 2\theta \cos \varphi - i\tilde{a} \sin \theta \sin \varphi \right),$$

$$f_{35} = -\frac{1}{\sqrt{2}} (a_0 + 3a_{4,\alpha_4}) \sin^2 \theta \sin 2\varphi,$$

$$f_{36} = \sqrt{2} \left[\frac{1}{\sqrt{3}} a_{9,\alpha_9} \left(\sin^2 \theta \sin^2 \varphi - 1 \right) + a_{10,\alpha_9} \left(1 - \sin^2 \theta \cos^2 \varphi \right) \right],$$

$$f_{37} = \sqrt{2} \left[\frac{1}{\sqrt{3}} a_{9,\alpha_9} \left(\sin^2 \theta \cos^2 \varphi - 1 \right) + a_{10,\alpha_9} \left(1 - \sin^2 \theta \sin^2 \varphi \right) \right],$$

$$f_{38} = b_0 \sin 2\theta \cos \varphi - i\tilde{b} \sin \theta \sin \varphi,$$

$$f_{39} = b_0 \sin 2\theta \sin \varphi + i\tilde{b} \sin \theta \cos \varphi,$$

$$f_{3,10} = b' \sin^2 \theta + b'' \left(1 + \cos^2 \theta \right),$$

$$a_0 = a_{12,\alpha_{10}} - a_{11,\alpha_{10}} - 2a_{4,\alpha_4}, \qquad \tilde{a} = 3a_{11,\alpha_{10}} + 2a_{12,\alpha_{10}},$$

$$b_0 = \frac{1}{2} (b_{4,\beta_3} - b_{3,\beta_3} - b_{6,\beta_3} - b_{8,\beta_3}), \qquad \tilde{b} = b_{3,\beta_3} - b_{4,\beta_3} - b_{6,\beta_3} - b_{8,\beta_3},$$

$$b' = b_{3,\beta_3} + 2b_{4,\beta_3}, \qquad b'' = b_{8,\beta_3} - b_{6,\beta_3}.$$

Appendix F
The Wave Functions $\Phi_M^J(ll')$

$$\Phi_1^1(01) = \frac{1}{2}\sqrt{\frac{3}{5}}[(+-)-(-+)]\left(\overset{+-}{(31)} - \frac{2}{\sqrt{3}}\overset{++}{(11)} + \overset{-+}{(13)}\right),$$

$$\Phi_0^1(01) = \frac{1}{2\sqrt{10}}[(+-)-(-+)](3\overset{+-}{(33)} + \overset{-+}{(33)} - \overset{+-}{(11)} - \overset{-+}{(11)}),$$

$$\Phi_3^3(03) = \frac{1}{\sqrt{2}}[(+-)-(-+)]\overset{++}{(33)},$$

$$\Phi_2^3(03) = \frac{1}{2}[(+-)-(-+)](\overset{++}{(31)} + \overset{++}{(13)}),$$

$$\Phi_1^3(03) = \frac{1}{\sqrt{10}}[(+-)-(-+)](\overset{++}{(31)} + \sqrt{3}\overset{++}{(11)} + \overset{-+}{(13)}),$$

$$\Phi_0^3(03) = \frac{1}{2\sqrt{10}}[(+-)-(-+)][\overset{+-}{(33)} + \overset{-+}{(33)} + 3(\overset{+-}{(11)} + \overset{-+}{(11)})],$$

$$\Phi_1^1(10) = \frac{1}{2}(++)(\overset{+-}{(33)} - \overset{+-}{(11)} + \overset{-+}{(11)} - \overset{-+}{(33)}),$$

$$\Phi_0^1(10) = \frac{1}{2\sqrt{2}}[(+-)-(-+)](\overset{+-}{(33)} - \overset{+-}{(11)} + \overset{-+}{(11)} - \overset{-+}{(33)}),$$

$$\Phi_0^0(11) = \frac{1}{\sqrt{10}}\left\{(++)\left(\overset{+-}{(13)} - \frac{2}{\sqrt{3}}\overset{--}{(11)} + \overset{-+}{(31)}\right) - \frac{1}{2\sqrt{3}}[(+-)+(-+)]\right.$$

$$\times [3(\overset{+-}{(33)} + \overset{-+}{(33)}) - \overset{+-}{(11)} - \overset{-+}{(11)}]$$

$$\left. + (--)\left(\overset{+-}{(31)} - \frac{2}{\sqrt{3}}\overset{++}{(11)} + \overset{-+}{(13)}\right)\right\},$$

I. Geru, D. Suter, *Resonance Effects of Excitons and Electrons*,
Lecture Notes in Physics 869, DOI 10.1007/978-3-642-35807-4,
© Springer-Verlag Berlin Heidelberg 2013

$$\Phi_1^1(11) = \frac{1}{2\sqrt{10}}\left\{(++)[3((\overset{+-}{33})+(\overset{-+}{33}))-(\overset{+-}{11})-(\overset{-+}{11})]\right.$$

$$\left. - \sqrt{3}[(+-)+(-+)]\left((\overset{+-}{31})-\frac{2}{\sqrt{3}}(\overset{++}{11})+(\overset{-+}{13})\right)\right\},$$

$$\Phi_0^1(11) = \frac{1}{2}\sqrt{\frac{3}{5}}\left\{(++)\left((\overset{+-}{13})-\frac{2}{\sqrt{3}}(\overset{--}{11})+(\overset{-+}{31})\right)\right.$$

$$\left. -(--)\left((\overset{+-}{31})-\frac{2}{\sqrt{3}}(\overset{++}{11})+(\overset{-+}{13})\right)\right\},$$

$$\Phi_2^2(11) = \sqrt{\frac{3}{10}}(++)\left((\overset{+-}{31})-\frac{2}{\sqrt{3}}(\overset{++}{11})+(\overset{-+}{13})\right),$$

$$\Phi_1^2(11) = \frac{1}{2\sqrt{10}}\left\{(++)[3((\overset{+-}{33})+(\overset{-+}{33}))-(\overset{+-}{11})-(\overset{-+}{11})]\right.$$

$$\left. + \sqrt{3}[(+-)+(-+)]\left((\overset{+-}{31})-\frac{2}{\sqrt{3}}(\overset{++}{11})+(\overset{-+}{13})\right)\right\},$$

$$\Phi_0^2(11) = \frac{1}{2\sqrt{5}}\left\{(++)\left((\overset{+-}{13})-\frac{2}{\sqrt{3}}(\overset{--}{11})+(\overset{-+}{31})\right)+(--)\left((\overset{+-}{31})\right.\right.$$

$$\left. -\frac{2}{\sqrt{3}}(\overset{++}{11})+(\overset{-+}{13})\right)+\frac{1}{\sqrt{3}}[(+-)+(-+)]$$

$$\left. \times[3((\overset{+-}{33})+(\overset{-+}{33}))-(\overset{+-}{11})-(\overset{-+}{11})]\right\},$$

$$\Phi_1^1(12) = \sqrt{\frac{3}{10}}(--)((\overset{++}{31})-(\overset{++}{13}))-\frac{1}{2}\sqrt{\frac{3}{10}}[(+-)+(-+)]$$

$$\times((\overset{+-}{31})-(\overset{-+}{13}))+\frac{1}{2\sqrt{10}}(++)((\overset{+-}{33})+(\overset{+-}{11})-(\overset{-+}{11})-(\overset{-+}{33})),$$

$$\Phi_0^1(12) = \frac{1}{2}\sqrt{\frac{3}{5}}(--)((\overset{+-}{31})-(\overset{-+}{13}))-\frac{1}{2\sqrt{5}}[(+-)-(-+)]$$

$$\times((\overset{+-}{33})+(\overset{+-}{11})-(\overset{-+}{11})-(\overset{-+}{33}))+\frac{1}{2}\sqrt{\frac{3}{5}}(++)((\overset{+-}{13})-(\overset{-+}{31})),$$

$$\Phi_2^2(12) = \frac{1}{\sqrt{6}}\{[(+-)+(-+)]((\overset{++}{31})-(\overset{++}{13}))-(++)((\overset{+-}{31})-(\overset{-+}{13}))\},$$

$$\Phi_1^2(12) = \frac{1}{\sqrt{6}}(--)((\overset{++}{3}1) - (\overset{++}{1}3)) + \frac{1}{\sqrt{6}}[(+-)+(-+)]$$
$$\times ((\overset{+-}{3}1) - (\overset{-+}{1}3)) - \frac{1}{2\sqrt{2}}(++)((\overset{+-}{3}3) + (\overset{+-}{1}1) - (\overset{-+}{1}1) - (\overset{-+}{3}3)),$$

$$\Phi_0^2(12) = \frac{1}{2}[(--)((\overset{+-}{3}1) - (\overset{-+}{1}3)) - (++)((\overset{+-}{1}3) - (\overset{-+}{3}1))],$$

$$\Phi_3^3(12) = \frac{1}{\sqrt{2}}(++)((\overset{++}{3}1) - (\overset{++}{1}3)),$$

$$\Phi_2^3(12) = \frac{1}{2\sqrt{3}}[(+-)-(-+)]((\overset{++}{3}1) - (\overset{++}{1}3)) + \frac{1}{\sqrt{3}}(++)((\overset{+-}{3}1) - (\overset{-+}{1}3)),$$

$$\Phi_1^3(12) = \frac{1}{\sqrt{30}}(--)((\overset{++}{3}1) - (\overset{++}{1}3)) + \sqrt{\frac{2}{15}}[(+-)+(-+)]$$
$$\times ((\overset{+-}{3}1) - (\overset{-+}{1}3)) + \frac{1}{\sqrt{10}}(++)((\overset{+-}{3}3) + (\overset{+-}{1}1) - (\overset{-+}{1}1) - (\overset{-+}{3}3)),$$

$$\Phi_0^3(12) = \frac{1}{\sqrt{10}}(--)((\overset{+-}{3}1) - (\overset{-+}{1}3)) + \frac{1}{2}\sqrt{\frac{3}{10}}[(+-)+(-+)]$$
$$\times ((\overset{+-}{3}3) + (\overset{+-}{1}1) - (\overset{-+}{1}1) - (\overset{-+}{3}3)) + \frac{1}{\sqrt{10}}(++)((\overset{+-}{1}3) - (\overset{-+}{3}1)),$$

$$\Phi_2^2(13) = \sqrt{\frac{5}{7}}(--)(\overset{++}{3}3) - \frac{1}{2}\sqrt{\frac{5}{21}}[(+-)+(-+)]$$
$$\times ((\overset{++}{3}1) + (\overset{++}{1}3)) + \frac{1}{\sqrt{105}}(++)((\overset{+-}{3}1) + \sqrt{3}(\overset{++}{1}1) + (\overset{-+}{1}3)),$$

$$\Phi_1^2(13) = \sqrt{\frac{5}{21}}(--)((\overset{++}{3}1) + (\overset{++}{1}3)) - \frac{2}{\sqrt{105}}[(+-)+(-+)](\overset{+-}{3}1)$$
$$+ \sqrt{3}((\overset{++}{1}1) + (\overset{-+}{1}3)) + \frac{1}{2\sqrt{35}}(++)[(\overset{+-}{3}3) + (\overset{-+}{3}3) + 3((\overset{+-}{1}1) + (\overset{-+}{1}1))],$$

$$\Phi_0^2(13) = \frac{2}{\sqrt{35}}(--)((\overset{+-}{3}1) + \sqrt{3}(\overset{++}{1}1) + (\overset{-+}{1}3)) - \frac{1}{2}\sqrt{\frac{3}{70}}[(+-)+(-+)]$$
$$\times [(\overset{+-}{3}3) + (\overset{-+}{3}3) + 3((\overset{+-}{1}1) - (\overset{-+}{1}1))] + \frac{2}{35}(++)((\overset{+-}{1}3) + \sqrt{3}(\overset{--}{1}1 + (\overset{-+}{3}1)),$$

$$\Phi_3^3(13) = \frac{1}{2}\sqrt{\frac{3}{2}}[(+-)+(-+)]\overset{++}{(33)} - \frac{1}{2\sqrt{2}}(++)\overset{++}{(31)} + \overset{++}{(13)},$$

$$\Phi_2^3(13) = \frac{1}{2}(--)\overset{++}{(33)} + \frac{1}{2\sqrt{3}}[(+-)+(-+)](\overset{++}{(31)} + \overset{++}{(13)})$$

$$- \frac{1}{2\sqrt{3}}(++)(\overset{+-}{(31)} + \sqrt{3}\overset{++}{(11)} + \overset{-+}{(13)}),$$

$$\Phi_1^3(13) = \frac{1}{2}\sqrt{\frac{5}{6}}(--)(\overset{++}{(31)} + \overset{++}{(13)}) + \frac{1}{2\sqrt{30}}[(+-)+(-+)] \times (\overset{+-}{(31)}$$

$$+ \sqrt{3}\overset{++}{(11)} + \overset{-+}{(13)}) - \frac{1}{2\sqrt{10}}(++) \times [\overset{+-}{(33)} + \overset{-+}{(33)} + 3(\overset{+-}{(11)} + \overset{-+}{(11)})],$$

$$\Phi_0^3(13) = \frac{1}{\sqrt{10}}[(--)(\overset{+-}{(31)} + \sqrt{3}\overset{++}{(11)} + \overset{-+}{(13)}) - (++)(\overset{+-}{(13)} + \sqrt{3}\overset{--}{(11)} + \overset{-+}{(31)})],$$

$$\Phi_4^4(13) = (++)\overset{++}{(33)},$$

$$\Phi_3^4(13) = \frac{1}{2\sqrt{2}}[(+-)+(-+)]\overset{++}{(33)} + \frac{1}{2}\sqrt{\frac{3}{2}}(\overset{++}{(31)} - \overset{++}{(13)}),$$

$$\Phi_2^4(13) = \frac{1}{2\sqrt{7}}(--)\overset{++}{(33)} + \frac{1}{2}\sqrt{\frac{3}{7}}[(+-)+(-+)]$$

$$\times (\overset{++}{(31)} + \overset{++}{(13)}) + \frac{1}{2}\sqrt{\frac{3}{7}}(++)(\overset{+-}{(31)} + \sqrt{3}\overset{++}{(11)} + \overset{-+}{(13)}),$$

$$\Phi_1^4(13) = \frac{1}{2}\sqrt{\frac{3}{14}}(--)(\overset{++}{(31)} + \overset{++}{(13)}) + \frac{1}{2}\sqrt{\frac{3}{14}}[(+-)+(-+)] \times (\overset{+-}{(31)}$$

$$+ \sqrt{3}\overset{++}{(11)} + \overset{-+}{(13)}) + \frac{1}{2\sqrt{14}}(++)[\overset{+-}{(33)} + \overset{-+}{(33)} + 3(\overset{+-}{(11)} + \overset{-+}{(11)})],$$

$$\Phi_0^4(13) = \sqrt{\frac{3}{70}}(--)(\overset{+-}{(31)} + \sqrt{3}\overset{++}{(11)} + \overset{-+}{(13)}) + \frac{1}{\sqrt{70}}[(+-)+(-+)]$$

$$\times [\overset{+-}{(33)} + \overset{-+}{(33)} + 3(\overset{+-}{(11)} + \overset{-+}{(11)})]$$

$$+ \sqrt{\frac{3}{70}}(++)(\overset{+-}{(13)} + \sqrt{3}\overset{--}{(11)} + \overset{-+}{(31)}).$$

Appendix G
Integral $I(s, t)$

Consider the integral

$$I(s, t) = \int_0^\infty e^{-x} sh\{s\sqrt{x(x+t)}\}\theta(x+t)\,dx \tag{G.1}$$

for $0 < s < 1$, where $s = (1 + G/4\xi k_0 T)^{-1}$, $t = (4\xi/G + 1/k_0 T) \times [E(\Gamma) - E(\Gamma')]$, $\xi = 1/\nu$, ν, $(1/2)(\nu + 1/\nu)$.

The integral $I(s, t)$ from (G.1) can be transformed into

$$I(s, t) = \frac{se^{t\theta(-t)}}{1 - s^2} \int_0^\infty e^{-x} \zeta_{s,t}(x)\,dx, \tag{G.2}$$

where

$$\zeta_{s,t}(x) = \sqrt{\left(x + |t|/2\right)^2 - \left(1 - s^2\right)t^2/4 - s|t|/2}. \tag{G.3}$$

Taking into account that $\zeta_{0,t}(x) = \sqrt{x(x + |t|)}$ and $\zeta_{1,t}(x) = \zeta_{s,0}(x) = x$, the function (G.3) can be approximated by the expression

$$\zeta_{s,t}(x) = \sqrt{x\left[x + (1 - s)|t|\right]}. \tag{G.4}$$

Substituting (G.4) into (G.2), we find

$$I(s, t) \simeq \frac{s}{1 - s^2} e^{st\theta(-t)} F\left(-(1 - s)t\right),$$

where the function $F(x)$ which appears in the formula (6.5) is determined in Sect. 6.2 (see formula (6.5) and the definition of $F(x)$ after formula (6.7)).

I. Geru, D. Suter, *Resonance Effects of Excitons and Electrons*,
Lecture Notes in Physics 869, DOI 10.1007/978-3-642-35807-4,
© Springer-Verlag Berlin Heidelberg 2013

Appendix H
Unitarity of the Spinor Operators and Two-Boson Representation of the Angular Momentum [527]

The two-boson representation of the angular momentum was considered for the first time by Schwinger [406]. According to [406] the operators of the projections of the angular momentum may be represented as square forms built from the operators of creation and annihilation of two bosons of different types bound between themselves. As the energy spectrum of the harmonic oscillator is not limited from above, but the spectrum of the operator of the angular momentum is limited by the final number of states, this representation is true only for small numbers of boson fillings which correspond to the states in the limits of the eigenfunctions of the angular momentum operator.

A more complex representation of the angular momentum J by means of bound bosons of $2J$ type is proposed in [482]. However, the complication conditioned by the large number of types of bosons does not enable simple methods for the treatment of the angular momentum projection operators in the case of multilevel systems. Therefore, further analysis of a two-boson Schwinger representation for the angular momentum with the attraction of such fundamental concepts as unitarity is considered to be reasonable.

We shall demonstrate that one can obtain Schwinger's representation for the operators of arbitrary angular momentum projections on the basis of two-boson spinor operators of $2J$ rank under the condition that these spinor operators are unitary. The unitarity of two-boson spinor operators leads to the complete reduction of what are called operator loads (polynomials of $2J - 1$ degree on the operator of both types of bosons) to the spin projection operators for $S = 1/2$ in Schwinger's representation [527]. This unitarity property of two-boson spinor operators is the main one for the two-boson representation of the angular momentum.

The physical consequence of the unitarity of two-boson spinor operators is the exclusion for bosons of both types to occupy any of the states which do not belong to $2J$ the lowest excited levels of every two harmonic oscillators. Another important consequence of the unitarity of two-boson spinor operators is the existence of the invariant $N = a_1^+ a_1 + a_2^+ a_2$, where a_i^+ and a_i are the creation and annihilation operators of the boson of ith type. This invariant commutes with the Hamiltonian; therefore, during the calculation of the thermal averages the exclusion of nonphys-

I. Geru, D. Suter, *Resonance Effects of Excitons and Electrons*,
Lecture Notes in Physics 869, DOI 10.1007/978-3-642-35807-4,
© Springer-Verlag Berlin Heidelberg 2013

ical states can be performed by the method which is used for the grand canonical ensemble during the introduction of the chemical potential in quantum statistics [474].

Let us define the following two-boson spinor operator:

$$
U_J(a_1, a_2) = \begin{pmatrix} [(2J)!]^{-1/2}a_1^{2J} \\ [(2J-1)!]^{-1/2}a_1^{2J-1}a_2 \\ \vdots \\ [(J+M)!(J-M)!]^{-1/2}a_1^{J+M}a_2^{J-M} \\ \vdots \\ [(2J-1)!]^{-1/2}a_1a_2^{2J-1} \\ [(2J)!]^{-1/2}a_2^{2J} \end{pmatrix} \tag{H.1}
$$

and its corresponding Hermitian conjugate operator (M is the eigenvalue of the operator of the zth projection of the angular momentum). By means of these two-boson spinor operators we shall pass from the operators initially written in the spinor basis to the angular momentum projection operators in the representation of the coupled bosons:

$$
U_J^+(a_1^+, a_2^+)\hat{J}_\mu U(a_1, a_2) = \hat{J}_\mu(1,2)
$$

$$
\mu = z, +, -. \tag{H.2}
$$

Using the commutation relations for Bose operators, we shall present the angular momentum projection operators in the form

$$
J_z(1,2) = S_z(1,2)O_z^{(J)}(1,2),
$$

$$
J_\pm(1,2) = J_x(1,2) \pm iJ_y(1,2) = S_\pm(1,2)O_z^{(J)}(1,2), \tag{H.3}
$$

where

$$
S_z(1,2) = \frac{1}{2}(a_1^+ a_1 - a_2^+ a_2),
$$

$$
S_+(1,2) = a_1^+ a_2, \qquad S_-(1,2) = a_2^+ a_1. \tag{H.4}
$$

Here $S_z(1,2)$ and $S_\pm(1,2)$ are the operators of the projections of spin $S = 1/2$ in Schwinger's representation; $O_z^{(J)}$ and $O_\pm^{(J)}$ are called the operator loads of the angular momentum J with respect to the operators $S_z(1,2)$ and $S_\pm(1,2)$ for spin $S = 1/2$ in Schwinger's representation.

We shall require that the two-boson spinor operator $U_J(a_1, a_2)$ from (H.1) be a unitary operator,

$$
U_J^+(a_1^+, a_2^+)U_J(a_1, a_2) = 1. \tag{H.5}
$$

Then we may show that in particular cases $J = 1/2, 1, 3/2, 2, 5/2, \ldots$ the condition of unitarity (H.5) for the spinor operator $U_J(a_1, a_2)$ results in the following sequence of operator equations of degree $2J$ relative to the invariant N:

$$N = 1$$
$$N(N-1) = 2$$
$$N(N-1)(N-2) = 3!$$
$$N(N-1)(N-2)(N-3) = 4!$$
$$N(N-1)(N-2)(N-3)(N-4) = 5!$$

$$\vdots \quad \vdots \quad \vdots \quad \vdots \quad \vdots \quad \vdots \quad \vdots \quad \vdots \quad \vdots \quad \vdots \quad \vdots$$

It is easy to see that any term of this sequence of equations is determined by the expression

$$\mathscr{P}_{2J}(N) = (2J)!, \tag{H.7}$$

where

$$\mathscr{P}_{2J}(N) = N(N-1)(N-2)\cdots\left[N-(2J-1)\right]. \tag{H.8}$$

Here $\mathscr{P}_{2J}(N)$ is the polynomial of degree $2J$ relative to the operator N which depends on the polynomial of degree $2J-1$ according to the recurrent relation

$$\mathscr{P}_{2J}(N) = (N+1-2J)\mathscr{P}_{2J-1}(N). \tag{H.9}$$

We shall calculate the operator loads $O_z^{(J)}(1,2)$, $O_+^{(J)}(1,2)$, and $O_-^{(J)}(1,2)$ at different values of the angular momentum J. First, let us note that for a fixed value J the operator loads are equal between themselves:

$$O_z^J(1,2) = O_+^{(J)}(1,2) = O_-^{(J)}(1,2). \tag{H.10}$$

For the sequence of values $J = 1/2, 1, 3/2, 2, 5/2, \ldots$ the operator loads (H.10) have the form

$$1$$
$$N-1$$
$$\frac{1}{2}(N-1)(N-2)$$
$$\frac{1}{3!}(N-1)(N-2)(N-3)$$
$$\frac{1}{4!}(N-1)(N-2)(N-3)(N-4)$$

$$\cdot \quad \cdot \quad \cdot \quad \cdot \quad \cdot \quad \cdot \quad \cdot \quad \cdot \quad \cdot \quad \cdot \quad \cdot$$

Then any term of the sequence (H.11) is the operator

$$O_\alpha^{(J)}(1,2) = \left[(2J-1)!\right]^{-1}(N-1)(N-2)\cdots(N+1-2J),$$

$$\alpha = z, +, -. \tag{H.12}$$

Since $N = 2J$ is one of $2J$ solutions of every one of (H.5)–(H.8), then when $N = 2J$ the operator load $O_\alpha^{(J)}(1,2)$ becomes an identical transformation:

$$O_\alpha^{(J)}(1,2)|_{N=2J} = \left[(2J-1)!\right]^{-1}\left[(2J-1)!\right] = 1. \tag{H.13}$$

Expressions (H.1)–(H.13) prove the following theorem.

Theorem *If the two-boson spinor operator $U_J(a_1, a_2)$ of rank $2J$ of the type (H.1) is unitary, then complete reduction of the operator loads of the arbitrary angular momentum J in the two-boson representation takes place ($O_z^{(J)}(1,2) = O_\pm^{(J)}(1,2) = 1$).*

Corollary 1 *The angular momentum operators in the two-boson representation are invariant relative to the orthogonal addition or to the orthogonal reduction (up to a number of orths not less than two) of the basis vectors of the angular momentum space. The view of these operators does not depend on the value of the angular momentum.*

Corollary 2 *In the two-boson representation all the multilevel specificity of the system with angular momentum J is enclosed only in the wave functions $|(J + M)_1, (J - M)_2\rangle$, where $(J + M)_1$ and $(J - M)_2$ are the numbers of the excited levels of the oscillators of the first and second types.*

Glossary

Frenkel exciton an exciton with radius less than the parameters of a crystal lattice. The electron and hole of the exciton are localized on the same molecule. Frenkel excitons are characteristic for molecular crystals.

Wannier–Mott exciton an exciton with radius much bigger comparatively than the parameters of a crystal lattice. The center of gravity of such an exciton is translated through the crystal. Wannier–Mott excitons are characteristic for semiconducting crystals.

Bloch function the wave function of the electron (hole) in a crystal presenting the modulated plane wave

$$\Psi_{\mathbf{k}\alpha}(\mathbf{r}) = \frac{1}{\sqrt{V}} U_{\mathbf{k}\alpha}(\mathbf{r}) \exp(i\mathbf{k}\mathbf{r}),$$

where $U_{\mathbf{k}\alpha}(\mathbf{r}) = U_{\mathbf{k}\alpha}(\mathbf{r}+\mathbf{n})$ is a periodic function, $\mathbf{n}$ is the vector connecting two neighboring atoms of the periodic crystal lattice, $\mathbf{r}$ and $\mathbf{k}$ are the radius vector and wave vector of the electron (hole), α denotes the energy band, and V is the volume of the crystal.

Far-acting Coulomb interaction (exchange, resonance, or annihilation interaction) the exchange interaction between electrons in the valence band and in the conduction band, which can be considered as a result of virtual recombination and generation of excitons.

Quasi-particle a low-energy excitation of a system, possessing a set of quantum numbers and/or well-defined expectation values of certain operators (position, charge, momentum, angular momentum, energy) often associated with isolated particles.

Fermi operator an operator which obeys Fermi statistics.

EHD electron–hole drop.

EHL electron–hole liquid.

EPR electron paramagnetic resonance.

ESR electron spin resonance.

NMR nuclear magnetic resonance.

APR acoustic paramagnetic resonance.

I. Geru, D. Suter, *Resonance Effects of Excitons and Electrons*,
Lecture Notes in Physics 869, DOI 10.1007/978-3-642-35807-4,
© Springer-Verlag Berlin Heidelberg 2013

FMR ferromagnetic resonance.

AFMR antiferromagnetic resonance.

ENDOR electron–nuclear double resonance.

Acoustical ENDOR electron–nuclear double magnetoacoustic resonance.

ELDOR electron–electron double resonance.

EIC exciton–impurity complex.

MEIC multiple exciton–impurity complex.

$\Gamma_6 \otimes \Gamma_6 \otimes \Gamma_1$ **exciton** an exciton in the $1S$ ground state of the relative motion electron–hole in a crystal with Γ_6 conduction band and Γ_6 valence band.

$\Gamma_6 \otimes \Gamma_8 \otimes \Gamma_1$ **exciton** an exciton in the $1S$ ground state of the relative motion electron–hole in a crystal with Γ_6 conduction band and Γ_8 valence band.

Isoelectronic impurities substitutional impurities which have the same valence electron structures as the atoms they replace. Isoelectronic impurities are neutral impurities. They do not lead to doping of the semiconductor host crystal. However, the host atom to be substituted and the substitutional isoelectronic impurity necessarily differ in their electron core structure.

Isoelectronic traps isoelectronic impurities which produce localized states (bound exciton states) within the band gap.

CRH$_2$ solid hydrogen in the form of a molecular crystal.

CR$_{ex2}$ biexcitons crystallized in a "solid" biexciton lattice.

Zero-phonon lines for single excitons bound on isoelectronic traps transitions in the luminescence decay of the states of single excitons localized on isoelectronic impurities. In crystals with valence bands formed from p-type electron states, there are two zero-phonon lines. They appear at optical transitions from these states of the bound exciton with total angular momentum $I_t = 1$ and $I_t = 2$ to the ground state of the crystal. This reflects the presence of two-fold spin degeneration of the conduction band and four-fold degeneration of the upper valence sub-band arising as a result of big spin–orbital splitting of the valence band.

Interserial exciton transitions quantum transitions between exciton states belonging to different exciton series.

Intraserial exciton transitions quantum transitions between exciton states belonging to the same exciton series.

Threshold of generation the pump level at which the self-excitation of a quantum generator takes place.

EDDR exciton dipole–dipole reservoir.

PER paraelectric resonance.

ExPR exciton paraelectric resonance.

Isotopic shift of ExPR a shift of ExPR frequency at isotopic substitution of one part of a crystal's atoms.

Exciton Knight shift the shift of NMR lines due to interaction of excitons with the nuclei of atoms situated in nodes of the crystal lattice.

Bose–Einstein condensation of excitons second-order phase transition of an exciton gas to a Bose–Einstein condensate state with a large fraction of excitons occupying the lowest quantum state, at which point quantum effects become apparent on a macroscopic scale.

Orthobiexciton a molecule formed by two excitons with a resulting spin of four particles (two electrons and two holes) equal to one.

WAHUHA the known sequence of four electromagnetic pulses used for the first time in solid state NMR spectroscopy by J.S. **Wa**ugh, L.M. **Hu**ber, and U. **Ha**eberlen for narrowing solid state NMR lines.

FR Faraday rotation.

OMDR optico–magnetic double resonance.

MCDR magnetic circular dichroism of the reflection.

Antirotation rotation of a point or geometric figure with a subsequent change in the physical sense (for example, the transition from a wave function to a complex conjugate one, a change in the magnetic moment orientation, etc.).

Antireflection reflection of a point or geometric figure with a subsequent change in the physical sense (for example, the transition from a wave function to a complex conjugate one, a change in the magnetic moment orientation, etc.).

Two-boson Schwinger representation the representation of angular momentum projection operators by means of two different Bose field operators.

Exciton–elastic wave a "mixture" of strictly resonant hypersonic phonons with excitons in two different bands, the distance between which corresponds to the resonant phonons' energy, under conditions of the inverse difference of their populations.

HONDOR hole–nuclear double resonance.

HOPR hole paramagnetic resonance.

RODOR radio–optical double resonance.

QIP quantum information processing.

NV nitrogen–vacancy center.

N@C60 endohedral fullerene: nitrogen in C_{60}.

FID free induction decay.

References

1. E.K. Zavoisky, Paramagnetic absorption in the perpendicular and parallel fields for salts, solutions and metals, Doctoral thesis Phys.-Math. Sci., Moscow Inst. of Phys. of Acad. of Sci. of USSR, 1944
2. S.A. Altshuler, B.M. Kozyrev, To the history of discovery of electron paramagnetic resonance, in *Paramagnetic Resonance (1944–1969)* (Nauka, Moscow, 1971), pp. 25–31
3. S.A. Altshuler, B.M. Kozyrev, *Electron Paramagnetic Resonance of Compounds of Elements of the Intermediate Groups*, 2nd edn. (Nauka, Moscow, 1972), 672 pp., adv. ed. A.I. Rivkind
4. W. Low, *Paramagnetic Resonance in Solids* (Academic Press, San Diego, 1960), 212 pp.
5. A. Abragam, B. Bleaney, *Electron Paramagnetic Resonance of Transition Ions* (Oxford University Press, Oxford, 1970), 700 pp.
6. A. Carrington, A.D. McLechlan, *Introduction to Magnetic Resonance with Applications to Chemistry and Chemical Physics* (Chapman & Hall, New York, 1979), 266 pp.
7. I.N. Marov, N.A. Kostromina, *EPR and NMR in the Chemistry of Coordination Compounds* (Nauka, Moscow, 1979), 268 pp.
8. B.M. Kozyrev, I.B. Ovchinnikov, EPR of coordination compounds in liquid crystals, in *Problems of Magnetic Resonance*, ed.-in-chief A.M. Prohorov (Nauka, Moscow, 1978), pp. 49–65
9. D. Ingram, *Free Radicals as Studied by Electron Spin Resonance* (Butterworth, Stoneham, 1958), 274 pp.
10. Y.S. Lebedev, V.I. Muromtsev, *EPR and Relaxation of Trapped Radicals* (Chem., Moscow, 1972), 254 pp.
11. G.E. Pake, *Paramagnetic Resonance*, 2nd edn. (W.A. Benjamin, New York, 1962), 205 pp.
12. C. Poole, *The Technique of EPR Spectroscopy* (Mir, Moscow, 1970), 557 pp. [Russian translation]
13. J.S. Hyde, The paramagnetic resonance signals from color centers, in *NMR and EPR Spectroscopy*, ed. by L.L. Dekabrun (Mir, Moscow, 1964), pp. 235–245 [Russian translation]
14. C. Slichter, *Principles of Magnetic Resonance*, 3rd edn. (Springer, New York, 1990), 655 pp.
15. N.A. Penin (ed.), *Electron Spin Resonance in Semiconductors* (IL, Moscow, 1962), 380 pp. [Russian translation] collection of articles
16. G.W. Ludwig, H.H. Woodbury, *Electron Spin Resonance in Semiconductors* (Academic Press, New York, 1962), p. 148
17. S. Metfessel, D.C. Mattis, *Magnetic Semiconductors*. Handbuch der Physik (Springer, New York, 1968), 562 pp.
18. A.A. Samokhvalov, B.C. Babushkin, M.I. Simonova, T.I. Arbuzova, Fiz. Tverd. Tela **14**, 2174–2175 (1972)
19. G. Sperlich, Int. J. Magn. **5**, 125–128 (1973)
20. I.I. Geru, I.G. Lupea, K.G. Nikiforov, S.I. Radautsan, V.E. Tezlevan, Fiz. Tverd. Tela **20**, 1534–1535 (1978)

I. Geru, D. Suter, *Resonance Effects of Excitons and Electrons*,
Lecture Notes in Physics 869, DOI 10.1007/978-3-642-35807-4,
© Springer-Verlag Berlin Heidelberg 2013

21. K. Morigaki, Y. Sano, I. Harabyashi, Optical detected in hydrogenated amorphous silicon, Techn. Repts., ISSR (1982), A, N1232, 37 pp.

22. N.A. Goryunova, A.F. Bendersky, G.S. Kuzmenko, E.O. Osmanov, I.I. Geru, A.G. Cheban, Electron paramagnetic resonance of Mn^{2+} ions in glasses $CdGeAs_2$, in *Ternary Semiconductors $A^{II}B^{IV}C_2^V$ and $A^{II}B_2^{III}C_4^{VI}$*, ed.-in-chief S.I. Radautsan (Stiinta, Chisinau, 1972), pp. 117–119

23. J. Winter, *Magnetic Resonance in Metals* (Oxford University Press, New York, 1971), 206 pp.

24. L.L. Hirst, Adv. Phys. **21**(93), 759–782 (1972). doi:10.1080/00018737200101358

25. N.E. Alekseevsky, I.A. Garifullin, B.I. Kochelaev, E.G. Harahashyan, JETP Lett. **18**(5), 323–326 (1973)

26. D. Davidov, A. Chelkowski, C. Rettori, R. Orbach, M.B. Maple, Phys. Rev. B **7**(3), 1029–1038 (1973)

27. B.I. Kochelaev, M.G. Khusainov, J. Exp. Theor. Phys. **80**(4), 1480–1487 (1981)

28. L.J. Berliner (ed.), *Spin Labelling II: Theory and Applications* (Academic Press, New York, 1979), 357 pp.

29. A. Abragam, *The Principles of Nuclear Magnetism* (Clarendon Press, Oxford, 1994), 614 pp. Reprint edition

30. E.R. Andrew, *Nuclear Magnetic Resonance* (Cambridge University Press, Cambridge, 1955), 278 pp.

31. A. Leshe, *Nuclear Induction* (IL, Moscow, 1963), 683 pp. [Russian translation]

32. I.Y. Slonim, A.N. Lyubimov, *Nuclear Magnetic Resonance in Polymers* (Chem., Moscow, 1966), 339 pp.

33. C.D. Jeffries, *Dynamic Nuclear Orientation* (Interscience, New York, 1963), 177 pp.

34. P.M. Borodin, A.V. Melnikov, A.A. Morozov, *Nuclear Magnetic Resonance in the Earth's Field* (Leningrad, 1967), 387 pp.

35. V.S. Grechishkin, *Nuclear Quadrupole Interactions in Solids* (Nauka, Moscow, 1973), 263 pp.

36. S.A. Altshuler, Proc. Acad. Sci. USSR **85**(6), 1235–1238 (1952)

37. A.R. Kessel, *Nuclear Acoustic Resonance* (Nauka, Moscow, 1969), 214 pp.

38. V.A. Golenischev-Kutuzov, V.V. Samartsev, N.K. Solovarov, B.M. Khabibulin, *Magnetic Quantum Acoustics* (Nauka, Moscow, 1977), 197 pp.

39. J. Taner, B. Rampton, *Hypersound in Solid State Physics* (Mir, Moscow, 1975), 453 pp. [Russian translation]

40. W.P. Mason (ed.), *Applications to Quantum and Solid State Physics, Part A*. Physical Acoustics, vol. 4 (Mir, Moscow, 1969), 436 pp. [Russian translation]

41. U. Kuhn, F. Lüty, Solid State Commun. **3**(2), 31–33 (1965)

42. M.F. Deigen, M.D. Glinchuk, Usp. Fiz. Nauk **114**(2), 185–211 (1974)

43. J.M. Luttinger, Phys. Rev. **102**(4), 1030–1041 (1956)

44. G.L. Bir, G.E. Pikus, *Symmetry and Strain-Induced Effects in Semiconductors* (Nauka, Moscow, 1972), 584 pp.

45. K. Seeger, *Semiconductor Physics: An Introduction*, 9th edn. (Springer, Berlin, 2004), 548 pp.

46. G. Bemsky, Spin resonance of conduction electrons in InSb, in *Electron Spin Resonance in Semiconductors*, ed. by N.A. Penin (IL, Moscow, 1962), pp. 117–121 [Russian translation]

47. J.A. Bratashevsky, V.B. Tyutyunik, I.S. Aver'yanov, I.M. Nesmelov, Fiz. Tekh. Poluprovodn. **9**(1), 168–169 (1975)

48. A.A. Bugay, Y.S. Gromovoi, B.D. Shanina, Ukr. Fiz. Zh. **26**(2), 1826–1830 (1981)

49. E.I. Rashba, I.I. Boiko, Fiz. Tverd. Tela **3**(4), 1277–1289 (1960)

50. A.I. Akhiezer, V.G. Baryakhtar, S. Peletminskii, *Spin Waves* (Nauka, Moscow, 1967), 368 pp.

51. A.G. Gurevich, *Magnetic Resonance in Ferrites and Antiferromagnets* (Nauka, Moscow, 1973), 591 pp.

52. J.A. Monosov, *Nonlinear Ferromagnetic Resonance* (Nauka, Moscow, 1971), 376 pp.

53. G. Feher, The electronic structure of donors in silicon determined by the method of electron–nuclear double resonance, in *Electron Spin Resonance in Semiconductors*, ed. by N.A. Penin (IL, Moscow, 1962), pp. 13–97 [Russian translation]

54. M.F. Deigen, Investigation of the structure of energy bands of crystals, the hyperfine and spin–phonon interactions by electron–nuclear double resonance method, in *Paramagnetic Resonance (1944–1969)*, adv. ed. A.I. Rivkind (Nauka, Moscow, 1971), pp. 202–210

55. S.R. Hartmann, E.L. Hahn, Phys. Rev. **128**(5), 2042–2053 (1962)

56. D. Schmid, Nuclear magnetic double resonance—principles and applications in solid state physics, in *Springer Tracts in Modern Physics*, vol. 68 (Springer, Berlin, 1973), pp. 1–75

57. E.T. Lipmaa, J. Struct. Chem. **8**(4), 717–780 (1967)

58. V.A. Benderskii, L.A. Blumenfeld, P.A. Sturzas, E.A. Sokolov, Nature **220**(5165), 365–367 (1968)

59. M.F. Deigen, I.I. Geru, Solid State Phys. **9**(9), 2611–2618 (1967)

60. A.S. Davydov, *Solid State Theory* (Nauka, Moscow, 1976), 639 pp.

61. E.M. Purcell, H.C. Torrey, R.V. Pound, Phys. Rev. **69**, 37–38 (1946)

62. F. Bloch, W.W. Hansen, M. Packard, Phys. Rev. **69**, 127 (1946)

63. E.A. Turov, M.P. Petrov, *Nuclear Magnetic Resonance in Ferro- and Antiferromagnets* (Nauka, Moscow, 1969), 260 pp.

64. A. Narat, Nuclear magnetic resonance in magnetics and metals, in *Hyperfine Interactions in Solids*, ed. by E.A. Turov (Mir, Moscow, 1970), pp. 163–236 [Russian translation]

65. M.P. Petrov, V.P. Chekmarev, A.P. Paugurt, Nuclear magnetic resonance in ferro- and antiferromagnets, in *Problems of Magnetic Resonance*, ed.-in-chief A.M. Prohorov (Nauka, Moscow, 1978), pp. 289–309

66. G.A. Smolensky, V.V. Lemanov, G.M. Nedlin, M.P. Petrov, R.V. Pisarev, *Physics of Magnetic Dielectrics* (Nauka, Leningrad, 1974), 454 pp.

67. M.P. Petrov, V.F. Pashin, A.P. Paugurt, JETP Lett. **12**(7), 359–362 (1970)

68. J.S. Waugh, *New NMR Methods in Solid State Physics* (Mir, Moscow, 1978), 179 pp. [Russian translation]

69. U. Heberlen, M. Mehring, *High Resolution NMR in Solids* (Mir, Moscow, 1980), 504 pp. [Russian translation]

70. M.F. Deigen, S.I. Pekar, J. Exp. Theor. Phys. **34**(1), 684–687 (1958)

71. A.V. Komarov, C.M. Ryabchenko, O.B. Terletsky, I.I. Geru, R.D. Ivanchuk, J. Exp. Theor. Phys. **73**(2(8)), 608–618 (1977). [Sov. Phys. JETP **46**, 318–323 (1977), 1978 American Institute of Physics]

72. A.V. Komarov, S.M. Ryabchenko, O.V. Terletsky, I.I. Geru, R.D. Ivanchuk, in *Fifth All-Union Symposium on Spectroscopy of Crystals Activated by Rare Earths, and Elements of Iron* (Abstracts Book, Kazan, 1976), p. 88

73. A.V. Komarov, S.M. Ryabchenko, O.V. Terletsky, I.I. Geru, R.D. Ivanchuk, in *All-Union Conference on Physics of Magnetic Phenomena* (Abstracts Book, Donetsk, 1977), p. 164

74. K. Morigaki, P. Dawson, B.C. Cavenett, Solid State Commun. **28**(9), 829–834 (1978)

75. R. Landauer, Phys. Lett. A **217**, 188–193 (1996)

76. R.P. Feynman, Simulating physics with computers. Int. J. Theor. Phys. **21**, 467–488 (1982)

77. A. Bernstein, U. Vazirani, Quantum complexity theory, in *Proceedings 25th Annual ACM Symp. on Theory of Computing*, San Diego, CA (1993), pp. 11–20

78. L.K. Grover, Phys. Rev. Lett. **79**, 325–328 (1997)

79. D. Deutsch, R. Jozsa, Proc. R. Soc. Lond. A **439**, 553–558 (1992)

80. Y.I. Frenkel, *Collection of Selected Works*, vol. 2 (Publishing house Acad. Sci. USSR, Moscow-Leningrad, 1958), pp. 127–175

81. G.H. Wannier, Phys. Rev. **52**(3), 191–197 (1937)

82. N.F. Mott, Trans. Faraday Soc. **34**(203), 500–506 (1938)

83. A.S. Davydov, *The Theory of Light Absorption in Molecular Crystals* (Publisher house Acad. Sci. USSR, Kiev, 1951), 230 pp.

84. A.S. Davydov, *The Theory of Molecular Excitons* (Nauka, Moscow, 1968), 296 pp.

85. V.M. Agranovich, V.L. Ginzburg, *Crystal Optics with Spatial Dispersion and Theory of Excitons* (Nauka, Moscow, 1965), 374 pp.
86. R.S. Knox, *Theory of Excitons* (Mir, Moscow, 1966), 219 pp. [Russian translation]
87. V.M. Agranovich, *Theory of Excitons* (Nauka, Moscow, 1968), 382 pp.
88. V.M. Agranovich, M.D. Galanin, *Transfer of Electronic Excitation Energy in Condensed Media* (Nauka, Moscow, 1978), 383 pp.
89. V.L. Broude, E.I. Rashba, E.F. Sheka, *Spectroscopy of Molecular Excitons* (Energoizdat, Moscow, 1981), 248 pp.
90. E.F. Gross, Usp. Fiz. Nauk **63**(3), 575–611 (1957)
91. M. Hayashi, K. Katsuki, J. Phys. Soc. Jpn. **7**, 599–603 (1952). doi:10.1143/JPSJ.7.599
92. M.A. Kozhushner, *Excitons—Quasiparticles in Solids* (Znanie, Moscow, 1973), 64 pp.
93. A.W. Overhauser, Phys. Rev. **101**(6), 1702–1715 (1956)
94. A.S. Davydov, *Biology and Quantum Mechanics* (Naukova Dumka, Kiev, 1979), 296 pp.
95. L.A. Blumenfeld, *Fundamentals of Biological Physics*, 2nd edn. (Nauka, Moscow, 1977), 336 pp.
96. K. Colbow, R.P. Danyluk, Biochim. Biophys. Acta **440**, 107–121 (1976)
97. L.V. Keldysh, Collective properties of excitons in semiconductors, in *Excitons in Semiconductors*, adv. ed. B.M. Vul (Nauka, Moscow, 1971), pp. 5–18
98. S.A. Moskalenko, *Introduction to the Physics of High-Density Excitons* (Stiinta, Chisinau, 1983), 280 pp.
99. S.V. Tyablikov, *Methods of Quantum Theory of Magnetism* (Nauka, Moscow, 1974), 250 pp.
100. N.N. Bogolyubov, *Selected Works*, 2nd edn. (Naukova Dumka, Kiev, 1970), 522 pp.
101. V.L. Bonch-Bruevich (ed.), *The Problems of Semiconductor Physics* (IL, Moscow, 1957), pp. 515–539 [Russian translation]
102. S.I. Pekar, *Studies on the Electron Theory of Crystals* (Gostekhizdat, Moscow, 1951), 256 pp.
103. I.S. Gradshteyn, M. Ryzhik, *Tables of Integrals, Sums, Series and Products* (Fizmatgiz, Moscow, 1962), 1100 pp.
104. G.G. Hall, Proc. R. Soc. Lond. Ser. A **270**(1341), 285–294 (1962)
105. E.I. Rashba, J. Exp. Theor. Phys. **36**(6), 1703–1708 (1959)
106. S.I. Pekar, *Crystal Optics and Additional Light Waves* (Naukova Dumka, Kiev, 1982), 295 pp.
107. S.I. Pekar, J. Exp. Theor. Phys. **38**(6), 1786–1797 (1960)
108. G.E. Pikus, G.L. Bir, J. Exp. Theor. Phys. **60**(1), 195–207 (1971)
109. G.E. Pikus, G.L. Bir, J. Exp. Theor. Phys. **62**(1), 324–332 (1972)
110. U. Fano, Phys. Rev. **124**(6), 1866–1878 (1961)
111. Y. Onodera, Phys. Rev. B **4**(8), 2751–2757 (1971)
112. E.F. Gross, A.A. Kaplyansky, Fiz. Tverd. Tela **2**(2), 2968–2981 (1960)
113. T. Koda, D.W. Langer, Phys. Rev. Lett. **20**(2), 50–53 (1968)
114. T. Koda, D.W. Langer, P.H. Yuvema, Investigation of the influence of uniaxial strain on exciton levels in the compound of the type $A^{II}B^{VI}$ with the wurtzite structure, in *IX International Conference on Semiconductor Physics*, Moscow, 23–29 July 1968, vol. 1 (Nauka, Leningrad, 1969), pp. 256–261
115. G.L. Bir, G.E. Pikus, L.G. Suslina, D.L. Fedorov, Fiz. Tverd. Tela **12**(4), 1187–1198 (1970)
116. G.L. Bir, G.E. Pikus, L.G. Suslina, D.L. Fedorov, Fiz. Tverd. Tela **12**(2), 3218–3228 (1970)
117. G.L. Bir, G.E. Pikus, L.G. Suslina, D.L. Fedorov, E.B. Shadrin, Fiz. Tverd. Tela **13**(12), 3551–3565 (1971)
118. D.W. Langer, R.N. Euwema, K. Era, T. Koda, Phys. Rev. B **2**(10), 4005–4022 (1970)
119. S. Syga, K. Cho, P. Heisinger, T. Koda, J. Lumin. **12**(13), 109–117 (1976)
120. H. Venghaus, R. Lambrich, Solid State Commun. **25**(2), 109–112 (1978)
121. H. Venghaus, P.E. Simmonds, I. Lagois, P.I. Dean, D. Brimberg, Solid State Commun. **24**(1), 5–9 (1977)
122. K. Cho, S. Suga, W. Dreybrodt, F. Willmann, Phys. Rev. B **2**(4), 1512–1521 (1975)
123. J. Lagois, Phys. Rev. B **16**(4), 1699–1705 (1977)

124. R.J. Elliott, Phys. Rev. **124**(2), 340–345 (1961)
125. R.J. Elliott, Theory of excitons, in *Polarons and Excitons*, ed. by C.G. Kuper, G.D. Whitfield (Oliver and Boyd, Edinburgh, 1962), pp. 269–294
126. G.Y. Lyubarsky, *Group Theory and Its Application in Quantum Mechanics* (Fizmatgiz, Moscow, 1958), 354 pp.
127. V. Heine, *Group Theory in Quantum Mechanics* (IL, Moscow, 1963), 522 pp. [Russian translation]
128. M.I. Petrashen', E.A. Trifonov, *Application of Group Theory in Quantum Mechanics* (Nauka, Moscow, 1967), 308 pp.
129. M. Hamermesh, *Group Theory and Its Application to Physical Problems* (Mir, Moscow, 1966), 587 pp. [Russian translation]
130. L. Michelle, M. Schaaf, *Symmetry in Quantum Physics* (Mir, Moscow, 1974), 250 pp. [Russian translation]
131. A. Barut, R. Raczka, *Theory of Group Representations and Applications*, vol. 1 (Mir, Moscow, 1980), 455 pp. [Russian translation]
132. A. Barut, R. Raczka, *Theory of Group Representations and Applications*, vol. 2 (Mir, Moscow, 1980), 395 pp. [Russian translation]
133. G. Jones, *Theory of Brillouin Zones and Electron States in Crystals* (Mir, Moscow, 1968), 264 pp. [Russian translation]
134. R. Knox, A. Gold, *Symmetry in the Solid State* (Nauka, Moscow, 1970), 424 pp.
135. G. Shtraitvolf, *Theory of Groups in Solid State Physics* (Mir, Moscow, 1971), 262 pp. [Russian translation]
136. W.A. Wooster, *Tensors and Group Theory for the Physical Properties of Crystals* (Oxford Univ. Press, London, 1973), 344 pp.
137. J. Birman, *Spatial Symmetry, and Optical Properties of Solids*, vol. 1 (Mir, Moscow, 1978), 387 pp. [Russian translation]
138. J. Birman, *Spatial Symmetry, and Optical Properties of Solids*, vol. 2 (Mir, Moscow, 1978), 352 pp. [Russian translation]
139. I.I. Geru, Study of the hyperfine and spin–phonon interactions in the local electron centers by invariants method, Diss. of candidate in phys-math. sci., Chisinau, 1967, 168 pp.
140. V.A. Koptsik, I.N. Kotsev, J.N.M. Kojukeev, Belovsky color groups and classification of magnetic structures, Preprint/Joint Inst Nucl. Research, P4-7513, Dubna, 1973, 13 pp.
141. A.P. Lungu, Sov. Phys. Crystallogr. **25**(5), 1051–1053 (1980)
142. F.J. Dyson, A. Lenard, J. Math. Phys. **8**(3), 423–434 (1967)
143. A. Lenard, P.J. Dyson, J. Math. Phys. **9**(5), 698–711 (1968)
144. F. Dyson, E. Montroll, M. Katz, M. Fisher, *Stability and Phase Transitions* (Mir, Moscow, 1973), 373 pp. [Russian translation]
145. V. Zbeling, V. Creft, D. Kremp, *Theory of Bound States and the Equilibrium Ionized Plasma in Solids* (Mir, Moscow, 1979), 262 pp. [Russian translation]
146. M. Fisher, D. Ruelle, J. Math. Phys. **7**, 260 (1966)
147. S.A. Moskalenko, D.W. Snoke, *Bose–Einstein Condensation of Excitons and Biexcitons and Coherent Nonlinear Optics with Excitons* (Cambridge University Press, Cambridge, 2000), 415 pp.
148. C.J. Radford, W.E. Hagston, F.J. Bryant, J. Lumin. **5**(1), 47–56 (1972)
149. A.F. Dite, V.I. Revenko, V.B. Timofeev, Fiz. Tverd. Tela **16**(7), 1953–1957 (1974)
150. J. Shah, Phys. Rev. B **9**(2), 562–567 (1974)
151. R.M. Habiger, A. Compaan, Solid State Commun. **18**(11/12), 1531–1534 (1976)
152. C. Benôit à la Guillaume, J.M. Debever, F. Salvan, Phys. Rev. **177**(2), 567–580 (1969)
153. C. Benôit à la Guillaume, J.M. Debever, F. Salvan, Theoretical and experimental investigation of Auger effect on excitons leading to stimulated emission in cadmium sulphide, in *Proc. IX International Conference on the Physics of Semiconductors*, vol. 1 (Nauka, Leningrad, 1969), pp. 615–621
154. D. Magde, H. Mahr, Phys. Rev. B **2**(10), 4098–4103 (1970)
155. G.K. Vlasov, M.S. Brodin, A.V. Kritskii, Phys. Status Solidi (b) **71**(2), 787–795 (1975)

156. L.V. Keldysh, Usp. Fiz. Nauk **100**(3), 514–517 (1970)
157. Ya.E. Pokrovskii, Phys. Status Solidi (a) **11**(2), 385–410 (1972)
158. J.B. Grun, Nuovo Cimento B **39**(2), 579–592 (1977)
159. A. Quattropani, J.J. Forney, Nuovo Cimento B **39**(2), 569–578 (1977)
160. E. Hanamura, H. Hang, Phys. Rep. C **33**(4), 209–284 (1977)
161. E.A. Andryushin, A.P. Silin, Fiz. Nizk. Temp. **3**(11), 1365–1394 (1977)
162. B.M. Vul (ed.), *Excitons in Semiconductors* (Nauka, Moscow, 1971), 143 pp.
163. H. Haken, S. Nikitine (eds.), *Excitons at High Density*. Springer Tracts in Modern Physics, vol. 73 (Springer, Berlin, 1975), 303 pp.
164. M. Grosmann, S.G. Elcomoss, J. Ringeissen (eds.), *Molecular Spectroscopy of Dense Phases* (Elsevier, Amsterdam, 1976), 814 pp.
165. V.A. Moskalenko (ed.), *Intrinsic Semiconductors at High Excitation Levels* (Stiinta, Chisinau, 1978), 197 pp.
166. S.A. Moskalenko, A.I. Bobrysheva, A.V. Lelyakov, M.F. Migleia, P.I. Hadji, M.I. Shmiglyuk, *The Interaction of Excitons in Semiconductors* (Stiinta, Chisinau, 1974), 211 pp.
167. P.I. Hadji, *The Kinetics of the Recombination Radiation of Excitons and Biexcitons in Semiconductors* (Stiinta, Chisinau, 1977), 242 pp.
168. A.I. Bobrysheva, *Biexcitons in Semiconductors* (Stiinta, Chisinau, 1979), 181 pp.
169. S.A. Moskalenko, P. Hadji, A.H. Rotaru, *Solitons and Nutation in the Exciton Region of the Spectrum* (Stiinta, Chisinau, 1980). 194 pp.
170. I.I. Geru, *Low Frequency Resonance of Excitons and Impurity Centers* (Stiinta, Chisinau, 1976), 194 pp.
171. S.A. Moskalenko, *Bose–Einstein Condensation of Excitons and Biexcitons* (RIO AN MSSR, Chisinau, 1970), 167 pp.
172. M.I. Shmiglyuk, P.I. Bardetsky, *Laser Spectroscopy of Excitons in Semiconductors* (Stiinta, Chisinau, 1980), 123 pp.
173. S.A. Moskalenko, Opt. Spectrosc. **5**(2), 147–155 (1958)
174. M.A. Lampert, Phys. Rev. Lett. **1**(7), 450–453 (1958)
175. L.V. Keldysh, in *Proc. IX Intern. Conf. Phys. Semicond.*, Moscow, 23–29 July 1968 (Nauka, Leningrad, 1969), pp. 1384–1392
176. I.M. Rice, Theory of the electron–hole fluid, in *Proc. XII Intern. Conf. Phys. Semicond.*, Stuttgart, 15–19 July 1974 (Elsevier, Amsterdam, 1976), pp. 23–32
177. S.A. Moskalenko, Fiz. Tverd. Tela **4**(1), 276–284 (1962)
178. J.M. Blatt, K.W. Böer, W. Brandt, Phys. Rev. **126**(5), 1691–1692 (1962)
179. S.A. Moskalenko, J. Exp. Theor. Phys. **45**(4), 1159–1163 (1963)
180. R.C. Casella, J. Appl. Phys. **34**(6), 1703–1705 (1963)
181. L.V. Keldysh, A.N. Kozlov, JETP Lett. **5**(7), 238–242 (1967)
182. L.V. Keldysh, A.N. Kozlov, J. Exp. Theor. Phys. **54**(3), 978–993 (1968)
183. A.B. Lelyakov, S.A. Moskalenko, Fiz. Tverd. Tela **11**(4), 3260–3265 (1969)
184. L.V. Keldysh, Coherent states of excitons, in *Problems of Theoretical Physics*, ed. by V.I. Ritus (Nauka, Moscow, 1972), pp. 433–444
185. V.F. Elesin, Y.V. Kopaev, J. Exp. Theor. Phys. **63**(4), 1447–1453 (1972)
186. H. Haken, A. Schenzle, Phys. Lett. A **41**(5), 405–406 (1972)
187. S.A. Moskalenko, M.F. Miglei, M.I. Shmiglyuk et al., J. Exp. Theor. Phys. **64**(5), 1786–1799 (1973)
188. O. Akimoto, E. Hanamura, J. Phys. Soc. Jpn. **33**(6), 1537–1544 (1972)
189. E. Hanamura, in *Optical Properties of Solids. New Developments*, ed. by B.O. Seraphin (North-Holland, Amsterdam, 1976), pp. 81–142
190. J. Adamowski, S. Bednarek, M. Suffczynski, Solid State Commun. **9**(23), 2037–2038 (1971)
191. R.K. Wehner, Solid State Commun. **7**(5), 457–458 (1969)
192. A.I. Bobrysheva, M.F. Miglei, S.A. Moskalenko, M. Shmiglyuk, On the interaction of two excitons in the crystal, in *Compound Semiconductors and Their Physical Properties*, adv. ed. S.I. Radautsan (Stiinta, Chisinau, 1971), pp. 25–34
193. A.I. Bobrysheva, M.F. Miglei, M.I. Schmiglyuk, Phys. Status Solidi (b) **53**(1), 71–84 (1972)

194. P. Čulik, Czechoslov. J. Phys. B **16**(3), 194–206 (1966)
195. E. Hanamura, J. Phys. Soc. Jpn. **29**(1), 50–57 (1970)
196. A.I. Bobrysheva, S.A. Moskalenko, V.I. Vybornov, Phys. Status Solidi (b) **76**(1), K5I–K56 (1976)
197. R. Levy, A. Bivas, J.B. Grun, Phys. Lett. A **36**(3), 159–160 (1971)
198. Y. Kato, T. Goto, T. Fujii, M. Ueta, J. Phys. Soc. Jpn. **36**(1), 169–176 (1974)
199. H. Kuroda, S. Shionoya, J. Phys. Soc. Jpn. **36**(2), 476–484 (1974)
200. T. Ugumori, K. Masuda, S. Namba, J. Phys. Soc. Jpn. **41**, 1991–1995 (1976). doi:10.1143/JPSJ.41.1991
201. R. Baltramiejunas, V. Narkevicius, E. Skaistys et al., Il Nuovo Cimento B **38**(2), 603–609 (1977). doi:10.1007/BF02723537
202. I.A. Carp, S.A. Moskalenko, Fiz. Tekh. Poluprovodn. **8**(2), 285–288 (1974)
203. A.I. Bobrysheva, I.A. Carp, S.A. Moskalenko, Fiz. Tekh. Poluprovodn. **9**(3), 605–607 (1975)
204. S. Nikitine, Properties of biexcitons, in *Excitons at High Density*, ed. by H. Haken, S. Nikitine. Springer Tracts in Modern Physics, vol. 73 (Springer, Berlin, 1975), pp. 18–42
205. S. Nikitine, Physical chemistry of excitons, in *Molecular Spectroscopy of Dense Phases*, ed. by M. Grosmann, S.G. Elkomoss, J. Ringeissen. Proceedings of the 12th European Congress on Molecular Spectroscopy, Strasbourg, France, 1–4 July 1975 (Elsevier, Amsterdam, 1976), pp. 3–15
206. W. Ekardt, M.I. Sheboul, Phys. Status Solidi (b) **74**(2), 523–529 (1976)
207. W. Ekardt, M.I. Sheboul, Phys. Status Solidi (b) **76**(2), K89–K91 (1976)
208. O. Akimoto, Excitonic molecules II. The case of anisotropic effective mass, Techn. Report. ISSP A, N592, 1973, 23 pp.
209. O. Akimoto, J. Phys. Soc. Jpn. **35**(4), 973–979 (1973)
210. W.F. Brinkman, T.M. Rice, B. Bell, Phys. Rev. B **8**(4), 1570–1580 (1973)
211. E.D. Gutljansky, V.E. Khartsiev, Solid State Commun. **12**(11), 1087–1090 (1973)
212. E.D. Gutljansky, V.E. Khartsiev, J. Exp. Theor. Phys. **71**(2), 472–477 (1976)
213. T. Tosatti, Phys. Rev. Lett. **33**(18), 1092–1094 (1974)
214. P.P. Schmidt, J. Phys. C **2**(5), 785–795 (1969)
215. J. Pollmann, H. Büttner, Solid State Commun. **12**(11), 1105–1108 (1973)
216. A.I. Bobrysheva, V.I. Vybornov, Fiz. Tekh. Poluprovodn. **10**(3), 447–451 (1976)
217. A.I. Bobrysheva, V.I. Vybornov, Phys. Status Solidi (b) **69**(1), 267–273 (1975)
218. E. Ostertag, R. Levy, J.B. Grun, Phys. Status Solidi (b) **69**(2), 629–638 (1975)
219. R. Levy, A. Bivas, C. Comte, E. Ostertag, Luminescence of highly excited CuBr, in *Molecular Spectroscopy of Dense Phases*, ed. by M. Grosmann, S.G. Elkomoss, J. Ringeissen. Proc. XII European Congress on Molecular Spectroscopy, Strasbourg, France, 1–4 July 1975 (Elsevier, Amsterdam, 1976), pp. 97–100
220. R. Planel, C. Benôit à la Guillaume, Phys. Rev. B **15**(2), 1192–1201 (1977)
221. J.B. Grun, C. Comte, R. Levy, E. Ostertag, J. Lumin. **12–13**, 581–586 (1976). doi:10.1016/0022-2313(76)90144-7
222. I. Balslev, Phys. Rev. B **20**(2), 648–653 (1979)
223. A. Maruani, D.S. Chemla, Phys. Rev. B **23**(2), 841–860 (1981)
224. A. Mysyrowicz, J.B. Grun, R. Levy, A. Bivas, S. Nikitine, Phys. Lett. A **26**(12), 615–616 (1968)
225. S. Nikitine, A. Mysyrowicz, J.B. Grun, Helv. Phys. Acta **41**(6–7), 1058–1063 (1968)
226. T. Goto, H. Souma, M. Ueta, J. Lumin. **1–2**, 231–240 (1970). doi:10.1016/0022-2313(70)90038-4
227. V.D. Kulakovsky, V.B. Timofeev, V.M. Eidelstein, J. Exp. Theor. Phys. **47**(1), 372–383 (1978)
228. P.L. Gourley, J.P. Wolfe, Phys. Rev. Lett. **40**(8), 526–530 (1978)
229. M.L.W. Thewalt, J.A. Rostworowski, Solid State Commun. **25**(12), 991–993 (1978)
230. J.S.-Y. Wang, C. Kittel, Phys. Lett. A **42**(3), 189–190 (1972)
231. J.R. Haynes, Phys. Rev. Lett. **17**(16), 860–862 (1966)
232. V.M. Asnin, A.A. Rogachev, J. Exp. Theor. Phys. **7**(12), 464–467 (1968)

233. Ya.E. Pokrovsky, K.I. Svistunova, J. Exp. Theor. Phys. **9**(7), 435–438 (1969)
234. V.M. Asnin, A.A. Rogachiov, J. Exp. Theor. Phys. **9**(7), 415–419 (1969)
235. V.S. Vavilov, V.A. Zayats, V.N. Murzin, JETP Lett. **10**(7), 304–309 (1969)
236. V.S. Bagaev, T.I. Galkina, O.V. Gogolin, L.V. Keldysh, J. Exp. Theor. Phys. **10**(7), 309–313 (1969)
237. A.S. Kaminsky, Ya.E. Pokrovsky, N.B. Alkeev, J. Exp. Theor. Phys. **59**(6), 1937–1947 (1970)
238. V.M. Asnin, A.A. Rogachev, N.I. Sablina, J. Exp. Theor. Phys. **11**(3), 162–165 (1970)
239. C. Benôit à la Guillaume, M. Voos, F. Salvan, J.-H. Laurat, A. Bonnot, C. R. Acad. Sci. **272**(3), 236–239 (1971)
240. Ya.E. Pokrovskii, K.I. Svistunova, J. Exp. Theor. Phys. **13**(6), 297–301 (1971)
241. V.S. Bagaev, L.V. Keldysh, N.N. Sibel'din, V.A. Tsvetkov, J. Exp. Theor. Phys. **70**(2), 702–716 (1976)
242. G.O. Muller, H.H. Weber, V.G. Lusenko, V.I. Revenko, V.B. Timofeev, Solid State Commun. **21**(2), 217–219 (1977)
243. D. Hulin, M. Combescot, N. Botemps, A. Musurowicz, Phys. Lett. A **61**(5), 349–352 (1977)
244. D. Bimberg, M.S. Skolnick, W.J. Choyke, Phys. Rev. Lett. **40**(1), 56–60 (1978)
245. W.F. Brinkman, T.M. Rice, P.W. Anderson, S.T. Chui, Phys. Rev. Lett. **28**(15), 961–964 (1972)
246. M. Combescot, P. Nozieres, J. Phys. C **5**(17), 2369–2391 (1972)
247. W.F. Brinkman, T.M. Rice, Phys. Rev. B **7**(4), 1508–1523 (1973)
248. P. Vashishta, P. Bhattacharyya, K.S. Singwi, Phys. Rev. Lett. **30**(25), 1248–1251 (1973)
249. P. Vashishta, S.G. Das, K.S. Singwi, Phys. Rev. Lett. **33**(15), 911–914 (1974)
250. T.M. Rice, J.C. Hensel, T.G. Phillips, G.A. Thomas, *Electron–Hole Liquid in Semiconductors* (Academic Press, New York, 1977), 349 pp.
251. I.H. Akopyan, E.F. Gross, B.S. Razbirin, J. Exp. Theor. Phys. **12**(7), 366–371 (1970)
252. H. Kuroda, S. Shionoya, Solid State Commun. **13**(8), 1195–1196 (1973)
253. H. Kuroda, S. Shionoya, J. Phys. Soc. Jpn. **36**(2), 476–484 (1974)
254. H. Kuroda, S. Shionoya, H. Saito, E. Hanamura, J. Phys. Soc. Jpn. **35**(2), 534–542 (1973)
255. N. Nagasawa, N. Nakata, Y. Doi, M. Yeta, J. Phys. Soc. Jpn. **38**(2), 593 (1975)
256. N. Nagasawa, N. Nakata, Y. Doi, M. Yeta, J. Phys. Soc. Jpn. **39**(4), 987–993 (1975)
257. L.L. Chase, N. Peyghambarian, G. Grymberg, A. Musyrowicz, Phys. Rev. Lett. **42**(18), 1231–1234 (1979)
258. N. Nagasawa, N. Nakata, Y. Doi, M. Yeta, J. Phys. Soc. Jpn. **38**(3), 903 (1975)
259. W.D. Johnston, K.L. Shaklee, Solid State Commun. **15**(1), 73–75 (1974)
260. M.G. Matsko, J. Exp. Theor. Phys. **21**(5), 281–285 (1975)
261. M.S. Brodin, M.G. Matsko, Solid State Commun. **25**, 789–792 (1978)
262. M.S. Brodin, M.G. Matsko, Izv. Akad. Nauk SSSR, Ser. Fiz. **45**(8), 1567–1579 (1981)
263. L.L. Chase, N. Peyghambarian, G. Grynberg, A. Musyrowicz, Opt. Commun. **28**(2), 189–192 (1979)
264. S. Nikitine, Opt. Commun. **35**(3), 377–383 (1980)
265. J.R. Haynes, Phys. Rev. Lett. **4**(7), 361–363 (1960)
266. R.A. Faulkner, Phys. Rev. **175**(3), 991–1009 (1968)
267. D.G. Thomas, M. Gershenzon, Phys. Rev. **131**(6), 2397–2404 (1963)
268. Y. Yafet, D.G. Thomas, Phys. Rev. **131**(6), 2405–2408 (1963)
269. J.C. Phillips, Phys. Rev. Lett. **22**(7), 285–287 (1969)
270. J.J. Hopfield, P.J. Dean, D.G. Thomas, Phys. Rev. **158**(3), 748–755 (1967)
271. R.A. Faulkner, J.L. Merz, P.J. Dean, Solid State Commun. **7**(11), 831–835 (1969)
272. J.L. Merz, R.A. Faulkner, P.J. Dean, Phys. Rev. **188**(3), 1228–1239 (1969)
273. A.S. Kaminsky, Ya.E. Pokrovsky, J. Exp. Theor. Phys. **11**(8), 381–384 (1970)
274. V.D. Kulakovsky, G.E. Pikus, V.B. Timofeev, Usp. Fiz. Nauk **135**(2), 237–284 (1981)
275. O.V. Konstantinov, V.I. Perel', J. Exp. Theor. Phys. **39**(1(7)), 197–208 (1960)
276. L.K. Aminov, J. Exp. Theor. Phys. **48**(5), 1398–1406 (1965)

277. I.I. Geru, Stimulated intraserial and interserial exciton transitions, in *Physics of Semiconductors and Dielectrics*, adv. ed. V.P. Mushinsky (Stiinta, Chisinau, 1982), pp. 34–42
278. I.I. Geru, Coherent magnons in magnetic semiconductors, in *Magnetic Semiconductors of the Type CdCr₂Se₄*, adv. ed. S.I. Radautsan (Stiinta, Chisinau, 1978), pp. 109–117
279. I.I. Geru, S.A. Moskalenko, M.I. Shmiglyuk, Paraelectric resonance on excitons in semiconductors, in *Paramagnetic Resonance 1944–1969, Part I "Electron Paramagnetic Resonance"*, adv. ed. A.I. Rivkind (Kazan Univ. Publish. House, Kazan, 1971), pp. 126–129
280. I.I. Geru, S.A. Moskalenko, M.I. Shmiglyuk, in *All-Union Jubilar Conf. on Paramagn. Resonance*, Kazan, 24–29 June 1969, Abstracts Book (Kazan, 1969), p. 54
281. I.I. Geru, S.A. Moskalenko, M.I. Shmiglyuk, Fiz. Tekh. Poluprovodn. **6**, 1532–1537 (1972)
282. I.I. Geru, S.A. Moskalenko, M.I. Shmiglyuk, Paraelectric resonance on excitons in semiconductors, in *2nd All-Union Conf. on Solid State Theory*, Moscow, 14–21 December 1969. Abstracts Book (Nauka, Moscow, 1969), pp. 65–66
283. H. Sternlicht, H.M. McConnel, J. Chem. Phys. **35**(5), 1793–1800 (1961)
284. N.I. Botoshan, S.A. Moskalenko, in *All-Union Conf. on Low Temperature Physics Dedicated to the Research of Condensed Systems at Low Temperatures by Resonance Methods*, Kazan, 25–29 June 1965. Abstracts Book (Kazan, 1965), pp. 18–19
285. L.K. Aminov, The diagram technique applied to problems of paramagnetic relaxation theory, in *Paramagnetic Resonance*, ed. by S.A. Al'tshuler (Kazan Univ. Publish. House, Kazan, 1968), pp. 3–26
286. D.N. Zubarev, *Nonequilibrium Statistical Thermodynamics* (Nauka, Moscow, 1971), 415 pp.
287. A.I. Ahiezer, S.V. Peletminskii, *Methods of Statistical Physics* (Nauka, Moscow, 1977), 367 pp.
288. N. March, Y. Yang, S. Sampanthar, *Many Body Problem in Quantum Mechanics* (Mir, Moscow, 1969), 496 pp. [Russian translation]
289. A.A. Andronov, V.A. Kozlov, L.S. Mazov, V.N. Shastin, J. Exp. Theor. Phys. **30**(9), 585–589 (1979)
290. I.I. Vosilyus, I.B. Levinson, J. Exp. Theor. Phys. **52**(4), 1013–1024 (1967)
291. V.A. Kozlov, L.S. Mazov, I.M. Nefedov, Population inversion of the subband of light holes and far-infrared radiation of hot carriers, in *Proceedings of the All-Union Conf. on the Phys. of Semicond*, vol. 1, Baku, 12–14 October (1982), pp. 195–196
292. Yu.K. Pozhela, E.V. Starikov, P.N. Shiktorov, Optimal conditions for the generation of IR radiation in p-Ge in crossed electric and magnetic fields, in *Proceedings of the All-Union Conf. on the Phys. of Semicond*, vol. 2, Baku, 12–14 October (1982), pp. 261–262
293. E.M. Gershenzon, A.P. Mel'nikov, R.I. Rabinovich, The capture of electrons into small neutral impurities in semiconductors and the possibility of creating an inverse distribution function, in *Proceedings of the All-Union Conf. on the Phys. of Semicond*, vol. 1, Baku, 12–14 October (1982), pp. 136–137
294. S.K. Avetisian, E.S. Kazarian, A.O. Melikian, H.R. Minasian, Resonance parametric generation of submillimeter radiation in semiconductors, in *Abstracts of Intern. Conf. and School "Lasers and Applications"*, Bucharest, 30 August–11 September 1982. Abstracts Book, vol. 2 (CIPAP, Bucharest, 1982), pp. 35–36
295. Yu.I. Ravich, B.A. Efimova, I.A. Smirnov, *Methods of Study of Semiconductors with Applications to Plumbum Chalcogenides PbTe, PbSe and PbS* (Nauka, Moscow, 1968), 383 pp.
296. H. Haken, H. Sauermann, Z. Phys. **176**(1), 47–62 (1963)
297. H. Haken, H. Sauermann, Z. Phys. **176**(3), 261–275 (1963)
298. E.G. Pestov, R.M. Lapshin, *Quantum Electronics* (Voenizdat MO SSSR, Moscow, 1972), 334 pp.
299. H.N. Spector, Solid State Commun. **6**(11), 811–813 (1968)
300. V.L. Gurevich, Fiz. Tverd. Tela **4**(4), 909–917 (1962)
301. M.D.C. Filho, L.C.M. Miranda, S.M. Rezende, Phys. Status Solidi (b) **57**(1), 85–91 (1973)
302. T. Kasuya, New aspect on the electronic properties of magnetic semiconductors, in *Proc. XI Intern. Conf. Phys. Semicond*, Warsaw (1972), pp. 141–154
303. E.L. Nagaev, *Physics of Magnetic Semiconductors* (Nauka, Moscow, 1979), 430 pp.

304. E.L. Nagaev, Usp. Fiz. Nauk **117**(3), 437–492 (1975)
305. L.L. Golik, Z.E. Kun'kova, T.G. Aminov, About nature of circular dichroism and Faraday effect in the absorption edge $CdCr_2Se_4$, in *All-Union Conference "Ternary Semiconductors and Their Applications"*, Chisinau, 8–10 October 1982. Abstracts Book (Stiinta, Chisinau, 1982), pp. 58–61
306. M. Krawczyk, H. Szymczak, W. Zbieranowski, Acta Phys. Pol. A **44**(3), 455–464 (1973)
307. A.S. Boruhovich, M.S. Marunya, N.I. Lobachevskaya, V.G. Bamburov, P.V. Gel'd, Fiz. Tverd. Tela **16**(7), 2084–2086 (1974)
308. H.L. Pinch, S.B. Berger, J. Phys. Chem. Solids **29**(12), 2091–2099 (1968)
309. W.E. Bron, R.W. Dreyfus, Phys. Rev. Lett. **16**(5), 165–168 (1966)
310. G. Hocherl, H.C. Wolf, Phys. Lett. A **27**(3), 133–134 (1968)
311. I. Shepherd, G. Feher, Phys. Rev. Lett. **15**(5), 194–198 (1965)
312. S. Nikitine, Excitons, in *Optical Properties of Solids*, ed. by S. Nudelman, S.S. Mitra (Plenum Press, New York, 1969), pp. 197–237, Chap. 9
313. A. Daunois, J.L. Diess, S. Nikitine, C. R. Acad. Sci. Paris B **268**(14), 977–980 (1969)
314. E.I. Rashba, Fiz. Tverd. Tela **1**(3), 407–421 (1959)
315. D.G. Thomas, J.J. Hopfield, Phys. Rev. **124**, 657–665 (1961). doi:10.1103/PhysRev.124.657
316. M.I. Shmiglyuk, Fiz. Tverd. Tela **14**(3), 816–822 (1972)
317. J.L. Deiss, A. Daunois, S. Nikitine, Solid State Commun. **8**(7), 521–525 (1970)
318. U.M. Grassano, G. Margaritondo, R. Rosei, Phys. Rev. B **2**(8), 3319–3322 (1970)
319. S.A. Moskalenko, A.I. Bobrysheva, Fiz. Tverd. Tela **4**(8), 1994–2004 (1962)
320. E.F. Gross, B.P. Zaharchenya, Proc. Acad. Sci. USSR **111**(3), 564–567 (1956)
321. E.F. Gross, V.T. Agekyan, JETP Lett. **8**(11), 605–610 (1968)
322. F.I. Kreingol'd, K.F. Lider, L.E. Solov'ev, JETP Lett. **23**(12), 679–681 (1976)
323. F.I. Kreyngold, K.F. Lider, V.F. Sapega, Fiz. Tverd. Tela **19**(10), 3158–3160 (1977)
324. F.I. Kreyngold, K.F. Lider, in *Exciton–Phonon Interaction in Crystals. XIV-th All-Union Seminar "Excitons in Crystals"*, Lviv, 1–3 October, 1979. Abstracts Book (Lviv, 1980), pp. 47–48
325. A.I. Bobrysheva, I.I. Geru, S.A. Moskalenko, Phys. Status Solidi (b) **113**(2), 439–445 (1982)
326. A.I. Bobrysheva, I.I. Geru, S.A. Moskalenko, The isotopic shift of exciton series in Cu_2O crystals, in *XXII-th All-Union Conference on Low Temperature Physics*, Chisinau, 20–23 October 1982. Abstracts Book, Part 1 (Stiinta, Chisinau, 1982), pp. 168–169
327. A.I. Bobrysheva, I.I. Geru, S.A. Moskalenko, in *Excitons in Semiconductors-82, All-Union Meeting with the Scientists of the Socialist Countries*, Leningrad, 22–25 November 1982. Abstracts Book (Leningrad, 1982), p. 64
328. K.G. Waters, F.H. Pollak, R.H. Bruce, H.Z. Cummins, Phys. Rev. B **21**(4), 1665–1675 (1980)
329. Ch. Uihlein, D. Fröhlich, R. Kenklies, Phys. Rev. B **23**(6), 2731–2740 (1981)
330. H.R. Trebin, H.Z. Cummins, J.L. Birman, Phys. Rev. B **23**(2), 597–606 (1981)
331. E.F. Gross, Usp. Fiz. Nauk **76**(3), 433–466 (1962)
332. V.A. Margulis, Vl.A. Margulis, A.D. Margulis, Fiz. Tekh. Poluprovodn. **19**(3), 787–790 (1977)
333. L.E. Gurevich, B.I. Shklovsky, J. Exp. Theor. Phys. **53**(5(11)), 1726–1734 (1967)
334. W.P. Mason, T.B. Bateman, Phys. Rev. A **134**(5), 1387–1396 (1964)
335. M. Pomerantz, Phys. Rev. B **1**(10), 4029–4036 (1970)
336. A.K. Srivastava, G.S. Verma, Phys. Rev. B **1**(2), 776–778 (1970)
337. G.L. Merkulov, L.A. Yakovlev, Acoust. J. **6**(2), 244–251 (1960)
338. I.I. Geru, P.I. Bardetsky, Fiz. Tekh. Poluprovod. **9**, 1995–1999 (1975)
339. I.I. Geru, Instability in the system of excitons and hypersound phonons, in *Semiconductors at High Levels of Excitation*, ed. by V.A. Moskalenko (Stiinta, Chisinau, 1978), pp. 136–145
340. I.I. Geru, P.I. Bardetsky, Izv. Akad. Nauk SSSR **37**(10), 2159–2165 (1973)
341. I.I. Geru, P.I. Bardetsky, Elastic scattering of hypersound by excitonic molecules in semiconductors A_2B_6, in *Abstracts of the International Meeting of Photoelectric and Optical Phenomena in Solid State*, Varna, 24–25 May (1974), p. 27
342. C. Herring, Phys. Rev. **95**, 954 (1954)

343. R. Nava, R. Azrt, I. Ciccarello, K. Dransfeld, Phys. Rev. A **134**(3), 581–589 (1964)
344. L.E. Gurevich, B.I. Shklovsky, Fiz. Tverd. Tela **9**(2), 526–534 (1967)
345. S. Simons, Proc. Phys. Soc. **82**(3), 401–405 (1963)
346. J. Klerk, Phys. Rev. A **139**(5), 1635–1639 (1965)
347. E.M. Ganapol'sky, A.N. Chernets, J. Exp. Theor. Phys. **51**(2(8)), 383–393 (1966)
348. D.H.R. Price, J. Phys. Ser. G (Solid State Phys.) **4**(9), 966–970 (1971)
349. G. Weinreich, Phys. Rev. **104**(2), 321–327 (1956)
350. G. Weinreich, T.M. Sanders, H.G. White, Phys. Rev. **114**(1), 33–44 (1959)
351. E.I. Blount, Phys. Rev. **114**(2), 418–436 (1959)
352. A.R. Hutson, D.L. White, J. Appl. Phys. **33**(1), 40–47 (1962)
353. D.L. White, J. Appl. Phys. **33**(8), 2547–2554 (1962)
354. V.L. Gurevich, V.D. Kagan, Fiz. Tverd. Tela **4**(9), 2441–2446 (1962)
355. I. Uchida, T. Ishiguro, Y. Sasaki, T. Suzuki, J. Phys. Soc. Jpn. **19**(5), 674–680 (1964)
356. P.D. Southgate, H.N. Spector, J. Appl. Phys. **36**(12), 3728–3734 (1965)
357. A.E. Glauberman, M.A. Ruvinsky, Fiz. Tverd. Tela **8**(11), 3335–3338 (1966)
358. A.I. Ansel'm, Yu.A. Firsov, J. Exp. Theor. Phys. **30**(4), 719–723 (1956)
359. V.L. Gurevich, Fiz. Tekh. Poluprovodn. **2**(11), 1557–1592 (1968)
360. J.A. Deverin, Helv. Phys. Acta **42**(3), 397–419 (1969)
361. A.A. Lipnick, Fiz. Tekh. Poluprovodn. **6**(5), 878–885 (1972)
362. V.D. Kagan, Fiz. Tverd. Tela **16**(7), 2022–2026 (1974)
363. D.C. Reynolds, C.W. Litton, T.C. Collins, Phys. Status Solidi (b) **12**(3), 645–648 (1965)
364. R.G. Wheeler, J.O. Dimmock, Phys. Rev. **125**(6), 1805–1815 (1962)
365. E.P. Pokotilov, Polaron and dissipative-relaxation processes due to lattice imperfection, Diss. Doctor habilit. phys.-mat. sciences, Chisinau, 1971, 315 pp.
366. J.H. McFee, Propagation and amplification of acoustic waves in piezoelectric semiconductors, in *Physical Acoustics, Vol. IV, Part A. Applications of Physical Acoustics in Quantum Physics and Solid State Physics*, ed. by W.P. Mason (Moscow, Mir, 1969), 436 pp. [Russian translation]
367. J.E. Rowe, M. Cardona, F.H. Pollak, Solid State Commun. **6**(4), 239–242 (1968)
368. A.M. Fedorchenko, N.Ya. Kotsarenko, *The Absolute and Convective Instability in the Plasma and Solids* (Nauka, Moscow, 1981), 176 pp.
369. P.A. Stworock, Phys. Rev. **112**(5), 1488–1503 (1958)
370. K.F. Renk, J. Deisenhofer, Phys. Rev. Lett. **26**(13), 764–766 (1971)
371. R.S. Grandall, Solid State Commun. **7**(16), 1109–1111 (1969)
372. E.H. Jacobsen, J. Phys. Suppl. **33**(11–12), 25–31 (1973)
373. U.H. Kopvillem, V.R. Nagibarov, Fiz. Tverd. Tela **10**(3), 750–753 (1968)
374. A.A. Lipnik, M.A. Ruvinsky, Phys. Status Solidi (b) **28**(1), 75–81 (1968)
375. S.L. Korolyuk, Fiz. Tverd. Tela **5**(1), 352–353 (1963)
376. P.D. Altukhov, V.I. Revenko, V.B. Timofeev, Fiz. Tverd. Tela **15**(4), 974–979 (1973)
377. E. Gross, S. Permogorov, V. Travnikov, A. Selkin, J. Phys. Chem. Solids **31**(12), 2595–2606 (1970)
378. I.I. Geru, A.H. Rotaru, Fiz. Tverd. Tela **18**(9), 2766–2771 (1976)
379. A.C. Alekseev, T.I. Galkina, V.N. Maslennikov, R.G. Khakimov, E.P. Schebnev, JETP Lett. **21**(10), 578–582 (1975)
380. L.V. Keldysh, S.G. Tikhodeev, JETP Lett. **21**(10), 582–585 (1975)
381. I.I. Geru, Ukr. Fiz. Zh. **7**, 726–733 (1965)
382. I.I. Geru, Symmetry in hyperfine interaction, in *Solid State Radiospectroscopy*, ed. by S.A. Al'tshuler et al. (Atomizdat, Moscow, 1967), pp. 14–16
383. G.E. Pikus, J. Exp. Theor. Phys. **41**(4(10)), 1258–1273 (1961)
384. G.E. Pikus, J. Exp. Theor. Phys. **41**(5(11)), 1507–1521 (1961)
385. I.I. Geru, Infringement of the equivalency of inversion connected nuclei at electron–nuclear–magneto–acoustic double resonance, in *Magnetic Resonance and Related Phenomena*, ed. by I. Ursu. Proceedings of the XVIth Congress AMPERE, Bucharest, 1–5 September

1970 (Publish. house of the Academy of Socialist Republic of Romania, Bucharest, 1971), pp. 286–287

386. I.I. Geru, Hole–nuclear double resonance in the localized biexciton, in *Magnetic Resonance and Related Phenomena*, ed. by V. Hovi. Proceedings of the XVIIth Congress AMPERE, Turku, August 1972 (North-Holland, Amsterdam, 1973), pp. 524–526

387. I.I. Geru, P.I. Bardetsky, Interaction between bound excitons in semiconductors, in *New Semiconducting Compounds and Their Properties*, adv. ed. S.I. Radautsan (Stiinta, Chisinau, 1975), pp. 198–211

388. I.I. Geru, A.H. Rotaru, Izv. Akad. Nauk SSSR **40**(9), 1893–1896 (1976)

389. M.F. Deigen, I.I. Geru, Fiz. Tverd. Tela **9**(6), 1679–1689 (1967)

390. M.I. Vladimir, I.I. Geru, To Stevens' operator-equivalent method, in *Quantum Theory of Multiparticle Systems*, ed. by V.A. Moskalenko (Red. Izd. Otdel AN MSSR, 1970), pp. 51–61

391. I.I. Geru, About time-reversal operator for high spin systems, in *Quantum Theory of Multiparticle Systems*, ed. by V.A. Moskalenko (Red. Izd. Otdel AN MSSR, 1970), pp. 66–80

392. I.I. Geru, Proc. Acad. Sci. USSR **286**, 1392–1394 (1983)

393. I.I. Geru, Influence of the polarization of electron shells on dynamic of the hyperfine interaction in the local electron centers, in *Complex Semiconductors and Their Physical Properties*, ed.-in-chief S.I. Radautsan (Stiinta, Chisinau, 1971), pp. 92–99

394. K. Steven, Matrix elements and equivalent operators associated with the magnetic properties of rare-earth ions, in *Symmetry in the Solid*, ed. by R. Knox, A. Gold (Nauka, Moscow, 1970), pp. 322–332 [Russian translation]

395. G.F. Koster, H. Statz, Phys. Rev. **113**(2), 445–454 (1959)

396. N.I. Deryugina, A.B. Roitsin, Ukr. Fiz. Zh. **11**(6), 594–603 (1966)

397. A.M. Leushin, *Tables of Functions that Are Transformed According to the Irreducible Representations of Crystallographic Point Groups* (Nauka, Moscow, 1968), 141 pp.

398. E. Wigner, *Group Theory and Its Application to Quantum-Mechanics Theory of Atomic Spectra* (Springer, Berlin, 1961), 443 pp. [Russian translation]

399. I.S. Zheludev, *Symmetry and Its Applications* (Atomizdat, Moscow, 1976), 286 pp.

400. Y.I. Sirotin, M.P. Shaskol'skaya, *The Basis of Crystallophysics* (Nauka, Moscow, 1979), 639 pp.

401. J.S. Hyde, A.H. Maki, J. Chem. Phys. **40**(10), 3117–3118 (1964)

402. M.F. Deigen, Yu.S. Gromovoi, I.I. Geru, G.A. Popovich, Fiz. Tverd. Tela **17**(10), 3075–3076 (1975)

403. J.M. Baker, F.I.B. Williams, Proc. R. Soc. Lond. A **267**, 283–294 (1962)

404. V.A. Golenischev-Kutuzov, U.H. Kopvillem, N.A. Shamukov, JETP Lett. **10**(6), 240–244 (1969)

405. G.F. Koster, J.O. Dimmock, R.G. Wheeler, H. Statz, *Properties of the Thirty-Two Point Groups* (MIT Press, Cambridge, 1963), 104 pp.

406. J. Schwinger, On angular momentum, in *Quantum Theory of the Angular Momentum*, ed. by L.C. Biedenharn, H. Van Dam (Academic Press, New York, 1965), pp. 229–279

407. E.B. Aleksandrov, A.P. Sokolov, Opt. Spectrosc. **31**(3), 329–331 (1971)

408. J.P. Barrat, C. Cohen-Tannodji, J. Phys. Radium **22**(6), 329–336 (1961). doi:10.1051/jphysrad:01961002206032900

409. J.P. Barrat, C. Cohen-Tannodji, J. Phys. Radium **22**(7), 443–450 (1961)

410. C. Cohen-Tannodji, Ann. Phys. **7**(9–10), 469–504 (1962)

411. C. Landre, C. Cohen-Tannodji, J. Dupont-Roc, S. Haroche, J. Phys. **31**(11–12), 971–983 (1970)

412. J. Dupont-Roc, J. Phys. **32**(2–3), 135–144 (1971)

413. S. Haroche, Ann. Phys. **6**(4–5), 189–326 (1971)

414. K. Morigaki, P. Dawson, B.C. Cavenett, Solid State Commun. **28**(9), 829–834 (1978)

415. W. Hayes, M. Yamaga, D.J. Robbing, B. Cockayne, J. Phys. C **13**(34), L1011–L1015 (1980)

416. B.C. Cavenett, J. Phys. Soc. Jpn. Suppl. A **49**, 611–618 (1980)

417. I.I. Geru, A.H. Rotaru, Fiz. Tverd. Tela **18**, 2993–2996 (1976)

418. I.I. Geru, A.H. Rotaru, Spin–lattice relaxations of nuclei via Wannier–Mott excitons, in *Crystalline and Vitreous Semiconductors*, ed. by S.I. Radautsan et al. (Stiinta, Chisinau, 1977), pp. 46–53

419. A.H. Rotaru, I.I. Geru, Spin–lattice relaxation of nuclei via orthobiexcitons, in *Crystalline and Vitreous Semiconductors*, ed. by S.I. Radautsan et al. (Stiinta, Chisinau, 1977), pp. 53–61

420. I.I. Geru, A.H. Rotaru, Relaxation of nuclear spin and Knight shift due to the Wannier–Mott excitons, in *Abstracts of the XVIIIth Congress AMPERE "Magnetic Resonance and Related Phenomena"*, Nottingham, England, 9–14 September (1974), p. 152

421. V.A. Piragas, Fiz. Tverd. Tela **13**(7), 2013–2017 (1971)

422. P.I. Hadji, Fiz. Tverd. Tela **15**(6), 1718–1726 (1973)

423. A. Bivas, R. Levy, S. Nikitine, J.B. Grun, J. Phys. **31**(2–3), 227–234 (1970)

424. A.P. Prudnikov, Yu.A. Brychkov, O.I. Marichev, *Integrals and Series* (Nauka, Moscow, 1981), 798 pp.

425. T. Ugumori, K. Masuda, S. Namba, J. Phys. Soc. Jpn. **41**(6), 1991–1995 (1976)

426. T. Ugumori, K. Masuda, S. Kamba, J. Phys. Soc. Jpn. **43**(1), 151–156 (1977)

427. M.S. Brodin, I.I. Geru, V.P. Karperko, M.G. Matsko, Ukr. Fiz. Zh. **26**(5), 867–869 (1981)

428. A. Balzarotti, M. Grandolfo, F. Somma, P. Vecchia, Phys. Status Solidi (b) **44**(2), 713–716 (1971)

429. M.S. Brodin, V.M. Bandura, I.I. Geru, M.G. Matzko, Fiz. Tverd. Tela **24**(10), 3133–3136 (1982)

430. I.I. Geru, J. Phys., Conf. Ser. **324**(1), 012023 (2011). doi:10.1088/1742-6596/324/1/012023

431. L.L. Buishvili, M.G. Menadbe, J. Exp. Theor. Phys. **77**(6), 2435–2442 (1979)

432. B.P. Provotorov, E.B. Fel'dman, J. Exp. Theor. Phys. **79**(6), 2206–2217 (1980)

433. I.I. Geru, Yu.G. Semenov, Fiz. Tverd. Tela **23**(5), 1506–1508 (1981)

434. I.I. Geru, Yu.G. Semenov, Fiz. Tekh. Poluprovodn. **15**(9), 1711–1716 (1981)

435. A.A. Barsuk, I.I. Geru, A.H. Rotaru, V.I. Vybornov, Indirect exchange of the paramagnetic centers via excitons in semiconductors, in *The IXth Meeting on the Theory of Semiconductors*, Tbilisi, 24–26 October (1978), pp. 41–42

436. D. Pines, D. Bardin, C. Slichter, Nuclear polarization and spin relaxation processes of impurity states in silicon, in *Electron Spin Resonance in Semiconductors*, ed. by N.A. Penin (IL, Moscow, 1962), pp. 125–156 [Russian translation]

437. M.F. Deigen, V.Ya. Bratus, B.E. Vugmeister, I.M. Zaritsky, A.A. Konchits, A.A. Zolotuhin, L.S. Milevsky, J. Exp. Theor. Phys. **69**(6), 2110–2117 (1975)

438. V.S. Vihnin, M.F. Deigen, Yu.G. Semenov, B.D. Shanina, Fiz. Tverd. Tela **18**(8), 2222–2228 (1976)

439. Yu.G. Semenov, Fiz. Tverd. Tela **22**(10), 3190–3192 (1980)

440. A.V. Komarov, S.M. Ryabchenko, Yu.G. Semenov, B.D. Shanina, N.I. Vitrihovsky, J. Exp. Theor. Phys. **79**(4(10)), 1554–1560 (1980)

441. Yu.G. Semenov, B.D. Shanina, Phys. Status Solidi (b) **104**(2), 631–639 (1981)

442. G.L. Bir, A.G. Aronov, G.E. Pikus, J. Exp. Theor. Phys. **69**(4), 1382–1397 (1975)

443. S.V. Vonsovsky, *Magnetism* (Nauka, Moscow, 1971), 1032 pp.

444. C. Kittel, *Quantum Theory of Solids* (Nauka, Moscow, 1967), 491 pp.

445. R.M. White, *Quantum Theory of Magnetism* (Mir, Moscow, 1972), 306 pp. [Russian translation]

446. R. Watson, Distribution of the charge and spin density of conduction electrons in metals with impurities, in *Hyperfine Interactions in Solids*, ed. by E.A. Titov (Mir, Moscow, 1970), pp. 237–287 [Russian translation]

447. B.V. Karpenko, A.A. Berdyshev, Fiz. Tverd. Tela **5**(10), 3026–3028 (1963)

448. B.I. Kochelaev, Fiz. Tverd. Tela **7**(9), 2859–2860 (1965)

449. L.K. Aminov, B.I. Kochelaev, J. Exp. Theor. Phys. **42**(5), 1303–1306 (1962)

450. R. Orbach, M. Tachiki, Phys. Rev. **158**(2), 524–529 (1967)

451. M.F. Deigen, N.I. Kashirina, L.A. Suslin, J. Exp. Theor. Phys. **75**(1(7)), 149–152 (1978)

452. N.I. Kashirina, L.A. Suslin, Phys. Scr. **20**(6), 669–672 (1979)

453. N.I. Kashirina, L.A. Suslin, Ukr. Fiz. Zh. **26**(1), 6–13 (1981)
454. N.I. Kashirina, Theor. Exp. Chem. **18**(5), 549–558 (1982)
455. N.I. Kashirina, The exchange interaction of paramagnetic centers due to elementary excitations of the crystals, Diss. of candidate in phys-math. sci, Kiev, 1983, 147 pp. (in Russian)
456. K. Cho, W. Dreybrodt, P. Hiesinger, S. Suga, F. Willmann, Magneto–optics of free and bound excitons in CdTe, in *Proc. of XIIth Int. Conf. Phys. Semicond*, Stuttgart, 15–19 July (1974), pp. 945–952
457. G. Bastard, C. Rigaux, A.M. Mycielski, Phys. Status Solidi (b) **79**(2), 585–593 (1977)
458. J.A. Gaj, R.R. Galazka, M. Nawrocki, Solid State Commun. **25**(3), 193–195 (1978)
459. J.A. Gaj, J. Ginter, R.R. Galazka, Phys. Status Solidi (b) **89**(2), 655–662 (1978)
460. J.A. Gaj, R. Planel, G. Fishman, Solid State Commun. **29**(5), 435–438 (1979)
461. A.V. Komarov, S.M. Ryabchenko, N.I. Vitrihovsky, JETP Lett. **27**(8), 441–445 (1978)
462. A.V. Komarov, S.M. Ryabchenko, N.I. Vitrihovsky, JETP Lett. **28**(3), 119–123 (1978)
463. A.V. Komarov, S.M. Ryabchenko, O.V. Terletsky, Phys. Status Solidi (b) **102**(2), 603–609 (1980)
464. S.I. Gubarev, J. Exp. Theor. Phys. **80**(3), 1174–1185 (1981)
465. A.V. Komarov, S.M. Ryabchenko, Yu.G. Semenov, B.D. Shanina, N.I. Vitrihovsky, J. Exp. Theor. Phys. **79**(4), 1554–1560 (1980)
466. A. Twardowski, J. Ginter, Phys. Status Solidi (b) **110**(1), 47–55 (1982)
467. A. Twardowski, J. Ginter, Phys. Status Solidi (b) **114**(2), 331–336 (1982)
468. J. Kossut, J.K. Furdyna (eds.), *Diluted Magnetic Semiconductors*. Semiconductors and Semimetals, vol. 25 (Academic Press, San Diego, 1987)
469. J. Kossut, Acta Phys. Pol. A Suppl. **100**, 111–138 (2001)
470. J. Kossut, J.A. Gaj, *Introduction to Physics of Diluted Magnetic Semiconductors* (Springer, Heidelberg, 2009)
471. Ya.B. Zel'dovich, J. Exp. Theor. Phys. **51**(5(12)), 1492–1495 (1966)
472. V.I. Ritus, J. Exp. Theor. Phys. **51**(5(11)), 1544–1549 (1966)
473. V.M. Galitsky, S.P. Goreslavsky, V.F. Elesin, J. Exp. Theor. Phys. **57**(1(7)), 207–217 (1969)
474. L.D. Landau, E.M. Lifshitz, *Quantum Mechanics. Non-relativistic Theory*, 2nd edn. (Fizmatgiz, Moscow, 1963), 702 pp. (in Russian)
475. A. Allen, J. Eberly, *Optical Resonance and Two-Level Atoms* (Mir, Moscow, 1978), 222 pp. [Russian translation]
476. E.Yu. Perlin, V.A. Kovarsky, Fiz. Tverd. Tela **12**(11), 3105–3112 (1970)
477. I.I. Geru, Time-reversal for systems with quasi-energy spectrum, in *Physics of Semiconductors and Dielectrics*, adv. ed. V.P. Mushinsky (Stiinta, Chisinau, 1982), pp. 54–59
478. I.I. Geru, in *VIII-th Conference on Theory of Semiconductors*, Kiev, October 1975. Abstracts Book (Naukova Dumka, Kiev, 1975), p. 68
479. I.I. Geru, Electron–nuclear double resonance under conditions of strong saturation, in *V-th All-Union Conference "Physical and Mathematical Methods in Coordination Chemistry"*, Chisinau, 6–8 June 1974. Abstracts Book (Stiinta, Chisinau, 1974), pp. 76–77
480. I.I. Geru, in *IV-th International Congress in Quantum Chemistry*, Uppsala, Sweden, 13–20 June 1982. Abstracts (Univ. of Uppsala, Quantum Chem. Inst., 1982), p. J-3
481. H. Sambe, Phys. Rev. A **7**(6), 2203–2213 (1973)
482. V.V. Mikhailov, Teor. Mat. Fiz. **15**(3), 367–374 (1973)
483. G.E. Moore, Electronics **38**, 114–117 (1965)
484. J.D. Meindl, Q. Chen, J.A. Davis, Science **293**(5537), 2044–2049 (2001). doi:10.1126/science.293.5537.2044
485. C.H. Bennett, R. Landauer, Sci. Am. **253**, 48–56 (1985). doi:10.1038/scientificamerican0785-48
486. S. Lloyd, Nature **406**, 1047–1054 (2000)
487. P.S. Peercy, Nature **406**, 1023–1026 (2000). doi:10.1038/35023223
488. L.B. Kish, Phys. Lett. A **305**, 144–149 (2002). doi:10.1016/S0375-9601(02)01365-8
489. R. Landauer, Phys. Today **May**, 23–29 (1991)
490. R. Landauer, Phys. Lett. A **217**, 188–193 (1996). doi:10.1016/0375-9601(96)00453-7

491. A. Church, Am. J. Math. **58**(2), 345–363 (1936)
492. A.M. Turing, Proc. Lond. Math. Soc. **s2-42**(1), 230–265 (1937). doi:10.1112/plms/s2-42. 1.230
493. M.A. Nielsen, I.L. Chuang, *Quantum Computation and Quantum Information* (Cambridge Univ. Press, Cambridge, 2001), 676 pp.
494. J. Stolze, D. Suter, *Quantum Computing: A Short Course from Theory to Experiment*, 2nd edn. (Wiley-VCH, Berlin, 2007), 255 pp.
495. D. Suter, T.S. Mahesh, J. Chem. Phys. **128**, 052206 (2008). doi:10.1063/1.2838166
496. X.-H. Peng, D. Suter, Front. Phys. China **5**(1), 1–25 (2010). doi:10.1007/s11467-009-0067-x
497. R.P. Feynman, Int. J. Theor. Phys. **21**, 467–488 (1982)
498. R. Feynman, Found. Phys. **16**(6), 507–531 (1986). doi:10.1007/BF01886518
499. P. Benioff, J. Stat. Phys. **29**(3), 515–546 (1982). doi:10.1007/BF01342185
500. E. Bernstein, U. Vazirani, Quantum complexity theory, in *Proc. 25th Annual ACM Symposium on Theory of Computing*, San Diego, CA, USA, 16–18 May 1993 (ACM, New York, 1993), pp. 11–20
501. D. Coppersmith, An approximate Fourier transform useful in quantum factoring, IBM Research Division, 1994. arXiv:quant-ph/0201067
502. P.W. Shor, Polynomial-time algorithms for prime factorization and discrete logarithms on a quantum computer, in *Proc. of 35th Annual Symposium on Foundations of Computer Science*, Santa Fe, NM, 20–22 November 1994
503. P. Zoller, Th. Beth, D. Binosi et al., Eur. Phys. J. D, At. Mol. Opt. Phys. **36**(2), 203–228 (2005). doi:10.1140/epjd/e2005-00251-1
504. D.P. DiVincenzo, Fortschr. Phys. **48**, 771–783 (2000)
505. N.A. Gershenfeld, I.L. Chuang, Science **275**(5298), 350–356 (1997). doi:10.1126/science. 275.5298.350
506. D.G. Cory, A.F. Fahmy, T.F. Havel, Proc. Natl. Acad. Sci. USA **94**, 1634–1639 (1997)
507. L.M.K. Vandersypen, I.L. Chuang, D. Suter, Encyclopedia of magnetic resonance, in *Liquid-State NMR Quantum Computing*, ed. by D.M. Grant, R.K. Harris (Wiley, New York, 2010)
508. J. Preskill, Quantum computation (available via DIALOG, 1997). http://theory.caltech.edu/~preskill/ph229/. Cited 25 May 2012
509. G. Chen, D.A. Church, B.G. Englert et al., *Quantum Computing Devices, Principles, Designs and Analysis* (Chapman & Hall CRC, Boca Raton, 2007), 560 pp.
510. E. Farhi, J. Goldstone, S. Gutmann et al., Science **292**(5516), 472–475 (2001). doi:10.1126/ science.1057726
511. X. Peng, Z. Liao, N. Xu et al., Phys. Rev. Lett. **101**, 220405 (2008). doi:10.1103/ PhysRevLett.101.220405
512. R. Raussendorf, D.E. Browne, H.J. Briegel, Phys. Rev. A **68**, 022312 (2003). doi:10.1103/ PhysRevA.68.022312
513. L.M.K. Vandersypen, I.L. Chuang, Rev. Mod. Phys. **76**(4), 1037–1069 (2005). doi:10.1103/ RevModPhys.76.1037
514. H. Kampermann, W.S. Veeman, J. Chem. Phys. **122**(21), 214108 (2005). doi:10.1063/ 1.1904595
515. V.F. Krotov, *Global Methods in Optimal Control* (Marcel Dekker, New York, 1996), 384 pp.
516. N.C. Nielsen, C. Kehlet, S.J. Glaser, N. Khaneja, Optimal control methods in NMR spectroscopy, in *Encyclopedia of Magnetic Resonance* (Wiley, New York, 2013). doi:10.1002/ 9780470034590.emrstm1043
517. J.P. Palao, R. Kosloff, Phys. Rev. Lett. **89**(18), 188301 (2002). doi:10.1103/PhysRevLett.89. 188301
518. G.L. Long, H.Y. Yan, Y.S. Li et al., Phys. Lett. A **286**(2–3), 121–126 (2001). doi:10.1016/ S0375-9601(01)00416-9
519. E.M. Fortunato, M.A. Pravia, N. Boulant et al., J. Chem. Phys. **116**(17), 7599 (2002). doi:10. 1063/1.1465412
520. R. Das, T.S. Mahesh, A. Kumar, Phys. Rev. A **67**(6), 062304 (2003). doi:10.1103/ PhysRevA.67.062304

521. W.K. Wooters, W.H. Zurek, Nature **299**, 802–803 (1982). doi:10.1038/299802a0
522. D. Dieks, Phys. Lett. A **92**(6), 271–272 (1982). doi:10.1016/0375-9601(82)90084-6
523. N. Gisin, G. Ribordy, W. Tittel, H. Zbinden, Rev. Mod. Phys. **74**(1), 145–195 (2002). doi:10.1103/RevModPhys.74.145
524. N. Gisin, R. Thew, Nat. Photonics **1**, 165–171 (2007). doi:10.1038/nphoton.2007.22
525. C.H. Bennett, G. Brassard, Quantum cryptography: public key distribution and coin tossing, in *Proceedings of the IEEE International Conference on Computers, Systems, and Signal Processing*, Bangalore, India, 10–12 December (1984)
526. H.-J. Briegel, W. Dür, J.I. Cirac, P. Zoller, Phys. Rev. Lett. **81**(26), 5932–5935 (1998). doi:10.1103/PhysRevLett.81.5932
527. I. Geru, V. Geru, Rom. J. Phys. **44**, 97–115 (1999)
528. V. Geru, Rom. J. Phys. **44**, 85–95 (1999)
529. D. Ruelle, *Statistical Mechanics: Rigorous Results* (World Scientific, Singapore, 1999), 219 pp.
530. J. Preskill, Proc. R. Soc. Lond. A **454**(1969), 385–410 (1998). doi:10.1098/rspa.1998.0167
531. R. Hanson, L.H. Willems van Beveren, I.T. Vink et al., Phys. Rev. Lett. **94**(19), 196802 (2005). doi:10.1103/PhysRevLett.94.196802
532. E. Von Bauer, Cryst. Mater. **110**(1–6), 372–394 (1958). doi:10.1524/zkri.1958.110.1-6.372
533. R. Hanson, L.P. Kouwenhoven, J.R. Petta et al., Rev. Mod. Phys. **79**(4), 1217–1265 (2007). doi:10.1103/RevModPhys.79.1217
534. A. Zrenner, E. Beham, S. Stufler et al., Nature **418**, 612–614 (2002)
535. P. Borri, W. Langbein, S. Schneider, Phys. Rev. B **66**(8), 081306(R) (2002). doi:10.1103/PhysRevB.66.081306
536. J. Berezovsky, M.H. Mikkelsen, O. Gywat et al., Science **314**(5807), 1916–1920 (2006). doi:10.1126/science.1133862
537. P. Borri, W. Langbein, U. Woggon et al., Phys. Rev. Lett. **91**(26), 267401 (2003). doi:10.1103/PhysRevLett.91.267401
538. A.W. Holleitner, R.H. Blick, A.K. Hüttel et al., Science **297**(5578), 70–72 (2002). doi:10.1126/science.1071215
539. G. Schedelbeck, W. Wegscheider, M. Bichler, G. Abstreiter, Science **278**(5344), 1792–1795 (1997). doi:10.1126/science.278.5344.1792
540. T. Fujisawa, T.H. Oosterkamp, W.G. van der Wiel et al., Science **282**(5390), 932–935 (1998). doi:10.1126/science.282.5390.932
541. R.I. Dzhioev, V.L. Korenev, B.P. Zakharchenya et al., Phys. Rev. B **66**(15), 153409 (2002). doi:10.1103/PhysRevB.66.153409
542. A. Shabaev, Al.L. Efros, D. Gammon, I.A. Merkulov, Phys. Rev. B **68**(20), 201305(R) (2003). doi:10.1103/PhysRevB.68.201305
543. R. de Sousa, S. Das Sarma, Phys. Rev. B **67**(3), 033301 (2003). doi:10.1103/PhysRevB.67.033301
544. A.V. Khaetskii, D. Loss, L. Glazman, Phys. Rev. Lett. **88**(18), 186802 (2002). doi:10.1103/PhysRevLett.88.186802
545. M.W. Doherty, N.B. Manson, P. Delaney, L.C.L. Hollenberg, New J. Phys. **13**, 025019 (2011). doi:10.1088/1367-2630/13/2/025019
546. F. Jelezko, J. Wrachtrup, Phys. Status Solidi (a) **203**(13), 3207–3225 (2006). doi:10.1002/pssa.200671403
547. H.J. Kimble, M. Dagenais, L. Mandel, Phys. Rev. Lett. **39**(11), 691–695 (1977). doi:10.1103/PhysRevLett.39.691
548. B. Smeltzer, L. Childress, A. Gali, New J. Phys. **13**, 025021 (2011). doi:10.1088/1367-2630/13/2/025021
549. H.Y. Carr, E.M. Purcell, Phys. Rev. **94**(3), 630–638 (1954)
550. S. Meiboom, D. Gill, Rev. Sci. Instrum. **29**, 688 (1958). doi:10.1063/1.1716296
551. J.R. Maze, P.L. Stanwix, J.S. Hodges et al., Nature **455**, 644–647 (2008). doi:10.1038/nature07279

552. L. Childress, M.V. Gurudev Dutt, J.M. Taylor et al., Science **314**(5797), 281–285 (2006). doi:10.1126/science.1131871
553. W.F. Koehl, B.B. Buckley, F.J. Heremans et al., Nature **479**(7371), 84–87 (2011). doi:10.1038/nature10562
554. G. Feher, E.A. Gere, Phys. Rev. **114**(5), 1245–1256 (1959). doi:10.1103/PhysRev.114.1245
555. B.E. Kane, Nature **393**, 133–137 (1998). doi:10.1038/30156
556. A.M. Tyryshkin, J.J.L. Morton, S.C. Benjamin et al., J. Phys., Condens. Matter **18**, S783 (2006). doi:10.1088/0953-8984/18/21/S06
557. A.R. Stegner, C. Boehme, H. Huebl et al., Nat. Phys. **2**, 835–838 (2006). doi:10.1038/nphys465
558. M. Xiao, I. Martin, E. Yablonovitch, H.W. Jiang, Nature **430**, 435–439 (2004). doi:10.1038/nature02727
559. S. Simmons, R.M. Brown, H. Riemann et al., Nature **470**, 69–72 (2011). doi:10.1038/nature09696
560. A. Morello, J.J. Pla, F.A. Zwanenburg et al., Nature **467**, 687–691 (2010). doi:10.1038/nature09392
561. R. Vrijen, E. Yablonovitch, K. Wang et al., Phys. Rev. A **62**(1), 012306 (2000). doi:10.1103/PhysRevA.62.012306
562. M. Friesen, P. Rugheimer, D.E. Savage et al., Phys. Rev. B **67**(12), 121301(R) (2003). doi:10.1103/PhysRevB.67.121301
563. W. Harneit, C. Meyer, A. Weidinger et al., Phys. Status Solidi (b) **233**(3), 453–461 (2002). doi:10.1002/1521-3951(200210)233:3<453::AID-PSSB453>3.0.CO;2-N
564. M. Waiblinger, Untersuchungen der endohedralen Fullerene mit eingeschlossenen Stickstoff- und Phosphor-Atomen, Dissertation, 2001
565. M.J. Butcher, F.H. Jones, P. Moriarty et al., Appl. Phys. Lett. **75**(8), 1074 (1999). doi:10.1063/1.124601
566. M. Mehring, W. Scherer, A. Weidinger, Phys. Rev. Lett. **93**(20), 206603 (2004). doi:10.1103/PhysRevLett.93.206603
567. D. Suter, K. Lim, Phys. Rev. A **65**(5), 052309 (2002). doi:10.1103/PhysRevA.65.052309
568. C. Ju, D. Suter, J. Du, Phys. Lett. A **375**(12), 1441–1444 (2011). doi:10.1016/j.physleta.2011.02.031
569. J.E. Grose, E.S. Tam, C. Timm et al., Nat. Mater. **7**, 884–889 (2008). doi:10.1038/nmat2300
570. J.H. van Vleck, J. Phys. Chem. **41**(1), 67–80 (1937). doi:10.1021/j150379a006
571. H. Lin, T. Wang, T.W. Mossberg, Opt. Lett. **20**(15), 1658–1660 (1995). doi:10.1364/OL.20.001658
572. T.L. Harris, K.D. Merkel et al., Appl. Opt. **45**(2), 343–352 (2006). doi:10.1364/AO.45.000343
573. N. Timoney, I. Baumgart, M. Johanning et al., Nature **476**, 185–188 (2011). doi:10.1038/nature10319
574. J.J. Longdell, M.J. Sellars, N.B. Manson, Phys. Rev. Lett. **93**(13), 130503 (2004). doi:10.1103/PhysRevLett.93.130503
575. W. Tittel, M. Afzelius, T. Chaneliere et al., Laser Photonics Rev. **4**(2), 244–267 (2010). doi:10.1002/lpor.200810056
576. J.J. Longdell, E. Fraval, M.J. Sellars, N.B. Manson, Phys. Rev. Lett. **95**, 063601 (2005). doi:10.1103/PhysRevLett.95.063601
577. E. Fraval, M.J. Sellars, J.J. Longdell, Phys. Rev. Lett. **95**, 030506 (2005). doi:10.1103/PhysRevLett.95.030506
578. S.E. Harris, Phys. Today **50**(7), 36 (1997). doi:10.1063/1.881806
579. B. Kraus, W. Tittel, N. Gisin et al., Phys. Rev. A **73**, 020302(R) (2006). doi:10.1103/PhysRevA.73.020302
580. N. Ohlsson, R.K. Mohan, S. Kroell, Opt. Commun. **201** (2002)
581. M.D. Lukin, Rev. Mod. Phys. **75**, 457–472 (2003). doi:10.1103/RevModPhys.75.457
582. C. Simon, M. Afzelius, J. Appel et al., Eur. Phys. J. D **58**(1), 1–22 (2010). doi:10.1140/epjd/e2010-00103-y

583. J.J. Longdell, G. Hetet, P.K. Lam, M.J. Sellars, Phys. Rev. A **78**, 032337 (2008). doi:10.1103/PhysRevA.78.032337

584. S.A. Moiseev, S. Kroell, Phys. Rev. Lett. **87**, 173601 (2001). doi:10.1103/PhysRevLett.87.173601

585. D.F. Phillips, A. Fleischhauer, A. Mair et al., Phys. Rev. Lett. **86**, 783–786 (2001). doi:10.1103/PhysRevLett.86.783

586. A.L. Alexander, J.J. Longdell, M.J. Sellars, N.B. Manson, Phys. Rev. Lett. **96**, 043602 (2006). doi:10.1103/PhysRevLett.96.043602

587. Th. Hannemann, D. Reiss, Ch. Balzer et al., Phys. Rev. A **65**, 050303(R) (2002). doi:10.1103/PhysRevA.65.050303

588. J.H. Wesenberg, K. Mølmer, L. Rippe, S. Kröll, Phys. Rev. A **75**, 012304 (2007). doi:10.1103/PhysRevA.75.012304

589. T. Böttger, C.W. Thiel, R.L. Cone, Y. Sun, Phys. Rev. B **79**, 115104 (2009). doi:10.1103/PhysRevB.79.115104

590. J.J. Longdell, A.L. Alexander, M.J. Sellars, Phys. Rev. B **74**, 195101 (2006). doi:10.1103/PhysRevB.74.195101

591. S.E. Beavan, E. Fraval, M.J. Sellars, J.J. Longdel, Phys. Rev. A **80**, 032308 (2009). doi:10.1103/PhysRevA.80.032308

592. M.N. Leuenberger, D. Loss, Lett. Nat. **410**, 789–793 (2001). doi:10.1038/35071024

593. J.R. Friedman, M.P. Sarachik et al., Phys. Rev. Lett. **76**, 3830–3833 (1996). doi:10.1103/PhysRevLett.76.3830

594. L. Thomas, F. Lionti, R. Ballou et al., Lett. Nat. **383**, 145–147 (1996). doi:10.1038/383145a0

595. C. Sangregorio, T. Ohm, C. Paulsen et al., Phys. Rev. Lett. **78**, 4645–4648 (1997). doi:10.1103/PhysRevLett.78.4645

596. M. Mannini, F. Pineider, C. Danieli et al., Nature **468**(7322) (2012)

597. V.A. Koptsik, I.N. Kotsev, Magnetic (spin) and magneto-electric point groups of P-symmetry, Preprint/Joint. Inst. Nuclear. Research, P4-8466, Dubna, 1974, 19 pp.

598. Yu.V. Rakitin, V.V. Volkov, V.T. Kalinnikov, Coord. Chem. **6**, 451–453 (1980)

599. R.J. Van Zee, C.A. Baumann, S.V. Bhat, W. Weltner Jr., J. Chem. Phys. **76**, 5636–5637 (1982). doi:10.1063/1.442871

600. A. Golnik, J.A. Gaj, M. Hawrocki, R. Planel et al., J. Phys. Soc. Jpn. Suppl. A **49**, 819–822 (1980)

Index

I. Geru, D. Suter, *Resonance Effects of Excitons and Electrons*,
Lecture Notes in Physics 869, DOI 10.1007/978-3-642-35807-4,
© Springer-Verlag Berlin Heidelberg 2013